A CENTURY OF CALCULUS
PART I

A CENTURY OF CALCULUS
PART I

*The Raymond W. Brink
Selected Mathematical Papers*

AMERICAN MATHEMATICAL MONTHLY
(Volumes 1–75)

and

MATHEMATICS MAGAZINE
(Volumes 1–40)

Selected and arranged by an editorial committee consisting of

TOM M. APOSTOL, Chairman
California Institute of Technology

HUBERT E. CHRESTENSON
Reed College

C. STANLEY OGILVY
Hamilton College

DONALD E. RICHMOND
Williams College

N. JAMES SCHOONMAKER
University of Vermont

Published by The Mathematical Association of America

Library of Congress Catalog Number 92-060987
ISBN 0-88385-205-5
Copyright © 1969, 1992 by
The Mathematical Association of America (Incorporated)
PRINTED IN THE UNITED STATES OF AMERICA

PREFACE

This book contains selected papers on calculus reprinted from THE AMERICAN MATHEMATICAL MONTHLY and the MATHEMATICS MAGAZINE. The selection was made by an editorial committee appointed by R. P. Boas during his tenure as Chairman of the Committee on Publications of the Mathematical Association of America. The charge to the editorial committee was to scan the back volumes of the MONTHLY and the MATHEMATICS MAGAZINE and to select articles that would be helpful to college teachers of freshman or sophomore calculus.

The editorial committee read more than 750 articles, of which 160 were finally selected for reproduction. The papers have been arranged into twelve categories, which are treated as chapters of the book. Some of the chapters are further divided into subcategories.

In each subcategory the selections from the MONTHLY appear in chronological order, followed by the selections from the MATHEMATICS MAGAZINE. A bibliography appears at the end of each subcategory listing further papers which the reader may wish to consult. Some of the bibliographic entries are accompanied by brief editorial comments describing in more detail the contents of the paper.

The original list of 750 articles contained 690 from the MONTHLY and 60 from the MATHEMATICS MAGAZINE. The list was divided chronologically into five equal parts, and each committee member was asked to prepare a brief abstract for each of the 150 articles on his list with a preliminary recommendation of *Yes*, *No*, or *Undecided*. The principal criterion at this stage was whether the reader felt the article would be helpful in some way to calculus teachers.

The abstracts were circulated to all committee members and all the papers were given a second reading by at least one more member of the committee. Those papers that received at least one vote of *Yes* or *Undecided* were given a third reading, and in some cases a fourth and fifth reading. There was remarkably good agreement on nearly all the papers that were finally selected.

The committee attempted to keep the volume to a reasonable size, while including a broad spectrum of ideas and maintaining some sort of balance among the various categories. Many of the papers contain material that can be used directly in the classroom. Others provide insights, background, or source material for special projects. Some papers discuss matters of controversy. Many of the bibliographic entries go beyond the level of elementary calculus but were included because they were considered valuable as reference material. The inclusion of any paper or bibliographic entry is not to be considered in any way an endorsement by the editorial committee of the point of view expressed by its author.

It is the committee's hope that there will be much of interest in this volume, both for the beginning teacher and the more experienced veteran, as well as for students.

THE EDITORIAL COMMITTEE

CONTENTS

PREFACE   v

1. HISTORY

Some Historical Notes on the Cycloid	E. A. WHITMAN	1
The Number e	J. L. COOLIDGE	8
The History of Calculus	ARTHUR ROSENTHAL	27
The Foremost Textbook of Modern Times	C. B. BOYER	32
The Unsatisfactory Story of Curvature	J. L. COOLIDGE	36
The Method of Archimedes	S. H. GOULD	41
Bibliographic Entries: History		44

2. PEDAGOGY

Objectives in Calculus	C. C. MACDUFFEE	45
Can We Teach Good Mathematics to Undergraduates?	R. G. HELSEL AND T. RADÓ	48
More Stress on General Formulations in Calculus Problems	LUISE LANGE	50
The Pupil's Advocate	RALPH BEATLEY	52
Parametric Equations and Mechanical Manipulation of Mathematical Symbols	A. D. FLESHLER	65
Parametric Equations and Proper Interpretation of Mathematical Symbols	A. ERDÉLYI	67
Student's Choice in Mathematics—Fundamental Reasoning or Blind Subservience to Rules	M. R. SPIEGEL	69
Bringing Calculus Up-to-Date	M. E. MUNROE	70

Concerning Domains of Real Functions	Hugh A. Thurston	80
The Area of the Ellipse	Roger Burr Kirchner	83
Calculus—A New Look	D. E. Richmond	84
Areas and Volumes Without Limit Processes	D. E. Richmond	93
Bibliographic Entries: Pedagogy		99

3. FUNCTIONS

(a) CONCEPTS AND NOTATION

A Kind of Problem that Effectively Tests Familiarity with Functional Relations	K. O. May	100
A Geometric Construction of Composite Functions	Kurt Kreith	101
Bibliographic Entires: Concepts and Notations of Functions		102

(b) TRIGONOMETRIC FUNCTIONS

Polar Coordinate Proofs of Trigonometric Formulas	Oswald Veblen	103
Relating to Finding Derivatives of Trigonometrical Functions	T. H. Hildebrandt	111
An Elementary Limit	M. S. Knebelman	113
Trigonometry from Differential Equations	D. E. Richmond	114
On the $\lim_{\theta \to 0} \cos \theta$	M. J. Pascual	117
On the Derivatives of Trigonometric Functions	M. R. Spiegel	119
Principal Values of Certain Inverse Trigonometric Functions	C. B. Read and Ferna Wrestler	122
A Classroom Proof of $\lim_{t \to 0}(\sin t)/t = 1$	Stephen Hoffman	124
Derivatives of $\sin \theta$ and $\cos \theta$	C. S. Ogilvy	126
Alternative Classroom Proof that $(\sin t)/t \to 1$ as $t \to 0$	Gary Perry	127
A Simple Way of Differentiating Trigonometric Functions and Their Inverses in an Elementary Calculus Course	H. A. Thurston	128
Bibliographic Entries: Trigonometric Functions		129

(c) LOGARITHMIC FUNCTIONS

An Existence Proof for Logarithms	F. E. Hohn	130
A Property of the Logarithm	D. S. Greenstein	132
Bibliographics Entries: Logarithmic Functions		132

(d) EXPONENTIAL AND HYPERBOLIC FUNCTIONS

Relating to the Exponential Function	Otto Dunkel	134
Editorial—The Proof of Euler's Equation	C. B. Allendoerfer	136
The Elementary Transcendental Functions	W. F. Eberlein	138
Bibliographic Entries: Exponential and Hyperbolic Functions		144

(e) FUNCTIONAL EQUATIONS AND INEQUALITIES

The Linear Functional Equation	G. S. Young	145
Bibliographic Entires: Functional Equations and Inequalities		146

4. CONTINUITY, ϵ AND δ, DISCONTINUITIES

A Type of Function with k Discontinuities	Raymond Garver	148
An Elementary Example of a Continuous Non-differentiable Function	Fred W. Perkins	150
A Peculiar Function	J. P. Ballantine	153
Two Discontinuous Functions	Marie M. Johnson	154
A Function with a Finite Discontinuity	J. A. Ward	155
An Everywhere Continuous Nowhere Differentiable Function	John McCarthy	156
A Note on δ and ϵ	Atherton H. Sprague	157
A Further Note on δ and ϵ	Allan Davis	158
Bibliographic Entries: Continuity, ϵ and δ, Discontinuities		159

5. DIFFERENTIATION

(a) THEORY

Derivatives of Implicit Functions	M. R. Spiegel	161
A Fundamental Theorem of the Calculus	Israel Halperin	163
On Proving the Chain Rule	David Gans	165
On the Derivative of x^c	S. Leader	167
A Simple Proof of a Basic Theorem of the Calculus	Mary Powderly	168

A Fallacy in Differentiability	ALBERT WILANSKY	169
Bibliographic Entries: Theory of Differentation		169

(b) APPLICATIONS TO GEOMETRY

On the Definition of Inflection Point	G. M. EWING	171
A Note on Envelopes	HOWARD EVES	174
Rulings	C. S. OGILVY	175
What Is a Point of Inflection?	A. W. WALKER	178
Bibliographic Entries: Applications to Geometry		179

(c) APPLICATIONS TO MECHANICS

Normal and Tangential Acceleration	R. C. YATES AND C. P. NICHOLAS	180
Inverse Square Orbits: A Simple Treatment	D. E. RICHMOND	182
Bibliographic Entries: Applications to Mechanics		184

(d) DIFFERENTIALS

Differentials	MARK KAC AND J. F. RANDOLPH	185
Differentials	ALONZO CHURCH	188
Bringing in Differentials Earlier	W. R. RANSOM	191
Differentials	M. K. FORT, JR.	193
The Relation of Differential and Delta Increments	C. G. PHIPPS	196
Toward Understanding Differentials	H. J. HAMILTON	199
Editorial [Differentials]	C. B. ALLENDOERFER	205
Bibliographic Entries: Differentials		207

(e) PARTIAL DERIVATIVES

Transformations of the Laplacian	R. P. AGNEW	208
A Proof of a Theorem on Jacobians	HIDEHIKO YAMABE	210
Bibliographic Entries: Partial Derivatives		211

6. MEAN VALUE THEOREM FOR DERIVATIVES, INDETERMINATE FORMS

(a) MEAN VALUE THEOREM

Proof of the First Mean Value Theorem of the Integral Calculus	T. PUTNEY	212
An Application of the Mean Value Theorem	DAVID ZEITLIN	214

The Law of the Mean	R. C. Yates	215
The Extended Law of the Mean by a Translation-Rotation of Axes	Jacqueline P. Evans	217
Extended Laws of the Mean	Louis C. Barrett and Richard A. Jacobson	219
A Natural Auxiliary Function for the Mean Value Theorem	M. J. Poliferno	222
On Avoiding the Mean Value Theorem	Lipman Bers	224
Mean Value Theorems and Taylor Series	M. R. Spiegel	225
Bibliographic Entries: Mean Value Theorem		228

(b) INDETERMINATE FORMS

A Geometric Example of an Indeterminate Form	Arthur C. Lunn	229
A Note on Indeterminate Forms	Roger Osborn	231
A Note on Indeterminate Forms	L. J. Paige	232
L'Hôpital's Rule and Expansion of Functions in Power Series	M. R. Spiegel	234
Bibliographic Entries: Indeterminate Forms		236

7. POLYNOMIALS AND POLYNOMIAL APPROXIMATIONS

(a) TAYLOR POLYNOMIALS

Elementary Development of Certain Infinite Series	J. P. Ballantine	237
A Note on Taylor's Theorem	C. L. Seebeck, Jr.	241
A Connection Between Taylor's Theorem and Linear Differential Equations	D. C. Lewis	243
More on Taylor's Theorem in a First Course	C. P. Nicholas	245
A Proof of Taylor's Formula	James Wolfe	247
Taylor's Theorem and Newton's Method	F. D. Parker	248
Remainder Formulae in Taylor's Theorem	William J. Firey	249
Geometric Interpretations of Polynomial Approximations of the Cosine Function	E. R. Heineman	251
Bibliographic Entries: Taylor Polynomials		252

(b) OTHER POLYNOMIALS

A Proof of Weierstrass's Theorem	Dunham Jackson	254
Bibliographic Entries: Other Polynomials		258

8. MAXIMA AND MINIMA

Maximum Parcels Under the New Parcel Post Law	W. H. Bussey	259
The Shortest Circular Path from a Point to a Line	Arnold Dresden	261
An Example in Maxima and Minima	Elijah Swift	265
A Problem in Maxima and Minima	Roger A. Johnson	266
Two Rectangles in a Quarter-Circle	B. M. Stewart	268
Minimal Tangents	Irving Kaplansky	272
A Rigorous Treatment of the First Maximum Problem in the Calculus	J. L. Walsh	273
End-Point Maxima and Minima	C. O. Oakley	275
One-Sided Maxima and Minima	J. D. Mancill	278
Fermat's Principle and Certain Minimum Problems	A. V. Baez	283
The Problem of a Non-vanishing Girder Rounding a Corner	Norman Miller	284
On the Harmonic Mean of Two Numbers	G. Pólya	287
A Simple Endpoint Minimum	Frank Hawthorne	290
Relative Maxima and Minima of Functions of Two or More Variables	A. S. Hendler	291
A Calculus Problem with Overtones in Related Fields	C. S. Ogilvy	294
Exceptional Extremum Problems	C. S. Ogilvy	297
So-Called "Exceptional" Extremum Problems	Hugh A. Thurston	303
Bibliographic Entries: Maxima and Mimima		305

9. INTEGRATION

(a) THEORY

The Fundamental Theorem of the Calculus	Louis Brand	307
Integration	D. G. Mead	310
The Two Fundamental Theorems of Calculus	F. Cunningham, Jr.	315
Bibliographic Entries: Theory of Integration		316

(b) TECHNIQUES OF INTEGRATION

The Dirichlet Formula, and Integration by Parts	L. M. Graves	318
A Remark on Integration by Parts	J. R. Borman	319
Undetermined Coefficients in Integration	J. B. Reynolds	321
Integration by Parts	K. W. Folley	322
Remarks on Integration by Parts	C. D. Olds	324
Partial Fractions with Repeated Linear or Quadratic Factors		
	M. R. Spiegel	327
Transformation of Standard Integrals	V. Punga	329
Integrals of Inverse Functions	F. D. Parker	330
Note on Integration by Operators	Roger Osborn	331
Integration by Matrix Inversion	William Swartz	332
A Useful Integral Formula	Donald K. Pease	334
An Extension of Integration by Parts	James W. Brown	335
Use of Hyperbolic Substitution for Certain Trigonometric Integrals		
	William K. Viertel	336
The Integration of Inverse Functions	John H. Staib	340
Bibliographic Entries: Techniques of Integration		341

(c) SPECIAL INTEGRALS

Note on the Quadrature of the Parabola	Otto Dunkel	342
The Probability Integral	C. P. Nicholas and R. C. Yates	344
The Calculus of Absolute Values	Kenneth O. May	345
Another Look at the Probability Integral	C. P. Nicholas	348
$\int \sec \theta \, d\theta$	Norman Schaumberger	350
Note on $\int_a^x t^y \, dt$	M. J. Pascual	351
$\int \sec^3 x \, dx$	Joseph D. E. Konhauser	352
Bibliographic Entries: Special Integrals		352

(d) APPLICATIONS

A Note on Areas	R. A. Johnson	354
Some Remarks on Centroids	R. B. Deal and W. N. Huff	356
Simple Problems on Arc Lengths	H. S. Thurston	357
Bibliographic Entries: Applications		358

(e) MULTIPLE INTEGRALS AND LINE INTEGRALS

The Use of Models While Teaching Triple Integration	E. A. Whitman	360
A Simple Problem in Cylindrical Coordinates	F. B. Hildebrand	364
Fubini Implies Leibniz Implies $F_{yx} = F_{xy}$	R. T. Seeley	365
Bibliographic Entries: Multiple Integrals and Line Integrals		366

10. NUMERICAL, GRAPHICAL, AND MECHANICAL METHODS AND APPROXIMATIONS

The Remainder in Computing by Series	R. K. Morley	367
A Simple Proof of Stirling's Formula	A. J. Coleman	369
A Graphical Integration	M. S. Klamkin	372
The Midpoint Method of Numerical Integration	Preston C. Hammer	373
Bibliographic Entries: Numerical, Graphical, and Mechanical Methods and Approximations		375

11. INFINITE SERIES AND SEQUENCES

(a) THEORY

Summability of Series	C. N. Moore	377
On a Certain Transformation of Infinite Series	J. A. Shohat	386
An Example of Double Series	J. E. Brock	390
Reversion of Series with Applications	J. B. Reynolds	391
A Slowly Divergent Series	R. P. Agnew	393
Term-wise Differentiation of Power Series	T. M. Apostol	395
A Note on Alternating Series	Philip Calabrese	399
On a Convergence Test for Alternating Series	R. Lariviere	401
Bibliographic Entries: Theory of Infinite Series and Sequences		402

(b) TRIGONOMETRIC SERIES

A Simple Discussion of the Representation of Functions by Fourier Series	Philip Franklin	405
Note on the Convergence of Fourier Series	Dunham Jackson	409
Bibliographic Entries: Trigonometric Series		411

(c) SUMS OF SPECIAL SERIES

A Formula for the Sum of a Certain Type of Infinite Power Series	Elbert H. Clarke	412
A Simple Derivation of the Leibnitz-Gregory Series for $\pi/4$	D. K. Kazarinoff	417
Probability and Sums of Series	G. B. Thomas, Jr.	419
An Elementary Proof of the Formula $\sum_{k=1}^{\infty} 1/k^2 = \pi^2/6$	Yoshio Matsuoka	422
Bibliographic Entries: Sums of Special Series		423

12. SPECIAL NUMBERS

(a) e

Methods of Presenting e and π	Karl Menger	425
An Application of a Famous Inequality	N. S. Mendelsohn	431
The Limit for e	Richard Lyon and Morgan Ward	432
Elementary Proof that e Is Irrational	L. L. Pennisi	434
A Note on the Base of Natural Logarithms	Ernest Leach	435
Elementary Proof that e Is Not Quadratically Algebraic	S. Beatty	437
Bibliographic Entries: e		438

(b) π

A Proof of the Irrationality of π	Robert Breusch	439
Bibliographic Entries: π		440

(c) EULER'S CONSTANT

A Note on the Calculation of Euler's Constant	Goldie Horton	442
On the Sequence for Euler's Constant	S. K. Lakshmana Rao	444
Bibliographic Entries: Euler's Constant		445

AUTHOR INDEX	447
CONTENTS, PART II	453

1

HISTORY

SOME HISTORICAL NOTES ON THE CYCLOID* †

E. A. WHITMAN, Carnegie Institute of Technology

1. Introduction. In this paper our interest is not in a renowned mathematician, a celebrated school, or a famous problem, but in a curve, the cycloid. More particularly, our interest is to center around its relation to the mathematics of the seventeenth century, one of the great centuries in the history of the subject. This curve had the good fortune to appear at a time when mathematics was being developed very rapidly and perhaps mathematicians were fortunate that so useful a curve appeared at this time. A new and powerful tool for the study of curves was furnished by the analytic geometry, whose year of birth is commonly given as 1637. New methods for finding tangents to curves, the areas under curves, and the volumes of solids bounded by curved surfaces were being discovered at a rapid pace, and a new subject, the calculus, was in the making. In these developments the cycloid was the one curve used preeminently and nearly every mathematician of the time used it in a trial of some of his new theory, even to the extent that much of the early histories of analytic geometry, calculus, and the cyloid are closely interwoven.

In the history that follows we shall not be concerned with historical minutiae, but only with the broad outlines of the story of this curve.

2. Early history of the curve. The original discoverer of the cycloid appears to be unknown. Paul Tannery has discussed a passage by Iamblichus referring to double movement and has remarked that it is difficult to see how the cycloid could have escaped the notice of the ancients.‡ John Wallis in a letter of 1679, ascribed the discovery to Nicolas Cusanus in 1450 and also mentioned Bouelles as one who in 1500 advanced the study of this curve. In the case of Cusanus, however, historians are agreed that Wallis was mistaken unless, says Cantor, he had access to some manuscript now lost. Now Bouelles mentions that he had observed a rolling wheel yet he seems to have considered the generated arch as a part of a circle whose radius was five-fourths that of the generating circle. The history of the cycloid becomes more definite when we come to Galileo. This

---

* From AMERICAN MATHEMATICAL MONTHLY, vol. 50 (1943), pp. 309–315.

† Presented to the Allegheny Section of the Mathematical Association of America, October 25, 1941.

‡ Pour l'histoire des lignes et surfaces courbes dans l'antiquité; Bulletin des sciences mathèmatique, Paris, 1883, p. 284.

scientist and teacher, famed for his telescope and microscope and as the discoverer of the isochronism of the vibrations of a pendulum, this Galileo attempted the quadrature of a cycloidal arch in 1599, at least so writes his pupil Torricelli in a publication of 1644. We here learn that Galileo had sought to measure its area and for this purpose used a balance upon which he placed a material cycloidal arch and a generating circle of like material. Always the arch was about three times as heavy as the circle, wherefore Galileo had given up his experiment since he believed that an incommensurable ratio was in question. Cantor writes of Galileo that he was the first to make this curve well known and that it was he who gave it its name. The curve was also known as a roulette and as a trochoid.

3. The work of Roberval. The scene now shifts to France, to the activities of Gilles Persone de Roberval, and to the problem of the quadrature of the cycloid. Going up to Paris in 1628, Roberval soon became a member of that small group of scientists and mathematicians who were wont to gather twice a week, generally at the home of Père Marin Mersenne, to discuss matters of common interest. Now Mersenne had brought the cycloid to the attention of French mathematicians at various times and Roberval soon learned of this curve but could not immediately effect the quadrature. However, a new method of finding the areas under curves was made known in 1629 when Cavalieri submitted his notes on the theory of *indivisibles* to show his fitness for the chair of mathematics at the University of Bologna, where he was a candidate. This new theory, and its extensions later, exerted an enormous influence upon the subject of finding the areas under curves, hence on the development of the calculus. In this paper we are concerned only with one part of this theory which is known as Cavalieri's Theorem* and which says that if two areas are everywhere of the same width one to the other, then the areas are equal.

About 1634 Roberval effected the quadrature of the cycloid, or trochoid as he called this curve. The first publication of his proof seems to have been in 1693 when his *Traité des Indivisibles*† appeared. To explain the long delay in publication of this important discovery, it may be noted that the Chair of Ramus at the Collège Royale which Roberval had won in 1634, automatically became vacant every three years, to be filled again by open competition. As the incumbent set the questions it seems plausible that Roberval should conceal his methods. In this way he would have a set of questions whereby he should win the coming contests. Professor Walker states that the accident of occupying this chair caused Roberval to lose credit for many of his discoveries.

Roberval's quadrature depends upon a so-called cycloid companion curve and an application of Cavalieri's Theorem. Professor Walker gives a translation

---

* A translation of this theorem and its proof is given in Smith's Source Book in Mathematics, also by Professor G. W. Evans in the American Mathematical Monthly for December, 1917.

† A Study of Roberval's Traité des Indivisibles, Columbia University, 1932, by Professor Evelyn Walker, gives an extended account of Roberval's works and discusses at some length the subject of indivisibles.

of this quadrature, but we shall describe it only in a general way. This is among the very earliest of the quadratures.

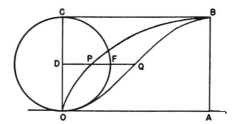

Let $OABP$ be the area under the half arch of the cycloid whose generating circle has the diameter OC. Take P any point on the cycloid and take PQ equal to DF. The locus of Q will be the companion curve to the cycloid. This curve OQB is the sine curve $y = a \sin x/a$ where a is the radius of the generating circle, if we take the origin at the midpoint of the arc OQB, and the x-axis parallel to OA. Now by Cavalieri's Theorem, the curve OQB divides the rectangle $OABC$ into two equal parts since to each line as DQ in $OQBC$, there corresponds an equal line in $OABQ$. The rectangle $OABC$ has its base and altitude equal respectively to the semicircumference and diameter of the generating circle, hence its area is twice that of the circle. Thus $OABQ$ has the same area as the generating circle. Also the area between the cycloid OPB and the curve OQB is equal to the area of the semicircle OFC since these two areas are everywhere of the same width one to the other. Hence the area under the half arch is one and one-half times the area of the generating circle, and the area under the arch is three times that of the generating circle.

4. Construction of the tangent. Early in 1638, Mersenne wrote to Fermat and Descartes presenting for their consideration the problem of the quadrature of the cycloid and the construction of a tangent to the curve. For a year or more previously Roberval and Fermat had been in correspondence, with Senator Carcavy as intermediary. The subjects discussed included tangents, cubatures, and centers of gravity. Mersenne's letters, however, brought to a focus the question of tangents for in August of this year Roberval, Fermat, and Descartes each gave Mersenne a method of drawing a tangent and each had a different method. In the ensuing dispute between Fermat and Descartes over the relative merits of their constructions, Roberval sided with Fermat. In turn Descartes wrote several letters to Mersenne bitterly ridiculing some of Roberval's tangent constructions which Mersenne had transmitted to him.

The question of priority in the matter of tangents we leave as unsettled and also unimportant, since each could not have borrowed from the others, so different were the methods. Part of the dispute over the relative merits of the constructions arose from different ideas as to the meaning of tangents to curves other than circles. The definition of a tangent as the limiting position of a secant had not yet been generally accepted.

We proceed to describe each of the three tangent constructions. Descartes' method is that which we now call instantaneous centers of curvature.

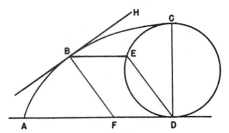

Let B be any point on the half arch of the cycloid ABC and let it be required to draw a tangent to the cycloid at B.

Draw BE parallel to the base AD cutting the circle at E. Draw BF parallel to ED and BH perpendicular to BF. BH is the required tangent.

The proof is based on the following considerations:

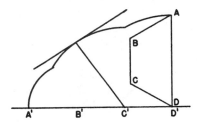

If a polygon $ABCD$ rolls on a straight line $A'D'$, any point A will describe a number of segments of circles whose centers will be at B', C', D', etc. The tangents to these segments will always be perpendicular to the line joining the point of tangency to the center of the circle. Consequently if the generating circle is considered as a polygon which has an infinite number of sides, the tangent at a given point will be the line perpendicular to the line joining this point to the point where the generating circle touches the base at the same instant it passes through the point.

Roberval's tangent construction makes use of the composition of forces and is easily understood in connection with his particular way of stating the definition of the cycloid.

Let the diameter AD of the circle move always parallel to its original position with A on the line AB until it takes the position BO with AB equal to a semicircumference. At the same time let the point A move on the semicircle ACD in such a way that the speed of AD along AB may be equal to the speed of A along the semicircle, thus allowing A to reach the point D at the same time AD reaches BO. The point A is carried along by two motions, its own on the semicircle and that of the diameter AD. The path of A due to these two motions is the half cycloid APO.

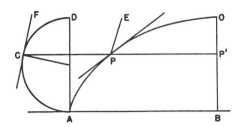

To construct a tangent at any point P on the cycloid, draw PP' parallel to AB cutting the semicircle at C. Then draw CF tangent to the semicircle and draw PE parallel to CF. The bisector of the angle EPP' is the required tangent since it is the resultant of two equal motions.

While finding the two components may be difficult for many curves, yet the cycloid is said to be the eleventh curve for which Roberval thus found tangents.

Fermat's construction is not unlike that of Descartes, but the proof appears to the casual reader to be quite as complicated as that of Descartes is simple. In the course of the proof a straight line is replaced by the arc of a circle. This is equivalent to assuming that an arc of a circle approaches coincidence with a certain straight line, making the method essentially one of limits. To one interested in the early approaches to the calculus, Fermat's method will be more interesting than that of Roberval or Descartes. The methods of the latter show what can be done in special cases and without the calculus. As Fermat's proof is quite long and is readily available elsewhere,* it will not be shown here.

With the area under the cycloidal arch and the tangent construction well mastered by his fellow Frenchmen, Mersenne announced these results to Galileo in 1638. Galileo, now old and blind, passed them on to his pupils Torricelli and Viviana, adding the suggestion that this curve would give a graceful form for the arch of the bridge that was projected for the nearby Arno River at Pisa. These pupils responded with a quadrature and a tangent. The interest thus kindled led Torricelli to a considerable study of the curve. In 1644 he made public his quadrature and a method of drawing a tangent. This was the earliest printed article on the cycloid.

Roberval was angered at seeing another print proofs that he considered his own discoveries. He wrote a letter to Torricelli charging plagiarism. More specifically, Roberval charged that a certain Frenchman had written out Fermat's method of maxima and minima and Roberval's propositions on the cycloid, that these papers had come into Torricelli's hands after the death of Galileo, and that Torricelli had published them as his own. This dispute was cut short by Torricelli's early death in 1647, a death caused, according to Cajori, by this charge of plagiarism.

---

* See Cantor's Vorlesungen über Geschichte der Mathematik, Volume II, pages 861–863, or Walker's Traité des Indivisibles, pages 132–134.

5. Pascal's mathematical contest. Our next episode in this history centers around Blaise Pascal, known for his *Pensées* and his *Lettres Provinciales* as well as for his mathematical works. After a brilliant early career in mathematics he had turned to theology. But suddenly the old mathematical propensity reasserted itself. Ball writes that Pascal was suffering from sleeplessness and toothache when the idea of an essay on the cycloid occurred to him. To his surprise the tooth ceased to ache. Regarding this as a divine intimation to proceed with the problem, he worked incessantly at it for eight days and completed a tolerably full account of the geometry of the cycloid. As certain questions about this curve had never been publicly answered, a prize was now offered by Pascal under the nom de plume of Amos Dettonville.

The year was 1658 when Newton was sixteen years old. The prizes were two in number, forty and twenty Spanish doubloons. The time allotted was June first to October first. Senator Carcavy was made recipient of the solutions offered and he, Pascal, and Roberval were the judges. The problems were as follows:

(1) The area and the center of gravity of that part of a cycloidal arch above a line parallel to the base.

(2) The volume and center of gravity of the volume generated when the above area is revolved about its base and also about its axis of symmetry.

(3) The center of gravity of the solids formed when each body is cut by a plane parallel to its axis of revolution.

Only two contestants, Wallis and Lalouvère, had submitted offerings when time was called. Ball says that Wallis did not submit solutions for the centers of gravity, and Cajori says that Wallis made many mistakes. Both historians agree that Lalouvère was quite unequal to the task. The judges declared that neither contestant was entitled to a prize.

At the time of this contest, Sir Christopher Wren sent to Pascal his proposition on the rectification of the cycloid, not, however, including any proof. When Pascal showed this to Roberval the latter is said to have proved the proposition immediately, claiming to have known it for many years. To Wren goes the credit for the first publication and its proof,* when Wallis published it as Wren's a year later in his *Tractatus duo*.

While the contest was on, Pascal published his *L'Histoire de la Roulette* and after the decision of the judges, his solution of the problems. With Pascal's and Wallis's publications at this time, the problems of quadrature, tangents, rectification, cubature, and centers of gravity are substantially completed in so far as the cycloid, or roulette as it was better known to the Frenchmen, was concerned. All this was accomplished in a period of about twenty-five years and before Newton's work in the calculus. The principle of *indivisibles*, or *infinites*, or whatever they had used, had in the hands of Roberval, Fermat, Torricelli, Wren, and Wallis led to important results. The cycloid curve was always being used; it was the pre-eminent curve, and its importance was to be seen later.

---

* The general character of the proof is given in Cantor, Volume II, page 904.

6. The brachistochrone problem.

In another fifteen years, Huygens was using the cyloidal pendulum in an attempt to get a better chronometer and made use of the property that the evolute of the cycloid is another equal cycloid. This same Huygens discovered that a heavy particle reached the bottom of an inverted cycloidal arch in the same length of time no matter from what point on the arch it began its descent. In 1686, Leibniz wrote the equation for the curve, thus showing the rapid progress that was being made in analytic geometry. This equation is given here as Leibniz wrote it since his form shows interesting variations from those employed at present:

$$y = \sqrt{2x - xx} + \int dx/\sqrt{2x - xx}.$$

In the decade following the publication of this equation, the Bernoulli brothers, Jacques and Jean, published several articles on the cycloid. But we shall hurry on to one final episode in the history of the curve.

In June, 1696, Jean Bernoulli proposed a new problem which mathematicians were invited to solve: If two points A and B are given in a vertical plane, to assign to a mobile particle M the path AMB along which, descending under its own weight, it passes from the point A to the point B in the briefest time. In later amplifying the problem Bernoulli says to choose such a curve that if the curve is replaced by a thin curve or groove and a small sphere placed in it and released, then this sphere will pass from one point to the other in the shortest possible time. Thus the famous brachistochrone problem appeared on the scene. The solution is the inverted cycloidal arch. An elaborate model of the brachistochrone formed a considerable part of the mathematics exhibit at the Golden Gate International Exposition in 1940, from which we may conclude that there is still considerable interest in the problem.

In giving out his solution,* Jean Bernoulli wrote that a new kind of maxima and minima is required. In this solution we see that mathematics had advanced at this time as far as the calculus of variations. In a few more years there began to appear articles on general methods for determining the nature of curves formed by other rolling circles and on curves of descent under activating forces other than gravity. As the cycloid thus loses its pre-eminence, this seems a proper place to close this recital of its history.

---

* See Smith's Source Book in Mathematics, pages 644–655.

THE NUMBER e *

J. L. COOLIDGE, Harvard University

1. The Greek beginning. The distinguished American mathematician, Benjamin Peirce, was wont to find all of analysis in the equation

$$i^{-i} = \sqrt{e^\pi}.$$

In fact, he had his picture taken in front of a blackboard on which this mystic formula, in somewhat different shape, was inscribed. He would say to his hearers, "Gentlemen, we have not the slightest idea of what this equation means, but we may be sure that it means something very important."

With regard to the symbols which appear in this charm, there is a vast literature connected with π; and i, when written $\sqrt{-1}$, leads into the broad field of analysis in the complex domain; but it seems surprisingly difficult to find a connected account of e.

I think we may make a fair beginning with the twelfth proposition of the Second Book of Apollonius, *Conics*, which tells us that if from a point on a hyperbola lines be drawn in given directions to meet the asymptotes, the product of the two distances is independent of the position of the point chosen on the curve. This theorem is more general than we shall need to arrive at the number e, and it is not original with Apollonius. Let us confine ourselves to the very special case where the hyperbola is rectangular, and we draw to each asymptote a line parallel to the other. When x and y are distances, we may write

(1)	$$xy = 1.$$

It is intriguing to inquire who first discovered the theorem which leads to this equation. In the commentary of Eutocius on the Sphere and Cylinder of Archimedes [1], we come to a discussion of the classical problem of inserting two mean proportionals between two given lengths. In one solution, which he labels "ut Menaechmus," we have what amounts to the equations

(2)
$$a/x = x/y = y/b;$$
$$y^2 = bx; \quad xy = ab.$$

He goes on to seek the intersection of a parabola and a hyperbola.

Eutocius' statement would place the theorem very early in the history of the conics, for Menaechmus is usually regarded as the discoverer or inventor of these curves, although this ascription is by no means certain. Allman writes [2], "It is much to be regretted that the two solutions of Menaechmus have not been transmitted to us in their original form. That they have been altered either by Eutocius or by some author whom he followed appears not only in the employment in these solutions of the terms parabola and hyperbola, as has frequently been pointed out, but more from the fact that the language used in

---

* From AMERICAN MATHEMATICAL MONTHLY, vol. 57 (1950), pp. 591–602.

them is, in character, altogether that of Apollonius." A similar doubt is shown in Loria [3]. On the other hand, Heath is perfectly definite on this point; he states, "This property in the particular case of the rectangular hyperbola was known to Menaechmus" [4].

But there is another reason for doubting the ascription to Menaechmus, aside from the linguistic objection. The classical Greek discussion of the conics always corresponds to our analysis when the axes are a tangent and the diameter through the point of contact, and with these data proofs are not simple. Heath, following Zeuthen, shows the fact that the hyperbola can be written immediately in the form (1) if we start with a technique like ours, that is, when the axes are a pair of conjugate diameters [5]. That is perfectly true, but the Greeks made surprisingly little study of the conics when expressed in this form more familiar to us; Apollonius comes to it quite late. It seems to me altogether doubtful that the first discoverer of the curves should have been able to make the transition.

2. Grégoire de St. Vincent. If we grant that the Greek mathematicians, perhaps Menaechmus, were familiar with the fundamental property of the rectangular hyperbola expressed in (1), what has this to do with e? We must look ahead some two thousand years to that original writer whose name appears at the head of this paragraph. In 1647, he published his fundamental *Prologomena a Santo Vincento, Opus geometricum quadraturae circuli et sectionum coni*. This I have not seen in its original form, but the content is given at great length by Bopp in [6]. Here is the general scheme. We take the hyperbola

(1) $$xy = 1.$$

On the x axis we take n equivalent rectangles whose bases are

$$P_0P_1, P_1P_2, \cdots, P_{n-1}P_n,$$

while each has an upper vertex on the curve Q_i. Then,

$$P_0P_1 \cdot P_0Q_0 = P_1P_2 \cdot P_1Q_1 = P_2P_3 \cdot P_2Q_2 = \cdots,$$

and

(2) $$\frac{P_0Q_0}{P_1Q_1} = \frac{P_1P_2}{P_0P_1}; \quad \frac{P_1Q_1}{P_2Q_2} = \frac{P_2P_3}{P_1P_2}$$

but

$$OP_0 \cdot P_0Q_0 = OP_1 \cdot P_1Q_1 = OP_2 \cdot P_2Q_2 \cdots,$$

so that

$$\frac{OP_0}{OP_1} = \frac{P_1Q_1}{P_0Q_0} = \frac{P_0P_1}{P_1P_2} = \frac{OP_1}{OP_2}, \quad \text{by composition.}$$

If

(3) $$OP_1 = \rho OP_0 \quad \text{then} \quad OP_j = \rho^j OP_0.$$

St. Vincent even treats the case where OP_0 and OP_j are incommensurable, but we need not follow him here.

The importance of this equation was early recognized, because of its connection with logarithms which were based on the relation of arithmetical and geometrical series. There is a good deal to be said in favor of the thesis that the credit for relating the rectangular hyperbola with logarithms is due to Sarasa. I have not seen his work, but like Cantor, I rely on Kästner. In 1649, Sarasa published *Solutio Problematis a R. P. Marino Mersenno propositi*. This was concerned with the problem: Given three positive quantities and the logarithms of two of them, find the logarithm of the third. Kästner writes [7], "Zu ihrer Beanwortung brang Sarasa drey Saetze aus des Gregorius Buche von der Hyperbel bey, die betreffen Flaechen der Hyperbel an der Aysmptoten, Sarasa erinnert wie das mit Logarithmen zusammenhangt." Cantor's view is similar [8]; he states, "Mit andern Worten, Gregorius hatte das Auftreten von Logarithmen bei der erhähnten Flächenraumen erkannt, wen auch nicht mit Namen genannt. Letzteres that Sarasa, und darin liegt das wirkliche Verdienst seiner Stratschrift."

A contrary view is expressed by Charles Hutton [9] in the words, "As to the first remarks on the analogy between logarithms and hyperbolic spaces, it having been shown by Gregory St. Vincent . . . that if an asymptote be divided into parts in geometrical progression, and from the points of division ordinates be drawn parallel to the other asymptote, they will divide the space between the asymptote and the curve into equal portions, from hence it was shown by Mersenne, that by taking continual sums of these parts there would be obtained areas in arithmetical progression which therefore were analogous to a system of logarithms."

This may be true, but I must point out that whereas St. Vincent published the work referred to above in 1647, Mersenne died in the middle of 1648, and the dates of all of his mathematical writings which I have seen were much earlier. However, St. Vincent's work was certainly well observed. We find Wallis writing in 1658 to Lord Brouncker [10], "Sumptis (in Asymptoto) rectis NH, NI, NK, NQ, NL, NM geometrice proportionalibus, in punctis H, I, K, Q, L, M, ducantur rectae parallelae alteri Asymptoto, spatium Hyperbolicum $A\,B\,H\,M$ in quinque partes dividi ostendit Gregor de Sancto Vincento (si memini) decimo."

3. The introduction of logarithms. The actual word logarithm occurs again in an account of Gregory's *Vera circuli et hyperbolae quadratura*, which was published in Padua in 1667 and laid before the Royal Society [11]. Here we read, "And lastly by the same method he calculates both the logarithm of any natural number, and, vice versa, the natural number of any given logarithm." Perhaps the wisest word on the subject has been pronounced by the kindly old writer Montucla [12], "Au reste la découverte de cette propriété est revindiquée par divers autres géomètres." Among these I surely must mention Christian Huygens, who acknowledges the work of St. Vincent, even though he does not claim for himself the discovery of the relation between the hyperbola and logarithms. This is admirably set forth in [13], first in a French account, then Huygens' own Latin. He finds the areas bounded by the x axis, which is an asymptote, the curve and ordinates. Two such areas terminating by the same ordinate of 1 are

$$\frac{\text{Area } FGDE}{\text{Area } ABDE} = \frac{\log_e FG}{\log_e 10} = \log_{10} FG.$$

Huygens divides numerator and denominator by 32, which amounts to finding the 32nd root of each area, but this has the effect of so far closing up the figure that we may safely replace the hyperbola by a parabola whose outside area is known. He checks by finding a very good value for $\log_{10} 2$.

In the same year, 1661, Huygens finds another curve which he calls logarithmic but we should probably call it exponential. This curve has the property that the ordinate corresponding to the point mid-way between two given points of the x-axis is the mean proportional between their ordinates. The equation of the curve is $y = ka^x$. Huygens takes

(4) $$y = 2^{x/x_0}; \qquad x = \frac{\log y}{\log 2} x_0.$$

The constant subtangent is

(5) $$\frac{y\,dx}{dy} = \frac{x_0}{\log_e 2}.$$

Huygens takes

$$x_0 = 10^n \log_{10} 2.$$

This gives for the constant subtangent

$$\log_{10} e = .43429448190325180,$$

"qualium logarithmus binarij est"

$$.30102995663981195.$$

These numbers had long been known as they had appeared, for instance, in Briggs' *Arithmetica logarithmica* of 1624, pages 10 and 14. As a matter of fact, there appeared in 1618 a second edition of Wright's translation of Napier's *Mirifici Logarithmorum Canonis Descriptio* which contained an appendix, probably written by Oughtred, giving the natural logarithms of various numbers from 100,000 to 900,000. This is probably the earliest table of natural logarithms, although a very similar table by John Spidell appeared in 1619 [14].

The astonishing thing about all of those writers who connected logarithms with hyperbolic areas is their lack of interest in what we should call the base. Napier began by considering the relation between an arithmetical and a geometric series. A geometric series consists in successive powers of one number. What is that number? Or given a set of logarithms, what number has the logarithm 1? I mentioned that Briggs gave the logarithm of e, to the base 10 but I find no mention of e itself. Of course, we might write

$$10^n \log_{10} \frac{10^n + \Delta x}{10^n} = 10^n \log_e \left(1 + \frac{\Delta x}{10^n}\right) \log_{10} e = \Delta x \log_{10} e + \cdots,$$

but e itself does not appear. The fact is that there was no comprehension that a logarithm was essentially an exponent. Tropfke is very explicit in this point; he writes, "Freilich dürfen wir nicht an die moderne Erklärung der Logarithmen denken, die in ihnen Potenzexponenten einer bestimmten Grundzahl erkennt. Diese Auffassung machte sich erst um die Mitte des achtenten Jahrhunderts geltend" [15]. This is perhaps too strong a statement, for in a note on the same page he quotes James Gregory (whom he calls David Gregory) as saying in his *Exercitationes Geometricae* of 1684, p. 14, "Exponentes sunt ut logarithmi." I have not been able to verify this, but we find in [16], "Si seriei Termonorum in Progressione geometrica ab 1 continue proportionalium, puta

1, 2, 4, 8, 16, 32, 64, etc.

accomedetur series Indicum, sive Exponentium, in progressione ab o continue procedentium, puta

0, 1, 2, 3, 4, 5, 6, etc.

Hos exponentes appelabant Logarithmos." We could not well ask for anything clearer or more explicit.

If most writers did not look on logarithms as exponents, how did they consider them? I think we find the clue in St. Vincent's identification of logarithms with hyperbolic areas, remembering that these were the days of Cavalieiri and Roberval, when an area was looked upon as the same thing as an infinite number of line segments, a very helpful if dangerous definition. We find Halley writing [17], "They may more properly be said to be numeri rationum exponentes, wherein we consider ratio as a quantity sui generis, beginning from the ratio of equality, or 1 to 1 = 0, $\cdots$ and the rationes we suppose to be measured by the number of ratiunculae in each. Now these ratiunculae are in a continued scale of proportionals, infinite in number, between the two terms of the ratio, which infinite number of mean proportionals is to that infinite number of the like and equal ratiunculae between any other two terms as the logarithm of one ratio is to the logarithm of the other. Thus if we suppose there to be between 1 and 10 an infinite scale of mean proportionals whose number is 100000 ad infinitum, between 1 and 2 there shall be 30102 of said proportionals and between 1 and 3, 47712 of them which numbers therefore are the logarithms of the ratio of 1 to 10 1 to 2, and 1 to 3, and so properly called the logarithms of 10, 2, and 3."

It is hard to see how there could be a much worse explanation of logarithms for those who "make constant use of logarithms without having an adequate notion of them." The one certain thing seems to be that a logarithm is an infinite number. I suppose we might translate this into the form

$$\frac{b}{a} = \frac{a+r_1}{a} \cdot \frac{a+r_2}{a+r_1} \cdot \frac{a+r_3}{a+r_2}, \cdots, \frac{a+r_n}{a+r_{n-1}} \cdot \frac{b}{a+r_n}.$$

If

$$\frac{a+r_j}{a+r_{j-1}} = r, \qquad \frac{b}{a} = r^n.$$

Then n would be the logarithm.

4. Mercator, Newton, Leibniz. It is fair to say that such a definition of a logarithm was not original with Halley. We find Mercator writing in 1668, [18] "Est enim Logarithmus nihil aliud, quam numerus ratiuncularum contentarum in ratione quam absolutus quisque ad unitatem obtinet." I may mention also that this seems the first place where the words "logarithmus naturalis" are used. But the real significance of the article comes from the fact that instead of studying log x he takes up log $(1+x)$, which enables him to start from 0. The article is not clearly written, so I follow the much clearer exposition in Wallis [19], which was published in the same year.

We study the area under the curve (1) from $x=1$ to $x=1+X$. We divide the length on the x-axis into n equal parts, each of length Δx. The abscissas are 1, $1+\Delta x$, $1+2\Delta x$, $\cdots$, $1+X$ and the corresponding ordinates are

$$1, \frac{1}{1+\Delta x}, \frac{1}{1+2\Delta x}, \cdots, \frac{1}{1+(n-1)\Delta x}.$$

The infinitesimal, rectangular areas are

$$\Delta x, \Delta x[1 - \Delta x + \Delta x^2 - \Delta x^3 + \cdots],$$
$$\Delta x[1 - (2\Delta x) + (2\Delta x)^2 - (2\Delta x)^3 + \cdots], \cdots.$$

Such infinite expansions were common in Wallis' work. The sum of these rectangular areas may be written

$$\Delta x[1 + 1 + 1 \cdots] - \Delta x[\Delta x + 2\Delta x + 3\Delta x + \cdots]$$
$$+ \Delta x[(\Delta x)^2 + (2\Delta x)^2 + (3\Delta x)^2 + \cdots] - \cdots.$$

Now $n\Delta x = X$, so we have

(6) $$X - \Delta x^2[1 + 2 + 3 \cdots] + \Delta x^3[1^2 + 2^2 + 3^2 \cdots] - \cdots.$$

With regard to these sums, Wallace says [19], page 222, "quod ostendit ille prop XVI etque a me alibi demonstratum." A reference he makes to Mercator is not conclusive as the statement is sketchy; as to his own work I will follow [20], as I shall need that again. Here he is seeking the area under the curve $y = x^m$ from $x = 0$ to $x = X$. His method is not perfectly clear, as he seems merely to generalize by analogy from cases worked out earlier, but what he does is essentially the following:

We take N equal lengths from 0 to $N\Delta x = X$. We have a set of rectangles whose combined areas are

$$\Delta x[0^m + (\Delta x)^m + (2\Delta x)^m + (3\Delta x)^m + \cdots].$$

Let us assume that $0^m + 1^m + \cdots + (N-1)^m = \alpha N^{m+1} + \beta N^m + \gamma N^{m-1} + \cdots$. Replacing N by $N+1$, and subtracting, we obtain

$$N^m = (m + 1)\alpha N^m + bN^{m-1} + cN^{m-2} \cdots,$$

so

$$\alpha = \frac{1}{m + 1}.$$

Substituting, and remembering that $N\Delta x = X$, there results

$$\text{Area} = \frac{X^{m+1}}{m + 1} + \beta \Delta x X^m + \gamma \Delta x^2 X^{m-1}.$$

The limit of this as $N \to \infty$ is $X^{m+1}/m+1$, since $\Delta x \to 0$. We thus can substitute this result in (6), when $m = 1, 2, 3, \cdots$, to obtain Mercator's famous formula:

(7) $$\log(1 + X) = X - \frac{X^2}{2} + \frac{X^3}{3} - \frac{X^4}{4} + \cdots.$$

A good deal has been written about this series, as we see from Mazeres and elsewhere. The obvious way to obtain the equation is to apply the calculus, so we now turn to see how this instrument was brought to bear. In 1669, a year after Mercator had published his work on logarithms [18], Newton sent to Collins his article, *De Analysi per aequationes numero terminorum Infinitas* [21]. This represents his first studies of areas under curves, which he had been working at for a year or two, but had not published. In fact, publication did not occur for a goodly number of years to come; there is, however, no question of giving his results precedence over those of Mercator. It begins as shown below:

Curvarum simplicium Quadratura

Reg. 1 $\quad\quad$ Si $ax^{m/n} = y$, erit $\dfrac{an}{m+n} x^{(m+n)/n} = $ Area ABD.

I must speak further of this. In [22] we read on p. 176, "Dr. Wallis published in his Arithmetica infinitorum in the year 1655 and in the 59th Proposition of that Book, if the Abscissa of any curvilinear figure be called x and m and n be two Numbers, the ordinates erected at right Angles be $x^{n/m}$ the area of the Figure shall be $(n/m+n)x^{m+n/n}$. And this is assumed by Mr. Newton, upon which he founds his Quadrature of Curves. Dr. Wallis demonstrated this by steps in many particular Propositions and then connected all the Propositions into one by a Table of Cases. Mr. Newton reduced all Cases to One, with an indefinite Index, and at the end of his Compendium demonstrated it at once by his method of moments, he being the first who introduced indefinite Indices of Dignites into the Operations of Analysis." This is Newton's own statement of the case and must be taken as final. It is true that Wallis worked out a number of special cases in a manner not exactly like the method followed here, and did not use a literal exponent. The greater generality of Newton's formula is found by replacing x by $x^{1/n}$. Newton's proof by "the method of moments" we should call differentiation, and consisted in showing that if

$$z = \frac{n}{m+n} x^{(m+n)/n}; \quad \text{then} \quad \frac{dz}{dx} = x^{m/n}.$$

It is fair to say also that although he gives Mercator's formula, he gives it as the area under the hyperbola, with no mention of Mercator or of logarithms.

It is time to turn for a moment to the other inventor of the calculus, Gottfried Leibniz. We find him writing in 1677 or 1678 [23],

"In Hyperbol sit $AB = 1$, $BM = x$, $ML = \dfrac{1}{1+x}$,

$$CBMLC = \frac{1}{1} x - \frac{1}{2} x^2 + \frac{1}{3} x^3 - \frac{1}{4} x^4 \ldots ."$$

This is proved by the straight expansion of $1/1+x$, after which there is integration term by term. We find something more interesting a dozen years later, when he writes to Huygens, who is said never to have understood Leibniz calculus of differences [24], "Soit donc x l'abscisse et y l'ordonnée de la courbe, et l'équation comme je vous ay dit

$$\frac{x^3 y}{h} = b^{2xy}.$$

Je désignerai le logarithme de x par log x et nous aurons

$$3 \log x + \log y - \log h = 2xy.$$

supposant que le log de l'unite soit 0 et le log $b=1$. Donc par la quadrature de l'hyperbole nous aurons

$$3\int \frac{dx}{x} + \int \frac{dy}{y} - \log h = 2xy$$

$$3\frac{dx}{x} + \frac{dy}{y} = 2xdy + 2ydx,$$

dx sera à dy, on bien DB sera à y comme $2x^2y - x$ est à $3y - 2xy$ c'est à dire DB sera

$$\frac{2x^2y - xa^2}{3a^2 - 2xy}$$

comme vous le demandées, a estant l'unité."

5. Leonhard Euler. It is now time to turn to the man who pulled all this together and who put the number e definitely on the map, Leonhard Euler. This he did in [25], beginning in "Caput VII" with the base a. His argument is outlined below:

Since $a^0 = 1$, we may put $a^w = 1 + kw$; $w = \log(1 + kw)$. Assume w to be very small, and write

$$a^{iw} = (1 + kw)^i = 1 + \frac{i}{1}kw + \frac{i(i-1)}{1 \cdot 2}k^2w^2 + \frac{i(i-1)(i-2)}{1 \cdot 2 \cdot 3}k^3w^3 + \cdots.$$

Since w is infinitesimally small, and i is infinitely large, we write $iw = z$

$$a^z = \left(1 - \frac{kz}{i}\right)^i = 1 + kz + \frac{(i-1)}{i \cdot 1 \cdot 2}k^2z^2 + \frac{(i-1)(i-2)}{i \cdot 1 \cdot 2 \cdot 3}k^3k^3 + \cdots.$$

Since i is very large, we may assume $(i - n)/i = 1$, then

$$a^z = 1 + kz + \frac{k^2z^2}{1 \cdot 2} + \frac{k^3z^3}{1 \cdot 2 \cdot 3} + \cdots.$$

If $z = 1$,

$$a = 1 + k + \frac{k^2}{1 \cdot 2} + \frac{k^3}{1 \cdot 2 \cdot 3}.$$

If we take $a = 10$, the base in the logarithm system of Briggs, Euler gives

$$k = 2.30238, \quad \text{approximately.}$$

For a natural logarithm we take $k = 1$; $a = e$; and

(8) $$e = 1 + \frac{1}{1} + \frac{1}{1 \cdot 2} + \frac{1}{1 \cdot 2 \cdot 3} + \cdots.$$

Euler gives this value to 18 places, without naming the source, namely,

$$\tag{9} e = \lim_{i \to \infty}\left(1 + \frac{1}{i}\right)^i.$$

With regard to the use of the letter e, Euler had long employed it, for we find him writing [26], page 80, "scribitur pro numero cujus logarithmus est unitas e, qui est 2.7182817...." Note that this is Leibniz' b.

I pass to Ch. VII of the *Introductio*. Euler assumes for small values of z,

$$\sin z = z; \quad \cos z = 1.$$

He then, following DeMoivre, writes,

$$\cos nz = \frac{(\cos z + \sqrt{-1} \sin z)^n + (\cos z - \sqrt{-1} \sin z)^n}{2},$$

$$\sin nz = \frac{(\cos z + \sqrt{-1} \sin z)^n - (\cos z - \sqrt{-1} \sin z)^n}{2\sqrt{-1}}.$$

Putting $nz = v$, and remembering that z is small,

$$\cos v = 1 - \frac{v^2}{1 \cdot 2} + \frac{v^4}{1 \cdot 2 \cdot 3 \cdot 4} - \cdots,$$

$$\sin v = v - \frac{v^3}{1 \cdot 2 \cdot 3} + \frac{v^5}{1 \cdot 2 \cdot 3 \cdot 4 \cdot 5} - \cdots.$$

Comparing these with the value given previously for a^z, one obtains

$$\tag{10} \cos v = \frac{e^{v\sqrt{-1}} + e^{-v\sqrt{-1}}}{2}; \quad \sin v = \frac{e^{v\sqrt{-1}} - e^{-v\sqrt{-1}}}{2\sqrt{-1}};$$

and

$$\tag{11} v\sqrt{-1} = \log[\cos v + \sqrt{-1} \sin v].$$

This last formula was not, strictly speaking, original. Roger Cotes in [27] sought the area of an ellipsoid of revolution. When the rotation is about the minor axis there is no trouble, but when the motion is about the major axis we find him writing "Posset hujus etiam superficiei per Logometriam designari, sed modo inexplicabili ... arcus erit rationis inter

$$EX + XC\sqrt{-1} \text{ a}CE \text{ mensura ducta in } \sqrt{-1}."$$

I will leave Euler for a moment to speak of the numerical value of e. William Shanks, who, until quite recently, held the world's record of 707 places for π, had a try at e [28]. Glaisher found an error in this, but Shanks corrected it, and calculated a value which he was sure was right to 205 places. Glaisher verified 137 of them. Boorman [29] calculated e to 346 places. He acknowledged that he and Shanks agreed only up to 187 places. "One is wrong, which one?" Boorman gives the impression of being a rather amateurish mathematician. Adams [30] calculated $\log_{10} e$ to 272 places, probably all correct. Many years ago I knew a youthful teacher of mathematics who had the vaulting ambition to calculate e by long-hand methods to 1,000 places. I lost sight of him over fifty years ago, probably he died early of heart failure.

I return to Euler. In Caput XVIII of *De Fractionibus continuis* [25], he describes methods of expansion into a continued fraction. When it is a question of turning a rational fraction into a continued one, the process is essentially that of finding a highest common factor, and can be done in only one way. Euler writes

$$e = 2.718281828459\cdots,$$

$$\frac{e-1}{2} - 0.8591409142295.$$

He writes this in the form,

$$\frac{e-1}{2} = \cfrac{1}{1+\cfrac{1}{6+\cfrac{1}{10+\cfrac{1}{14+\cfrac{1}{18+\cfrac{1}{etc.}}}}}},$$

and remarks [25], page 388, "Cuius fractio ex Calculo infinitesimali dari potest."

Euler assumes that the quotients will increase by 4 each time, so that the fraction goes on indefinitely. Hence e is not a rational fraction.

As for finding this "ex Calculo infinitesimali" he returns to this very much later in life, "Summatio fractionis continuae cujus indices progressionem arithmeticam constituunt" in Vol. 23 of his Opera mentioned in [25]. The method consists in establishing contact with a Riccati differential equation. For a fuller discussion see [31]. Euler did not complete all the details with modern rigor, but what I have just shown is the first attempt to demonstrate the irrationality of e.

We must wait a whole century for anything really new and startling in this line. This came in 1874 with Hermite's proof that e is not an algebraic number [32], that is, not the root of any equation with integral coefficients. A much simpler demonstration is given by Klein in [33].

References

1. Archimedes, Opera omnia. Third ed. Heiberg, vol. III, 1915, p. 79.
2. G. J. Allman, Greek Geometry. Dublin, 1889.
3. Loria, Le Scienze esatte nella antica Grecia. 2nd Ed. Milan, 1914, p. 155.
4. Heath, The Works of Archimedes. Cambridge, 1897, p. 1.
5. Zeuthen, Die Lehre von den Kegelschnitten. Kopenhagen, 1886, p. 463.
6. Die Kegelschnitte des Gregorius a St. Vincento, Abhandlungen zur Geschichte der mathematische Wissenschaften. Vol. XIX, Part 2, Leipzig, 1907.
7. Abraham Gotthilf Kästner, Geschichte der Mathematik. Vol. 3, Göttingen, 1799.
8. Cantor, Geschichte der Mathematik. Vol. 2, Second Ed., Leipzig, 1900, p. 715.
9. Charles Hutton, Mathematical Tables. London, 1804, p. 80.
10. Wallis, Opera Mathematica. Oxford, 1693, Vol. 2.
11. Philosophical Transactions, Abridged. Vol. I, p. 232.
12. Montucla, Histoire des mathématiques. Vol. 2, Paris, 1800, p. 80.
13. Huygens, Oeuvres Complètes. Vol. 14, La Haye, 1920, pp. 433, 441, and 474.
14. Glaisher, The earliest use of the Radix Method for calculating logarithms, Quarterly Journal of Mathematics, Vol. 46, 1914–15, especially p. 174.
15. Tropfke, Geschichte der Elementarmathematik. 3rd Ed., Vol. 2, Berlin, 1933, p. 205.
16. Wallis, Algebra, Ch. XII, Opera, Vol. 2. Oxford, 1693, pp. 57, 58.
17. A most compendius and facile Method for constructing Logarithms exemplified and demonstrated from the Nature of Numbers, Philosophical Transactions, abridged, Vol. IV. 1695–1702. London, 1809, p. 19.
18. Logarithmo-technica Auctore Nicolao Mercatore. See Mazeres, Scriptores Logarithmici, Vol. I. London, 1791, p. 169.
19. Wallis, Logarithmo-technica Nicola Mercatoris, Philosophical Transactions, August, 1668. Mazeres cit. in 18, p. 221.
20. Wallis, Arithmetica Infinitorum. Oxford, 1656. Especially Prop. 59.
21. Newton Commercium Epistolicum Collinsii et aliorum, published by Biot and Leffort, Paris, 1856.
22. An Account of the Book entitled Commercium Epistolicum Collinsii et aliorum. Anonymously by Newton, Philosophical Transactions, Vol. XXIX, London.
23. Leibnizens Mathematische Schriften. Gerhardt Ed., Part 2, Vol. I. Halle, 1858.
24. *Ibid.*, Part I, Vol. 2.
25. Euler, Introductio in Analysin infinitorum. Lausanne, 1748, also his Opera Omnia Series prima, Opera mathematica, Vol. 8.
26. Euler, Meditatio in Experimenta explosione Opera Postuma. Petropoli, 1862.
27. Cotes, Harmonia Mensurarum. Cambridge, 1722, p. 28.
28. Proceedings of the Royal Society, Vol. 6, 1854.
29. Computation of the Naperian Base. Mathematical Magazine, Vol. I, 1884, p. 204.
30. Shanks, On the Modulus of Common Logarithms. Proceedings of the Royal Society, Vol. 43, 1887.
31. Pringsheim, Ueber die ersten Beweise der Irrationalität von e und Sitzungsberichte der K Akademie der Wissenschaften zu München, Vol. 28, 1898.
32. Charles Hermite, Sur la fonction exponentielle. Paris, 1874; Oeuvres, Vol. III, Paris, 1912.
33. Klein, Ausgewählte Fragen der Elementargeometrie. Leipzig, 1895, pp. 47 ff.

THE HISTORY OF CALCULUS*†

ARTHUR ROSENTHAL, Purdue University

Everyone knows that Newton and Leibniz are the founders of Calculus. Some may think it suffices to know just this one fact. But it is worthwhile, indeed, to go into more details and to study the history of the development of Calculus, in particular, up to the time of Newton and Leibniz.

In our courses on Calculus we usually begin with differentiation and then come later to integration. This is entirely justified, since differentiation is simpler and easier than integration. On the other hand, the historical development starts with integration; computing areas, volumes, or lengths of arcs were the first problems occurring in the history of Calculus. Such problems were discussed by ancient Greek mathematicians, especially by Archimedes, whose outstanding and penetrating achievements mark the peak of all ancient mathematics and also the very beginning of the theory of integration. The method applied by Archimedes for his proofs was the so-called method of exhaustion, that is, in the case of plane areas, the method of inscribed and circumscribed polygons with an increasing number of edges. This method was first rigorously applied, in the form of a double *reductio ad absurdum*, by the great Greek mathematician Eudoxus at the beginning of the fourth century B.C. He first proved the facts, previously stated by Democritus, that the volume of a pyramid equals one third of the corresponding prism and the volume of a cone equals one third of the corresponding cylinder. The same method was also used by Euclid and then with the greatest success by Archimedes (third century B.C.). It is well known that Archimedes was the first to determine the area and the length of the circle, that is, to give suitable approximate values of π, and moreover to determine the volume and the area of the surface of the sphere and of cylinders and cones. But he went far beyond this [1]; he found the areas of ellipses, of parabolic segments, and also of sectors of a spiral, the volumes of segments of the solids of revolution of the second degree, the centroids of segments of a parabola, of a cone, of a segment of the sphere, of right segments of a paraboloid of revolution and of a spheroid. These were amazing achievements, indeed. Archimedes proved his results in the classical manner, by the method of exhaustion. Sometimes the type of approximation is just the same as we would use. For instance, in order to obtain the volume of a solid of revolution of the second degree, Archimedes approximates the volume by a sum of cylindrical slabs. But the direct evaluation of the limit of such sums was cumbersome. Hence we may ask: what was the method used by Archimedes for finding his results?

There is an indication of his method in the beginning of his book on the quadrature of the parabola. But a full explanation of his procedure was given

---

* From AMERICAN MATHEMATICAL MONTHLY, vol. 58 (1951), pp. 75–86.

† Based on two addresses given at the Mathematics Club of Purdue University, June 2, 1949, and at the meeting of the Indiana Academy of Science, Crawfordsville, Ind., November 4, 1949.

by him in a work rediscovered as late as 1906. It is his *Method Concerning Mechanical Theorems, dedicated to Eratosthenes*, known as Archimedes' *Method* or ἔφοδος [2]. This manuscript was found in Istanbul as a so-called palimpsest. That is to say, in the 10th century A.D. the manuscript of the *Method* was written on this parchment; later, in the 13th century, since nobody there was interested in it any longer or could even understand it, the *Method* was washed off and a religious text of the orthodox church, a so-called euchologion, was written on the parchment. Fortunately most of the Archimedean text could be restored.

The method of Archimedes may be called a mechanical infinitesimal method. We must remember that Archimedes was also the founder of statics, and of hydrostatics too. Now his method of integration consists in an application of the principle of the lever to elementary parts of the figure. As an example, we shall see how Archimedes determines the area of a segment of a parabola; this indeed is the first example given in his ἔφοδος.

Let $\widehat{AB}$ be the given segment of the parabola, let R be the midpoint of the chord AB, and draw the diameter d (parallel to the axis) through R, intersecting the arc AB in U (cf. Figure 1). We draw the tangent of the parabola at B and the parallel to d through A. So we obtain $\triangle ABC$ and wish to compare the area

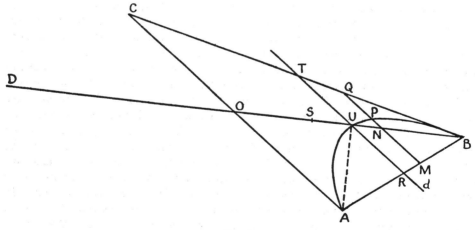

FIG. 1

of the segment $\widehat{AB}$ with the area of this triangle. Let T be the intersection of the tangent BC with the diameter d. From an elementary property of the parabola we obtain: $UT = UR$. Hence the line BU intersects AC in its midpoint O. The centroid S of the $\triangle ABC$ lies on OB with $OS = \frac{1}{3}OB$. Make $OD = OB$. Now through any point P of the arc AUB draw a parallel to the diameter, $MNPQ$. Archimedes proves as a property of the parabola that $MP:MQ = AM:AB$, hence $MP:MQ = ON:OB = ON:OD$. Therefore $MP \times OD = MQ \times ON$. That is, according to the principle of the lever: If DOB is considered to be the bar of a balance with the fulcrum in O (cf. Figure 2), then the line-segment PM suspended at D is in equilibrium with the line-segment MQ which remains, without any change, suspended at N. Since this holds for every line parallel to d, it follows that the total segment $\widehat{AB}$ of the parabola suspended at D is in equilibrium with the $\triangle ABC$ remaining unchanged or, what amounts to the same thing, the segment $\widehat{AB}$ suspended at D is in equilibrium with the $\triangle ABC$ suspended at its centroid S. Therefore, since $OS = \frac{1}{3}OD$, the segment $\widehat{AB} = \frac{1}{3} \triangle ABC$. Moreover, since $RU = \frac{1}{2}RT = \frac{1}{4}AC$, we have $\triangle ABC = 4 \triangle AUB$, and hence also segment $\widehat{AUB} = \frac{4}{3} \triangle AUB$.

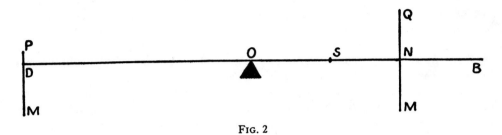

Fig. 2

By this ingenious method of statics, Archimedes found the final result. But he did not regard such reasoning as a proof. Having thus obtained the result, he was then able to give a rigorous formal proof by the method of exhaustion.

One should expect that the wonderful achievements of Archimedes would have become a great stimulus to the further development of Greek mathematics, similar to the great influence of Newton and Leibniz on the mathematical production of succeeding generations. But it is surprising that Archimedes found almost no successors to continue his work. In this connection only one of the subsequent mathematicians is to be mentioned, namely, Dionysodorus who found the volume of the torus. Of course, one has to remember that at the time of Archimedes there lived another outstanding Greek mathematician, Apollonius, about 25 years younger than Archimedes. Apollonius, in a masterly way, completed the Greek theory of conic sections. It is very strange, and I do not understand the reason for it, that soon after Archimedes and Apollonius Greek mathematics declined and that the later development essentially took a different direction. Under the influence of the needs of astronomy, a new branch of mathematics (the roots of which, however, go back also to Archimedes), namely trigonometry, was established and furthermore, much later, the theory of numbers was developed by the work of Diophantus. Original contributions in the direction of Archimedes' work were finally made by one of the latest Greek mathematicians, Pappus (end of the third century A.D.) who stated the important general theorems named after him, in particular, the theorem that the volume of a body of revolution equals the area of the revolving plane figure times the length of the path of the centroid of this area.

When Teutonic tribes, still barbarian at that time, invaded the Roman empire and conquered it, the interest in mathematics almost vanished there; mathematics receded to the Orient, to Byzantium, where at least valuable manuscripts were preserved, to Persia, and afterwards to the Arabian countries, where—on the basis of Greek tradition—mathematics flourished in the period about 800–1200 A.D. One of these mathematicians, the Mesopotamian Ibn Al Haitham (about 1000 A.D.), was able to compute the volume of a solid that is generated by rotation of a segment of a parabola about a line perpendicular to its axis.

Under the influence of the Orient, interest in mathematics was slowly awakened in Europe, in particular in the 12th and 13th centuries. As early as the 16th century great discoveries in algebra were made by Italian mathematicians, namely the solution of the algebraic equations of the third and fourth degree. Simultaneously Archimedes' works were studied and understood again.

Then about the beginning of the 17th century the further development of the ideas of Archimedes starts. This was the same great period in which modern science was first established by Galileo. The Flemish engineer Simon Stevin (as early as 1586) and the Italian mathematician Luca Valerio (1604) were the first ones who, by direct passage to the limit, tended to avoid the double *reductio ad absurdum* of the method of exhaustion. Valerio showed directly that the areas under certain curves can be approximated by sums of circumscribed and inscribed rectangles, whose difference can be made arbitrarily small.

Then, in particular, we have to mention the great German astronomer Johannes Kepler who, in 1615, published a book, *Nova stereometria doliorum vinariorum*, on determining the volumes of wine-casks. Somewhat earlier there had been a year of plenty and there was need of barrels for storing the great supply of wine; moreover, Kepler was puzzled by the rules which dealers applied to estimate the approximate contents of a barrel. So he discussed in a popular manner the volumes of various casks and, in particular, asked which cask has the most economic shape. He found that the Austrian barrel was the most economic one. Kepler used the results and methods of Archimedes, but also discussed quite a few new cases. Because of his popular purpose he replaced the rigorous proofs of Archimedes by an intuitive infinitesimal reasoning, in this way stressing the essential points.

Another mathematician of that time had a great influence on further progress; this was the Italian Bonaventura Cavalieri who published in 1635 an important book on the so-called indivisibles, entitled *Geometria indivisibilibus continuorum nova quadam ratione promota*. Indivisibles mean elements of a given dimension which by their motion generate figures of the next higher dimension. Thus a moving point generates a line, a moving line (parallel to a fixed line) generates a plane figure, a moving plane figure (parallel to a fixed plane) generates a solid. Cavalieri speaks, for instance, about "all lines of a plane figure" ("omnes lineae figurae"). Well known is Cavalieri's principle: Two solids (lying between two parallel planes) have the same volume if they intersect each intermediate parallel plane in two equal areas. Cavalieri's views, influenced by late medieval speculations, have somewhat of the spirit of Archimedes' Method, which, however, was not known at that time.

In connection with Cavalieri we must mention also the Swiss Paul Guldin who, besides criticizing Cavalieri, rediscovered Pappus' theorems on bodies of revolution, the Flemish mathematician Gregorius a St. Vincentio who was the first to observe (1647) that the area between a hyperbola and an asymptote behaves like a logarithm, and also the Italian mathematician and physicist Evangelista Torricelli [3] and the French mathematician Gil Persone de Roberval [4]. The important achievements of these last two men will be discussed presently.

About this time another outstanding event occurred in mathematics, the invention of Analytic Geometry by Descartes (1637) and, simultaneously and independently, by Fermat; this invention, of course, had great influence on the development of Calculus. Both Descartes and, in particular, Fermat also made valuable direct contributions to Calculus.

René Descartes, in his *Géométrie*, gave a method of finding the tangents, or rather the normals, to algebraic curves. He draws a circle with center on the x-axis, which cuts the given curve in two points. If these two points coincide, he obtains the normal. Hence the question is reduced to determining double roots of an algebraic equation. Somewhat later, in a letter, Descartes remarked that, instead of circles, intersecting straight lines could also be used for the same purpose.

Fermat's achievements in Calculus were even more important. In fact, he was the greatest mathematician of the first part of the 17th century, not only in general but particularly in the domain of Calculus. Pierre Fermat was a jurist, a councillor of the parliament at Toulouse in southern France. This position left him enough time for intensive mathematical activity. His outstanding work in the theory of numbers is well known. Now, what was *his* method of finding tangents? His procedure was first applied by him to the particular case of determining maxima and minima [5(a)]. He found this method as early as in 1629, communicated it to Descartes in 1638, and had it published in 1642. In order to find the maximum or minimum of an expression, one replaces the unknown A by $A+E$,* and both expressions obtained in this manner are considered approximately equal. One must cancel on both sides all that is possible to cancel. In this way only terms containing E are left. Now divide by E and then drop all terms still containing E. There remains an equation giving that value of A which yields the desired maximum or minimum. That means, if we write $F(A)$ for the given expression, we have to determine A from the equation

$$\left[\frac{F(A+E)-F(A)}{E}\right]_{E=0}=0.$$

---

* Fermat always used the letter A for the variable and the letter E for its increment.

This is just our usual method. Of course, the condition is only necessary, but not sufficient for the extreme, and the statement of Fermat yields the result only for polynomials F.

Fermat [5(a)] gave at the same time a general method for finding the tangent, in the form of determining the subtangent. Let PT (with T lying on the the x-axis) be the tangent line of the given curve $\mathfrak{C}$ at the point P (cf. Figure 3), let P_1 be a point of $\mathfrak{C}$ in the neighborhood of P, let Q and Q_1 be the projections

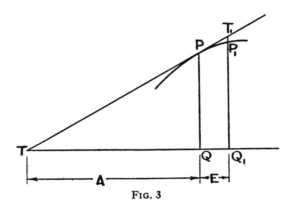

FIG. 3

on the x-axis of P and P_1, respectively, and let T_1 be the point on the tangent line whose projection on the x-axis is Q_1. In order to find the subtangent $A(=TQ)$, whose increment QQ_1 is again designated by E, Fermat uses the similarity of the triangles TQP and TQ_1T_1 and replaces T_1 approximately by P_1. Then he obtains approximately: $A:QP=E:(Q_1P_1-QP)$, that is, in our usual notation if we write the equation of the curve $\mathfrak{C}$ in the form $y=F(x)$,

$$A:F(x) = E:(F(x+E)-F(x));$$

hence
$$A = \frac{F(x)\cdot E}{F(x+E)-F(x)}.$$

Now again one divides the denominator by E and afterwards sets $E=0$.

It should be mentioned here that at this time an entirely different method of constructing tangents of curves, using the parallelogram of velocities, was invented also by Roberval and Torricelli, independently of each other; both of them published it in 1644. On the other hand, somewhat later (before 1659) two Netherlanders Johannes Hudde, for many years mayor of Amsterdam, and René François de Sluse advanced along the road opened up by Descartes and Fermat, giving quite explicit formal rules for finding extremes and subtangents of algebraic curves.

But let us return to Fermat. He had great success also in the theory of integration. He was the first who, by 1636 or earlier, had found and proved the power formula of integration for positive integral exponents n, i.e. a geometrical statement equivalent to the formula which we now write as

$$\int_0^a x^n dx = \frac{a^{n+1}}{n+1}.$$

Roberval also, at the suggestion of Fermat, then found and proved the same theorem. Afterwards Cavalieri discovered it independently,* and he was the first to publish it (1639, 1647); but he proved it explicitly only for the first few cases, including $n=4$, while, as he stated, the general proof which he published was communicated to him by a French mathematician Beaugrand, who quite probably had obtained it from Fermat. At that time Fermat, and then also Torricelli, had already generalized this power formula to rational exponents $n(\neq -1)$. Fermat determined areas under curves which he called "general parabolas" and "general hyperbolas," that is, curves

$$y^m = cx^n, \text{ which, in our notation, leads to } c\int_0^a x^{n/m} dx,$$

and

$$y = \frac{c}{x^m}, \text{ which, in our notation, leads to } c\int_a^{+\infty} \frac{dx}{x^m} \quad (m > 1).$$

It is remarkable that Fermat in this work does not use subdivisions into equal parts, but subdivisions according to a geometric progression. Other areas were reduced by him to areas under such general parabolas and hyperbolas.

Fermat did interesting work also on the rectification of curves, in a memoir published 1660 [5(b)], where he approximated the arc by segments of tangents, thus using a saw-like figure. At that time various mathematicians obtained rectifications of curves which are now considered classical. In 1645, Torricelli had rectified the logarithmic spiral. The semicubical parabola was rectified independently by the Englishman William Neil (1657), by the Hollander Hendrik van Heuraet, and by Fermat. The rectification of the cycloid was first achieved by the English mathematician and great architect Christopher Wren (1658), and then by Fermat and Roberval, after they had heard of his result. It is noteworthy that Fermat, Neil, van Heuraet, and also Wallis and Huygens, reduced rectifications of curves to the determination of areas of other curves.

Reviewing the achievements of Fermat we see that he was aware of the relation among various problems in differential calculus, and similarly for various problems about definite integrals. But he had not observed the general relation between differentiation and integration.

Another famous French mathematician of that time was Blaise Pascal (a younger friend of Fermat and Roberval), who may be considered a master of

---

* Much later, Blaise Pascal (1654) and Wallis (1656) rediscovered it also.

integration. Roberval was the first to integrate certain trigonometric functions. Pascal was able to integrate more such trigonometric functions as well as some algebraic functions. As an important means for some of his results Pascal used relations between integrals obtained by interchanging the order of integration in double integrals. Of course, he did this in a geometric form. That is to say, certain volumes were found by means of intersections parallel to one plane and also by intersections parallel to another plane. Since the volume is always the same, Pascal thus obtained a relation between two different integrals, which may be considered to be a kind of integration by parts.

At about the same time important contributions were made by the English mathematician John Wallis, whose book *Arithmetica infinitorum* of 1656 was written in contrast to the geometric work of most of his predecessors. He stressed the notion of the limit. On the other hand, he was very audacious with regard to generalizations and interpolations, but his strong power of intuition kept him on the right track. For instance, he stated the general power formula of integration for any real exponent $n(\neq -1)$.

The notion of limit was carefully considered at that time by the Italian Pietro Mengoli in his book *Geometria speciosa* (1659) [6(a)]. In particular, by modifying the procedure of Luca Valerio (see above), he gave in a precise way a representation of the area under certain special curves as limits of sums of rectangles [6(b)]. Later the same method was also employed by Newton [11, book 1, lemma 2].

One should now mention two other mathematicians who, like Wallis and Mengoli, were contemporary with Newton and Leibniz, but began their work earlier, so that they too are to be considered, at least partially, as predecessors of Newton and Leibniz. One of these men is the great Dutch physicist and mathematician Christiaan Huygens who, among other important results, introduced the notion of evolutes and involutes. It is remarkable that Huygens used to great extent the classical methods of Archimedes, and only for differentiation employed Fermat's method.

The second of these two mathematicians is the Scotsman James Gregory, who like Torricelli and Pascal died in the prime of life, when only 37 years old, and whose genius has found its full recognition only recently [7, 8]. He did excellent work in integration; for instance, in 1668 he published such a difficult result as the following (written in modern notation):

$$\int_0^a \sec x\, dx = -\log(\sec a - \tan a)$$

and other trigonometric integrals. Moreover, for example, he obtained Newton's interpolation formula, independently of Newton. But perhaps the most important achievements of Gregory belong to the theory of series.

The first great result in the theory of series is due to the German mathematician Nicolaus Mercator (1668), who found the logarithmic series. For this purpose, Mercator used term by term integration of a geometric series. This method was independently discovered, but not published, by Newton. Subsequent contributions were made by William Lord Brouncker, the first president of the Royal Society in London. Then the most outstanding results concerning infinite series were obtained by Newton and Gregory, who worked essentially independent of each other, though Gregory was influenced by the knowledge of some of Newton's statements, but not of his methods. Both discovered the binomial series and also many series for trigonometric and inverse trigonometric functions. In particular, Gregory found the series for arctan x; a special case of it is the series $\pi/4 = 1 - \frac{1}{3} + \frac{1}{5} - \frac{1}{7} + \cdots$, which was later found independently by Leibniz. Gregory obtained some more complicated series, *e.g.*, for tan x, sec x, log sec x, using processes of differentiation for determining the coefficients of these series, thus anticipating Brook Taylor by more than forty years. This fact has recently been demonstrated by H. W. Turnbull who studied and interpreted notes in Gregory's handwriting (*cf.* [7], pp. 168–176, 350–359). It is obvious that many of the contributions to the theory of series were closely connected with the growth of Calculus.

From all that has been discussed so far we have seen that there certainly was an extensive development of the theory of integration and differentiation in the period immediately before Newton and Leibniz, and that many mathematicans of various nations made great contributions. So we shall ask: What then was missing at that time? One very important point still missing was the general fact that differentiation and integration are inverse processes, that is, the so-called fundamental theorem of integral calculus. It is true that a few mathematicians had already come quite close to this knowledge. First in this connection we must mention Torricelli. In manuscripts left at his death (1647) and published as late as 1919 [cf. 3], he had obtained the distance $s(t)$ of a moving point by means of the quadrature of the velocity $v(t)$, while by the construction of the tangent of the curve $s(t)$ he could recover $v(t)$. It is very doubtful, however, whether he really conceived the significance of this relation. Secondly, as we have already stated, some of the mathematicians of the time knew that the rectification of one curve $\mathfrak{C}_1$ could be obtained by means of the area under another curve $\mathfrak{C}_2$. Fermat in his memoir [5(b)], published in 1660, found the relation between the slopes of these two curves. Moreover, James Gregory, in his book *Geometriae pars universalis* (1668), solved the following problem, which is inverse to the rectification of $\mathfrak{C}_1$, just mentioned: Given a certain curve $\mathfrak{C}_2$, find another curve $\mathfrak{C}_1$ whose length equals the area under $\mathfrak{C}_2$. In the solution of this problem Gregory used the quadrature of an auxiliary curve, which indeed corresponds to obtaining the primitive function of a derivative by means of integration [9]. In spite of this, one again may doubt whether, at that time, he perceived in general that differentiation and integration are inverse operations. The first who made this important discovery in full generality was Isaac Barrow, the teacher of Newton at Cambridge University. Barrow first was professor of

Greek language at Cambridge, then he was professor of mathematics first in London and then again at Cambridge. In 1669 he resigned his chair to his pupil Newton, whose superior genius he had recognized. Barrow then devoted the rest of his life to theology. His principal mathematical work was his *Lectiones Geometricae*, published in 1670, together with the second edition of his *Lectiones Opticae*. These geometrical lectures [10] contained important systematic contributions to the theory of differentiation and integration, almost all in purely geometric form. Here, for the first time, the inverse character of differentiation and integration was explicitly stated and proved.

At this remarkable point of the development of the theory we must again ask the same question as above: What more remained to be done? The answer is: What had to be created was just the *Calculus*, a general symbolic and systematic method of analytic operations, to be performed by strictly formal rules, independent of the geometric meaning. Now it is just this Calculus which was established by Newton and Leibniz, independent of each other and using different types of symbolism. Newton's first discoveries were made about ten years before those of Leibniz; on the other hand, Leibniz' publications preceded those of Newton, and—what is more important—Leibniz' symbolism, the same as that used by all mathematicians at present, is superior to that of Newton.

Isaac Newton, influenced by his teacher Barrow and also by the work of Wallis, started his "method of fluxions" in the years 1665–1666, his most creative period, at the age of 23 years. Some of his early manuscripts were known to friends of his and indications of his method were contained in some of his letters. He wrote his *Methodus fluxionum et serierum infinitarum* in 1670–1671; but it was not published until 1736, nine years after his death. In his profound work *Philosophiae naturalis principia mathematica* (1687) Newton avoided his method of fluxions, except for a few indications [11, book 2, lemma 2], and presented his great discoveries in the classical geometric form, though simplified by using the notion of limits. The first publication of both the method and the notation of Newton is found as late as 1693 in Wallis' works, where Wallis included two of Newton's letters to him. Newton himself then published an account of his method, entitled *Tractatus de quadratura curvarum*,* as an appendix to his *Optics* in 1704.

Newton, considering motions, took the time t as the independent variable, called the dependent variable x "fluent" and its velocity "fluxion," and wrote $\dot{x}$ for the fluxion, i.e. for the derivative with respect to t. The higher derivatives were then designated by $\ddot{x}$, $\dddot{x}$, etc. For the increment of the independent variable t Newton used the letter o and called $\dot{x}o$ (i.e., the differential of x) the "moment" of x. In the case of the inverse process, that is, if the variable x is given as a fluxion, he first designated the fluent (i.e., the antiderivative of x) by $\square\, x$ or $\boxed{x}$, later by $\acute{x}$, and used then for iterated integration $\acute{\acute{x}}$, $\acute{\acute{\acute{x}}}$, etc. It has to be stressed that Newton was the first to use systematically the results of differenti-

---

* Most of it was already written in 1676.

ation in order to obtain antiderivatives, and hence to evaluate integrals.

Our notations, now used generally in Calculus, are due to Gottfried Wilhelm Leibniz. He was a universal spirit, extremely versatile and interested in every kind of knowledge and scholarship, perhaps most famous as a philosopher. He started as a jurist, was soon active in diplomacy, then became librarian and historian at Hanover, and later (1700) founded the Berlin Academy of Sciences. In mathematics, his first publication, as a graduate student, concerned combinations and permutations; then he soon became interested in the theory of differences and in constructing a computing machine. At the suggestion of Huygens (1673), Leibniz thoroughly studied the works of previous mathematicians on integration and differentiation. In particular, he was much influenced by the work of Pascal. The story of Leibniz' own discoveries can be traced in all details, since his manuscripts with his dated sketches were found at the Hanover library (edited and published by C. I. Gerhardt in the middle of the 19th century) [12, 13, 14, 14a]. Leibniz' new notation was first introduced by him on October 29, 1675 [12, pp. 121–127]. On this day, as on preceding days, Leibniz discussed integrations using Cavalieri's "omnes lineae." Here he abbreviated "omnes" or "omnia" to "omn." and applied quite a few formal operations to this symbol. Then he remarked: "It will be useful to write $\int$ for omn., thus $\int l$ for omn. l, that is, the sum of those l's." Hence $\int$ is derived from the first letter of the word *summa*. Later in the same manuscript, he came to the "contrary calculus" and continued thus: "If $\int l \sqcap ya$" ($\sqcap$ is the equal sign of Leibniz), "then let us set $l \sqcap ya/d$. Certainly as $\int$ will increase the dimensions, so d will diminish them. $\int$, however, designates a sum, d a difference." Hence Leibniz first wrote the differential sign d into the denominator of the variable. But two or three weeks later he writes dx, dy, dx/dy, and the integrals $\int ydy$ or $\int ydx$. So he arrived at the notation, now classical.

In 1684, Leibniz first published his differential calculus in a paper (issued in the newly founded *Acta Eruditorum*) with the title *Nova methodus pro maximis, itemque tangentibus,* · · · . Here he deals with differentials, and it is noteworthy that he introduces dx as an arbitrary finite interval and then defines dy by the proportion $dy:dx=$ ordinate : subtangent. Then, in 1686, he published also a paper containing his notation of the integral. (The word "integral" was introduced by Jakob Bernoulli, 1690.)

Later Leibniz was accused by friends and followers of Newton of having plagiarized Newton's ideas. This unfortunate and undignified controversy started first in 1699 and became continuously more and more furious, so that Leibniz' last years were filled with bitterness. But considering Leibniz' manuscripts, nobody can doubt at present that Newton and Leibniz founded their Calculus independently.

The invention of Calculus stimulated an immense and energetic further development. On the English side Taylor and Maclaurin, on the continent the eminent Basel mathematicians, first the brothers Bernoulli, then the prodigious Euler, and later the Frenchman D'Alembert and the Italian Lagrange contributed greatly to this development. Moreover, other closely related parts of mathematics soon originated in connection with the Calculus, e.g., the theory of differential equations, the calculus of variations, differential geometry.

While Newton and Leibniz had rather reasonable (although not always consistent) ideas about the fundamentals of the new Calculus, the extremely rapid further development caused the basic concepts to be neglected or to be treated in a very unsatisfactory manner. In particular, Euler is an example of this tendency. A few mathematicians, among them D'Alembert, stressed the necessity of using the notion of the limit as foundation of the Calculus [15]. But it was Cauchy in the beginning of the 19th century who, in such a way, developed the Calculus systematically and consistently. So at last Cauchy and his many successors gave a solid basis to the Calculus.

References

1. T. L. Heath, The Works of Archimedes, Cambridge 1897.
2. The Method of Archimedes, edited by T. L. Heath, Cambridge 1912.
3. E. Bortolotti, I progressi del metodo infinitesimale nell' Opera geometrica di Evangelista Torricelli, Periodico di Matematiche (4) 8 (1928), pp. 19–59.
4. Evelyn Walker, A Study of the Traité des Indivisibles of Gilles Persone de Roberval, Teachers College, Columbia University, New York 1932.
5. Œuvres de Fermat, 4 volumes and supplements, Paris 1891–1922. (a) Methodus ad disquirendam maximam et minimam, Œuvres, vol. 1, pp. 133–136; see also pp. 136–179 (French translation: vol. 3, pp. 121–123 and 124–156); (b) De linearum curvarum cum lineis rectis comparatione, Œuvres, vol. 1, pp. 211–254 (French translation: vol. 3, pp. 181–215).
6. A. Agostini, Peridico di Matematiche (4) 5 (1925). (a) La teoria dei limiti in Pietro Mengoli, pp. 18–30; (b) Il concetto d'integrale definito in Pietro Mengoli, pp. 137–146.
7. James Gregory. Tercentenary Memorial Volume, edited by H. W. Turnbull, London 1939.
8. M. Dehn and E. D. Hellinger, Certain mathematical achievements of James Gregory, Am. Math. Monthly 50 (1943), pp. 149–163.
9. A. Prag, On James Gregory's Geometriae pars universalis, in [7], pp. 487–505.
10. J. M. Child, The Geometrical Lectures of Isaac Barrow, Chicago and London, 1916.
11. Sir Isaac Newton's Mathematical Principles of Natural Philosophy and His System of the World. Translated by A. Motte, revised by F. Cajori, University of California Press, Berkeley, Calif., 1946.
12. C. I. Gerhardt, Die Entdeckung der höheren Analysis, Halle 1855.
13. J. M. Child, The Early Mathematical Manuscripts of Leibniz, Chicago & London 1920.
14. D. Mahnke, Neue Einblicke in die Entdeckungsgeschichte der höheren Analysis, Abhandlungen der Preuss. Akad. der Wissenschaften, Phys.-Math. Klasse, 1925, No. 1.
14a. J. E. Hofmann, Die Entwicklungsgeschichte der Leibnizschen Mathematik während seines Aufenthalts in Paris (1672–1676), Munich 1949.
15. As for the long lasting struggle for a rigorous foundation of the Calculus see C. B. Boyer, The Concepts of the Calculus, New York 1939, 2nd printing 1949.

THE FOREMOST TEXTBOOK OF MODERN TIMES*†

C. B. BOYER, Brooklyn College

The most influential mathematics textbook of ancient times (or, for that matter, of all times) is easily named. The *Elements* of Euclid, appearing in over a thousand editions [1], has set the pattern for the teaching of elementary geometry ever since it was composed more than two and a quarter millenia ago. The medieval textbook which most strongly influenced mathematical development is not so easily selected; but a good case can be made out for *Al jabr wa'l muquabala* of Al-Khowarizmi [2], just about half as old as the *Elements*. From this Arabic work algebra took its name and, to a great extent, its origin. Is it possible to indicate a modern textbook of comparable influence and prestige? Some would mention the *Géométrie* of Descartes or the *Principia* of Newton or the *Disquisitiones* of Gauss; but in pedagogical significance these classics fell far short of a work less well known. The *Géométrie* was not strictly a textbook, and hence most mathematicians learned their analytic geometry from the works of other authors, such as Schooten, De Witt, Sluze, and Lahire. The *Principia*, the greatest of all works in the field of science, affected the course of pure mathematics only indirectly; few readers understood the elements of the calculus which it contained, and the effective teachers of the differential calculus were Leibniz, L'Hospital, and the Bernoullis. The *Disquisitiones*, a work of great profundity, was too specialized to make its influence widely felt except among ardent number-theorists. It is perhaps significant that none of these three modern works appeared in what may be considered the greatest age of textbooks in recent times. The eighteenth century often is characterized as a prosy age in the history of mathematics, for it contributed no single discovery which captured the imagination as had analytic geometry and the calculus. And yet the century was of capital importance in the consolidation of earlier work, a task which was facilitated by the appearance of outstanding textbook writers. At the opening of the century one finds the texts of L'Hospital dominating the fields of analytical conics and the calculus; at the close there were the textbooks of Lacroix which covered the whole elementary field and which appeared in dozens of editions [3], not to mention the Legendre *Euclid* of the same time. But over these well known textbooks there towers another, a work which appeared in the very middle of the great textbook age and to which virtually all later writers admitted indebtedness. This was the *Introductio in analysin infinitorum* of Euler, published in two volumes in 1748 [4]. Here in effect Euler accomplished for analysis what Euclid and Al-Khowarizmi had done for synthetic geometry and elementary algebra, respectively. The function concept and infinite processes had arisen by the seventeenth century, yet it was

---

* From AMERICAN MATHEMATICAL MONTHLY, vol. 58 (1951), pp. 223–226.

†Presented at the International Congress of Mathematicians, Cambridge, Mass., Sept. 1, 1950.

Euler's *Introductio* which fashioned these into the third member of the mathematical triumvirate comprising geometry, algebra, and analysis. From the point of view of leading textbooks, then, one might refer (with, of course, some oversimplification) to geometry as ancient, algebra as medieval, and analysis as modern.

Euler was not the first to use the word "analysis," or even to incorporate it into the title of a book; but he did give the word a new emphasis. Plato's analysis had reference to the logical order of steps in geometrical reasoning, and the analytic art of Viète was akin to our algebra; but the analysis of Euler comes close to the modern orthodox discipline, the study of functions by means of infinite processes, especially through infinite series. It is in this newer sense that Euler, especially after 1748, used the word; and it is for this reason that he has been referred to as "analysis incarnate" [5]. The word, analysis, took on a new lease of life. Euler himself used it in the titles of dozens of his published papers; and soon others were publishing books on "analytic optics" and "analytical mechanics," on "analytical trigonometry" and "analytic geometry." Euler avoided the phrase "analytic geometry," probably to obviate confusion with the older Platonic usage; yet the second volume of the *Introductio* has been referred to, not inappropriately, as "the first text on analytic geometry" [6], and this appeared more than a century after the publication of *La géométrie*! It is the earliest systematic graphical study of functions, not only of a single variable, but of two as well; and Euler's analysis is transcendental as well as algebraic. The notion of "elementary function" stems largely from the *Introductio*, but the book contains also such higher curves as $y = x^{\sqrt{2}}$ and $y^x = x^y$. Polar coordinates had appeared in at least half a dozen earlier works, including Newton's *Method of fluxions*; but the clarity and generality of Euler's treatment in the *Introductio* were such that most subsequent writers traced the use of polar coordinates back to this book. Here, in fact, the spiral of Archimedes appears in its dual form, probably for the first time. The use of parametric equations, implicit even in the work of Descartes, was first systemized in the *Introductio*; and here also one finds formalized for the first time the equations for the transformation of coordinates for two and three dimensions, the latter in a form still referred to as "Euler's equations." The *Introductio* was the first textbook to recognize the five proper general quadric surfaces as members of a single family, a century and a half after Kepler had done the same for the conics, and the names proposed were very similar to those now adopted [7]. In this same book Euler also did for plane quartic curves what Newton had done for cubics—he ordered them according to genus and species.

The word "first" is a hazardous one to use in the history of mathematics, and yet it has been applied freely to Euler's *Introductio*. The cases already cited by no means exhaust the respects in which this textbook was first. It contains the earliest algorithmic treatment of logarithms as exponents and of the trigonometric functions as numerical ratios. It was the first textbook to list systematically the multiple-angle formulas, calling attention to the periodicities of the functions; and it included the first general analytic treatment of these as infinite products, as well as their expansion into infinite series. The well-known "Euler identities," relating the trigonometric functions to imaginary exponentials, are also found here. The first volume contains as well an exposition of continued fractions and some excellent work of the zeta function and number theory [8].

In scope alone the *Introductio* ranks among the greatest of textbooks, for it is doubtful that any other essentially didactic work includes as large a portion of original material which survives in the college courses of today. Yet the book is outstanding also for its pedagogical lucidity. The immortal Gauss, a man not given to exaggerated expressions of flattery, held that "The study of Euler's works will remain the best school for the various fields of mathematics, and nothing can replace it" [9]. Written as it was more than two hundred years ago (a letter from Euler to D'Alembert indicates that it was completed by 1745), the *Introductio in analysin infinitorum* nevertheless can be read with comparative ease by the modern student—unlike the *Géométrie*, the *Principia*, or the *Disquisitiones*. Not only is the viewpoint quite similar to that of today; even the terminology and notation are almost modern—or perhaps, as Struik 1 is well written, "we should better say that our notation is almost Euler!" [10]. Under the circumstances one should expect that a textbook exhibiting the qualities of the *Introductio* would boast an impressive list of editions and translations; but the facts belie this. The work was not reprinted until the time of the author's death, thirty-five years later; and, including reprintings and incomplete translations, the fewer than a dozen editions are about equally distributed among the three languages Latin, French, and German [11]. No English translation has appeared, and a partial Russian translation apparently has not been published. However, that the worth of a textbook is not necessarily measured by the number of its editions is conclusively evidenced by the *Introductio*, the influence of which was unusually pervasive. Almost without exception the authors of the ubiquitous compendia of the second half of the eighteenth century refer to Euler as the source of their analysis. The *Introductio* became, in a sense, the prototype of modern textbooks. Is not imitation the sincerest form of flattery?

References

1. An excellent English edition is *The thirteen books of the Elements*, translated from the text of Heiberg with introduction and commentary by T. L. Heath, 3 vols., Cambridge, 1908.

2. See *Robert of Chester's Latin translation of the Algebra of Al-Khowarizmi*, with an introduction, critical notes, and an English version by L. C. Karpinski, New York, 1915.

3. In 1848, for example, there appeared at Paris the 20th edition of his *Traité élémentaire d'arithmétique* and the 16th edition of his *Éléments de géométrie*; and ten years later his *Éléments d'algèbre* appeared in a 20th edition.

4. Published at Lausanne.

5. See, e.g., E. T. Bell, *Men of mathematics* (New York, 1937), chapter 9.

6. D. J. Struik, *A concise history of mathematics* (2 vols., New York, 1948), II, 169.

7. His names were elliptoides, elliptico-hyperbolicae, hyperbolico-hyperbolicae, elliptico-parabolicae, and parabolico-hyperbolicae.

8. An extensive account of this work is found in the preface, by Andreas Speiser, to series 1, volume 9, of Euler's *Opera omnia* (Geneva, 1945).

9. *Ibid.*, p. viii. Speiser adds to the words of Gauss: "The *Introductio* in this connection may stand in first place."

10. Struik, *op. cit.*, II, 174.

11. Latin editions appeared in 1748, 1783, 1797, and in volumes 9 and 10 of Euler's *Opera*, series 1; French editions were published in 1785, 1796, and 1835; German editions appeared in 1788, 1835, and 1885. For bibliographic details on Euler's multitudinous works see Gustav Eneström, "Verzeichnis der Schriften Leonhard Eulers," *Jahresbericht der Deutschen Mathematiker-Vereinigung, Ergänzungsband* IV, Leipzig, 1910.. Cf. also P. H. Fuss, *Correspondance mathématique et physique de quelques célèbres géomètres du XVIIIème siècle. Précédée d'une notice sur les travaux de Léonard Euler* (2 vols., St. Pétersbourg, 1843); and Felix Müller, "Über bahnbrechende Arbeiten Leonhard Eulers aus der reinen Mathematik," *Abhandlungen zur Geschichte der mathematischen Wissenschaften* XXV (1907), 61–116.

THE UNSATISFACTORY STORY OF CURVATURE

J. L. COOLIDGE, Harvard University

1. The Preliminaries. The fact that plane loci are sometimes straight and sometimes are not would seem to be about as obvious as any property of such figures; it is therefore very surprising that this was not placed in the foreground by early writers who dealt with the matter. Euclid defines a straight line as one which lies evenly with the points on itself, which meant that the end points would completely cover those which lay in between or completely hide the latter points, and countless variations of this idea were evolved by the early writers of geometry, but the dynamic idea that a straight line was generated by a point which always moved in the same direction was not, as far as I know, stressed by anyone. Of course whoever admits direction as a primary idea of geometry is involving himself in a sea of trouble, and this may be the reason for its avoidance, but the total neglect of anything so fundamental is certainly striking.

The Greek writers were familiar with all sorts of specific curved loci, witness Proclus' famous summary of the history of geometry, and he says that Aristotle recognized three kinds of loci "*C'est aussi la raison pour laquelle il-y-a trois mouvements, l'une en direction de la ligne droite, l'autre circulaire, et le troisième mixte*" [1]. Proclus himself expands this idea in [1] p. 234, "*Tandis que d'autres ont affectuès la section au moyen des quadratices de Hippias et de Nicomède lesquels avaient aussi fait usage de lignes mixtes quadratices.*" The Greeks were familiar with curved loci, but singularly slow in pointing out the distinguishing characteristics.

The fundamental idea lay very close at hand. Some curves were straight, others were curved. Whatever curvature might be, a circle was everywhere equally curved. The greater the curvature, the less the radius, so it would be natural to take as the curvature a quantity inversely porportional to the radius of the circle, and for any curve, the curvature of the circle lying nearest to it. All this is so simple and natural, but it was strangely slow in coming to be realized. I am sure that a complete account can be found; it certainly is not in [2]. I write in the hope that someone will succeed where I have failed to find it.

The question of curvature is intimately connected with that of the **centre of curvature**, and Apollonius is perfectly aware that from certain points but one normal can be drawn to a conic, and the careful discussion of this question was earnestly pursued, but the writer did not go on to other loci, and it is curious that neither he nor Euclid in discussing optics took the quite obvious steps in this connection. On the contrary, the attention was diverted to the analogous problem of horn angles. A horn or cornicular angle is roughly the figure formed by the circumference of a circle and its tangent; a good discussion by Heath is

---

* From AMERICAN MATHEMATICAL MONTHLY, vol. 59 (1952), pp. 375–379.

found in [3]. We learn by Euclid III, 16 that the angle of the semicircle is greater and the remaining angle is less than any acute rectilineal angle. As long as a secant cuts a circumference twice, the angle which it makes with the curve is less than the horn angle, but the latter, though it can be indefinitely increased, by decreasing the radius of the circle, is nevertheless smaller than a right angle. This is very much contrary to the notion that the two have a definite ratio. Vieta in [4] takes up the question of whether a circumference and its tangent really can be said to make an angle, and points out that in any case this must be a different sort of object from other figures which cannot pass from greater to less without passing through equality. Wallis in [5] insists, in a withering attack on Clavius, that a horn angle is not an angle in Euclid's sense as it is not an inclination between two loci. A new word is needed "*Cui respondeo, mihi cum Clavio hactem convenire (et convenisse semper). Quod quem ille vocat Angulum Contactus, nil aliud est quod ego voco Gradum Curvitatis. Sunt utque curvitatum gradum semper proportionales longitudinibus Diametrorum Chordarium arcuum similibus.*"

I will not go further into the complicated question of horn angles but return to the larger question of definition of curvature. The first writer to give a hint of the definition of curvature was the fourteenth century writer Nicolas Oresme, whose work was called to my attention by Carl Boyer. We find Oresme saying in [6] "*Nunc restat de Curvitate dicendum.*" He assumes the existence of something which he calls *Curvitas*, and if we have two curves touching the same line at the same point, and on the same side, the smaller curve will have the greater curvature. He further states that the curvature of a circle is "*Uniformis*" and on p. 219 "*Sit circulus major cuius semidyameter sit AB et circulus minor cuius semidyameter sit AC, tunc si sit semidyameter AB duplo ad semidyanmeter AC, curvitas minoris circuli erit duplo intensior curvitate majoris, et ita proportionibus et curvitatibus.*" We could not have a clearer proof that Oresme conceived the curvature of a circle as inversely proportional to the radius; how did he find this out?

We apparently have to wait nearly three hundred years before finding anything further on the subject of curvature when we turn to the work of Kepler [7]. He undertakes to find the image of a certain brilliant point, the generalized problem of Al Hazen. He shows the standard approach but adds that instead of dealing with the curve itself it would be wiser to deal with its circle of curvature "*At verior ratio jubet invenire circulum, qui continett rationem curvitatis.*" The ratio he connects with a certain circle, the circle of curvature. His editor, Frischauf in a note on p. 403 to [7] adds "*Primum hic occurit circulus quem dicunt osculatorem, quem posteriores mathematici tum add lineas tum ad superficies curvas maximo commodos adhibebunt, cujus inventio Leibnitio huc usque tributa ets.*" We shall presently see that it was certainly a mistake to associate Leibniz with the invention of the circle of curvature, but we have a glimpse of the fact that the problem of the rightful ascription was not too simple.

A great step in advance is due to Huygens. I mention especially his *Horologium oscillatorium sive de motu pendulorum ad horologia aptato demon strationes geometricae* of 1673 [8]. Kepler starts curiously with the involutes of curves. We take a curve all of whose tangents are on one side. To this is attached a flexible string which is pulled taut and then unwound. The curve from which the string springs is called the evolute. He assumes that the string will always remain tangent to this evolute; the locus of a point fixed in the string will be the involute. He begins with a careful proof that the string will always be normal to the involute. Then comes the curious theorem that two curves with tangents on one side which have a common point cannot have the same set of normals. After this we see that if all tangents to one curve are normals to another they will be an involute and evolute as defined above. He next passes to certain specific curves, notably the cycloid, and in theorem XI he undertakes the problem of finding the radius of curvature of any given curve, defined as the distance up the normal from the foot to the point of contact with the evolute. This is defined as the limit of the intersection with an infinitely near normal. In our notation this would involve Δx and $\Delta [y(dy/dx)]$. He says in regard to this "*Illas vero dari in omnibus curvis geometricis.*" He indicates the method of doing this in general, but his ignorance of the calculus prevented him from attaining a satisfactory result in every case.

2. Sir Isaac Newton. The first writer to handle the question of curvature of a plane curve in what we should call today a thoroughly satisfactory manner was Sir Isaac Newton, no less. Huygens wrote the *Horologium oscillatorium* in 1673. Newton started thinking about the Calculus in around 1665 but published nothing on the subject till his letter to Collins of 1669; his ideas were well developed in 1671. The first systematic account appeared in [9] with the date of 1736. There is a long appendix due to Colson himself, but the bulk of the work purports to be a direct translation of Newton's own Latin. I shall return later to the possible relation of this to the *Horologium* of Huygens.

I begin by quoting Newton's own words, pp. 59–61 of [9]:

"The same Circle has everywhere the same curvature, and in different Circles it is reciprocally proportional to their Diameters.

"If a Circle touches any Curve on its concave side, in any given point, and if it be of such magnitude that no other tangent Circle be inscribed in the angle of contact of that Point, that Circle will be of the same Curvature as the Curve is of, in the Point of Contact.

"Therefore the Centre of Curvature to any Point of the Curve is the Centre of the Circle equally curved, and thus the Radius or Semi-diameter of Curvature is Part of the Perpendicular to the Curve which is terminated at the Centre.

"And the Proportion of Curvature at different Points will be known from the Proportion of Curvature of aequi-curve Circles, or from the reciprocal Proportion of the Radii of Curvature."

A further explanation comes presently:

"But there are several Symptoms or Properties of this Point C which may be used in its Determination:

1) That it is the Concourse of Perpendiculars to that Arc on each side at an infinitely little distance from DC.
2) If DC be conceived to move while it insists perpendicularly to the Curve, that point of it C (if you accept the motion of approaching to, or receding from the point of insistance C) will be the least moved, but will be its Centre of Motion.
3) If a Circle be described with the Centre C, and the Distance DC, no other Circle can be described that can lie between it and the Circle of Contact.
4) Lastly if the Centre H or h of any other touching Circle approaches by degrees to C the Centre of this, till at last it coincides with it, then any of the points in which the Circle shall cut the Curve, will coincide with the Point of Contact at D.
5) And each of these Properties may supply the means of solving this Problem in different ways. But we shall make choice of the first, as being the most simple."

What all this amounts to is the following. Newton assumes, as does Huygens, that if a point not an inflection is fixed on a curve, and a second point approaches it from either side, the intersection of the normals approaches a definite limiting position. This is the centre of curvature, the centre of a circle having the same curvature, and no other tangent circle can lie between this and the curve. All of these properties can be proved if one of them is assumed. He seeks the intersection of nearby normals, and if we take x as an independent variable so that $\dot{x}=1$ and $z=\dot{y}/\dot{x}$ he proves very simply

$$DC = \frac{(1+z^2)^{3/2}}{\dot{z}}.$$

This is very satisfactory and may be said to close the question of the discovery of curvature, but there remains a troublesome question of priority. Cantor in his *Geschichte der Mathematik*, Vol. 3, p. 17, raises the question whether between 1671 and 1736 Newton did not see Huygens' work of 1673 and find therein an excellent opportunity to make use of his own vastly superior methods. Newton made reference to flexible strings and pendulums and the preoccupation of both mathematicians with the cycloid troubles Cantor. On the other hand, it should be noticed that this was a time when a good many geometers were occupied with the cycloid where the determination of the radius was particularly easy. Newton pointed out various methods of finding the radius of curvature, and chooses the best. I cannot feel that we are justified in accusing Newton of plagiarism; it was not in his nature.

There remains the question of Leibniz. In [10] he discusses the angles of mutually tangent curves, and the contact with the *"Circulus osculans"* as if this were an already familiar figure. But the date is late, 1686, and he erroneously says that the osculating circle has four coincident intersections, a mistake promptly pointed out by James Bernoulli.

This represents the limit of my most incomplete knowledge. Where did Oresme get the idea that the curvature of a circle was inversely proportional to the radius? Where did Kepler find the circle of curvature? There are various intersecting questions still to be answered. More power to the persevering man who will answer them.

References

1. Les Commentaires sur le livre des Eléments d'Euclide par Paul Ver Ecke, Bruges, 1948.
2. August Haas, Untersuchung einer Darstellung der Geschichte des Krimmungs masses, Tübingen, 1881.
3. The thirteen books of Euclid's Elements by T. L. Heath. Especially vol. II, pp. 39–43.
4. Vieta, Opera, Ch. XIII, p. 386.
5. Wallis, Opera, II, p. 656, Oxford, 1693.
6. Wieleitner, Ueber den Funktionsbegriff und die graphische Darstellung bei Oresme Bibliotheca mathematica, vol. 14, part 1, Ch. 20, p. 217.
7. Kepler, Opera, Frischauf Ed., vol. 2, p. 175, Frankfort, 1869.
8. Huygens, Ouvre completes, vol. 18. The Hague, 1934.
9. The method of fluxions and infinite series with application to the geometry of curved lines by the inventor Sir Isaac Newton. Translated by Colson, 1736.
10. Meditation nova de nature anguli contactus et osculi Acta Eruditorum, 1686, Schriften (2), vol. 3, p. 326.

THE METHOD OF ARCHIMEDES*†

S. H. GOULD, Williams College

The works of Archimedes have come down to us in two streams of tradition, one of them continuous, the other broken by a gap of a thousand years between the tenth century and the year 1906, when the discovery of a manuscript in Constantinople brought to light an important work called the *Method*, on the subject of integration.

Newton and his contemporaries in the seventeenth century were much puzzled by one aspect of the integrations to be found in the continuous tradition. In the books on the *Sphere and Cylinder*, for example, it is clear that the somewhat complicated method employed there for finding the volume of a sphere represents merely a rigorous proof of the correctness of the result and gives no indication how Archimedes was led to it originally. The discovery of 1906 removes the veil, at least to some extent.

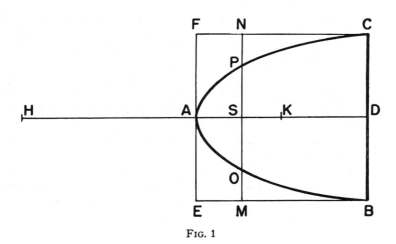

Fig. 1

The newly discovered *Method* consists of imagining the desired volume as cut up into a very large number of thin parallel slices or discs, which are then suspended at one end of an imaginary lever in such a way that they are in equilibrium with a solid whose volume and center of gravity are known. Thus, in Proposition 4 of the *Method*, Archimedes shows that the volume of a paraboloid of revolution is one-half of the volume of the circumscribing cylinder by slicing the two solids (see Figure 1 which represents a plane section through their common axis AD) at right angles to AD. For let us take HAD to be the bar of a

---

* From AMERICAN MATHEMATICAL MONTHLY, vol. 62 (1955), pp. 473–476.

† An address to the Mathematical Association of America at the 1953 Summer Meeting in Kingston, Ontario, Canada.

balance with $HA = AD$ and with the fulcrum at A, and imagine the circle PO to be removed from the paraboloid and suspended at H. Since $AD/AS = DB^2/SO^2$ in the parabola BAC, we have

$$\frac{HA}{AS} = \frac{AD}{AS} = \frac{MS^2}{SO^2} = \frac{\text{(circle in cylinder)}}{\text{(circle in paraboloid)}},$$

so that, by the law of the lever, the circle in the cylinder, remaining where it is, is in equilibrium with the circle from the paraboloid resting in its new position. If we deal in the same way with all the circles making up the paraboloid, we find that the cylinder, resting where it is with its center of gravity at the midpoint K of AD, is in equilibrium about A with the paraboloid placed with its center of gravity at H. Since $HA = AD = 2AK$, the volume of the paraboloid is therefore one-half of that of the cylinder, as desired.

Many accounts of the *Method* have been given since its discovery in 1906; for example, by T. L. Heath in his *Supplement to the Works of Archimedes*, Cambridge, 1912. In all of them, as in the original work of Archimedes himself, we are invited to *imagine* the lever and the objects suspended from it. But if we construct an *actual* lever and *actual* discs, the various figures, which may be spheres, cones, *etc.*, see below, will be observed to balance, slice by slice, as successive slices are added. The whole procedure then becomes a picturesque and effective illustration of the concept of an integral as the limit of a sum.

To find the volume of a sphere, a problem which Archimedes considered so important that he asked to have the result engraved on his tombstone, a cone and a sphere are together weighed against a cylinder (see Figure 2 and the accompanying sketch). Here the circle NM, resting where it is in the large cylinder $GLEF$ is in equilibrium about A with two circles placed at H, the one circle PO being taken from the given sphere and the other RQ from the cone FAE. For we have

$$OS^2 + QS^2 = OS^2 + AS^2 = AO^2 = CA \cdot AS = MS \cdot SQ$$

and therefore

$$\frac{HA}{AS} = \frac{MS}{SQ} = \frac{MS^2}{MS \cdot SQ} = \frac{MS^2}{OS^2 + QS^2}.$$

Thus, by the law of the lever as before,

one-half of cylinder equals cone plus sphere

from which, since the cone is one-third of the cylinder,

sphere equals one-sixth cylinder.

Thus the cylinder circumscribed about the sphere, being one-quarter as great as the large cylinder $GLEF$, is three-halves as great as the sphere, which is the result stated on the tombstone of Archimedes.

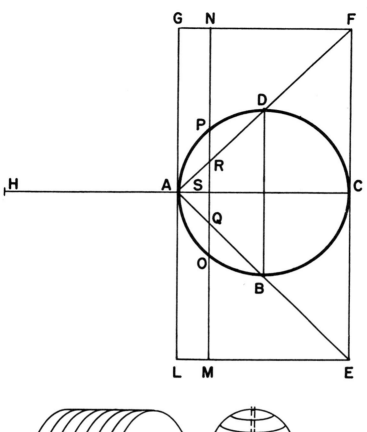

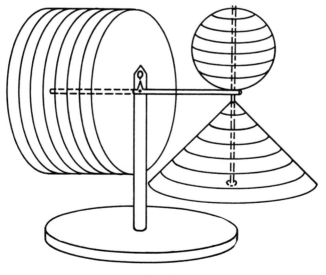

Fig. 2

If squares are substituted for the circles of cross-section in these figures, the argument remains unchanged and we have the solution of another famous problem (Proposition 15 in the *Method*), namely to find the volume common to two right circular cylinders intersecting at right angles.

The actual models were constructed by D. A. Eberle of the Psychology Workshop at Purdue University. The various slices were cut from a piece of white pine 1/2" thick and 7" wide. Thus the cylinder $GLEF$ is composed of seven slices, each with a diameter of 7". The seven slices for the cone, being first cut as stepwise increasing cylindrical discs with easily calculated radii, were placed all together on a mandrel passing through a 3/16" hole through their centers and were then shaped down on a lathe, a procedure found to be especially necessary for the square cross-sections in the problem of the intersecting cylinders. The lever itself is a piece of steel 9" by 1/2" by 1/32", placed so that its 1/2" face is vertical. In each disc a thin slit was cut with a fine hacksaw from edge to center so that the disc could be slipped onto the lever.

BIBLIOGRAPHIC ENTRIES: HISTORY

The references below are to the AMERICAN MATHEMATICAL MONTHLY.

1. Smith, D. E., Historical survey of the attempts at the computation and construction of π, vol. 2, p. 348.

   An English translation of Chapter II of Felix Klein's Vorträge über ausgewählte Fragen der Elementargeometrie. The entire book was later translated into English by W. W. Berman and David Eugene Smith, Famous Problems of Elementary Geometry, Ginn and Co., 1897.

2. Halsted, G. B., Pi in Asia, vol. 15, p. 84.

   Remarks concerning π in early Japanese mathematics.

3. Cajori, F., History of the exponential and logarithmic concepts, vol. 20, p. 5, continued on pp. 35, 75, 107, 148, 173, and 205.
4. Cajori, F., Discussion of fluxions: from Berkeley to Woodhouse, vol. 24, p. 145.
5. Club topic, Euler integrals and Euler's spiral . . . sometimes called Fresnel integrals and the clothoide or Cornu's spiral, vol. 25, p. 276.
6. Schoy, C., Al-Bîrûnî's computation of the value of π, vol. 33, p. 323.

   Remarks concerning π in early Arab mathematics.

7. Cairns, W. D., Napier's logarithms as he developed them, vol. 35, p. 64.
8. Wren, F. L. and Garrett, J. A., The development of the fundamental concepts of infinitesimal analysis, vol. 40, p. 269.
9. Coolidge, J. L., The story of tangents, vol. 58, p. 449.
10. Coolidge, J. L., The origin of polar coordinates, vol. 59, p. 78.
11. Coolidge, J. L., The lengths of curves, vol. 60, p. 89.
12. Davis, Philip J., Leonhard Euler's integral: A historical profile of the gamma function, vol. 66, p. 849.

2

PEDAGOGY

OBJECTIVES IN CALCULUS*

C. C. MacDUFFEE, University of Wisconsin

What should be the objectives in a beginning course in the calculus? That is a question which many college teachers ask themselves, and to which it is difficult to frame an answer. Calculus is the course for which the student has long been preparing through college algebra, trigonometry and analytics, and for many a student it is the last mathematics course which he will ever take. The amount of interesting and valuable material which is at that point open to him is large and beyond the capacity of the time available for its complete presentation. What gems shall be presented and which omitted is a problem which we all have to face.

Our problem is complicated by the fact that no two students have exactly the same backgrounds, interests, personalities or plans for the future. Education should be a very personal matter between Professor A and Student X. The ideal college would give to each student a tailor-made course fitted to his exact needs and capacities, allowing him to proceed as rapidly and deeply as his abilities allow. But such methods under competent teachers are too expensive for any modern college so that the student must be fitted as best he may with a ready-made suit of clothes. The problem is to design the suit so that it will fit the largest number of students reasonably well.

The first objective in a course in calculus has to be the basic techniques of differentiation and integration. Just as the fundamentals of spelling and grammar have to be learned before one can compose literature, so these techniques have to be acquired before one can use the calculus. Any student who can qualify to enter a class in calculus can with patience master these techniques. The pity of it is that so many students get not much else for their labors.

Beyond the fundamental techniques, there seems to be a difference of opinion as to the best procedure. There seems to be one school of thought which would have the student solve large numbers of problems in mechanics and physics without much attempt at incisive reasoning or rigor, apparently with the idea that after enough experience the student's subconscious mind will take over and set up the problem for him. This is known as "standard engineering practice."

---

* From AMERICAN MATHEMATICAL MONTHLY, vol. 54 (1947), pp. 335–337.

There is another school which tends to minimize the applications and to teach the calculus as a pure and lofty discipline of the mind. Most of the class time is spent on real variable theory, and a few students may even learn to throw ϵ's and δ's around with what the teacher believes (until the day of the final examination) to be a fair degree of intelligence. This method is in great favor with young Ph.D's. It is not in favor among deans.

The ideal method obviously lies in neither of these directions. The mathematician cannot afford to forget that the calculus was developed for the purpose of solving problems in mechanics and physics, and that its greatest glory even now is in connection with the applications. The engineer too must remember that Newton and Laplace and the Bernoullis were deep and incisive thinkers whose institutions were merely the manifestations of careful and rigorous thinking.

If we grant that the education of a scientist consists in the development of the power to do, we must admit that our proper goal in the calculus is to develop the student's ability to interpret the physical world in mathematical terminology. This presupposes of course that he shall be able to speak the language of mathematics, that he shall have a command of the techniques and also of the theory behind these techniques. But over and beyond this fluency, he must have achieved an intuitive feeling for the elementary concepts of mechanics and physics, and his thinking along these lines must be in the language of mathematics. This is a large order for a first course in calculus, and obviously incapable of complete achievement. But I believe it is a measuring rod upon which the success of such a course must be judged.

There is a pedagogical sequence for the presentation of ideas which seems to be inherent in the human animal, and which seems to be quite unrelated to what we consider the logical order. Successful teaching respects this pedagogical order. Thus the modern foundations of the calculus, which make the concept of limit a purely static concept without appeal to the notions of time or motion, is magnificent. Every graduate student should be required to master it. But it should not be the calculus teacher's one and only god.

One of the weaknesses of American universities is their intense departmentalization. Nature does not recognize the fine distinctions between what belongs to mathematics, what to physics, and what to chemistry. Neither did the great universities of Europe to the extent that we do. We sometimes teach calculus with no applications to physics, we frequently try to teach physics without using mathematics. What is perhaps even more demoralizing to the students, our mathematics teachers sometimes demonstrate an incompetence in physics and vice versa.

There have been attempts to coördinate the teaching of calculus and mechanics to the extent of preparing textbooks in the two courses which can be used in the two courses simultaneously, the calculus being available by the time it is needed in mechanics, and the problems in mechanics ready to illustrate the theorems of the calculus. Perhaps by the time our elective system has been modified so that the average student does not have an irregular program, this scheme will be more widely tried.

In these days of educational experimentation, various combinations of courses are being tried which have never been tried before. What could be more natural than a combination course of basic physics and calculus? This course would probably have to be spread over two years if it were to contain a complete course in both physics and calculus. It would have to be given by a man who is competent in and sympathetic toward both courses. He could not be a physicist who teaches a little mathematics as a "tool," nor a mathematician who "runs in a few illustrations from physics." But he would be able to develop physicists to whom mathematics is a mother tongue. Can you think of a better background for scientists of the present age?

Regardless of the framework in which it is taught, the first course in calculus must be handled with a fine sense of balance. It should be rigorous up to the capacity of the student to appreciate rigor, and this rigorous treatment must be extended to the problems, not merely confined to the proof of the existence of the definite integral. But the fundamental and basic problem is to develop the student's intuitions so that mathematics is to him a spoken language. Then and only then is he in a position to appreciate the meaning of rigor. For is rigor anything else than clarity?

CAN WE TEACH GOOD MATHEMATICS TO UNDERGRADUATES?*

R. G. HELSEL AND T. RADÓ, Ohio State University

We decided upon our topic shortly after Professor Birkhoff invited us to contribute to this symposium. Later Professor Pólya sent us detailed information concerning the nature of the symposium and we realized then that our topic was perhaps too general; however, we felt the question is of such import to teachers of college mathematics that it should not be put aside. One of the goals of our instruction certainly should be to show the student mathematics of the highest caliber, if we can.

Let us consider then the ingredients of good mathematics. First, it must be *relevant*. Fortunately, most of the subjects we teach are filled with pertinent topics. Calculus, for example, is overflowing with relevancy. Second, good mathematics must be *rigorous*. In other words, all of the reasons must be given. Some persons feel that to teach calculus rigorously the student must learn his ϵ's and δ's and be acquainted with the real number system as defined by Cantor or Dedekind. This is certainly not the case. In fact the sophomore regards any discussion of the real number system as being irrelevant, and it is, at that level. Finally, good mathematics must be *elegant*. If we fail to show our students elegant mathematics, we deprive them of the very thing which affords us our greatest pleasure.

Having listed the ingredients of good mathematics, there remains the question of whether it is possible to serve such a dish to the undergraduate. To answer this, consider the three parts of a theorem: *hypotheses*, *proof*, and *conclusion*. The conclusion, which we assume to be interesting and relevant, is fixed. There may be leeway in the method of proof, but the real freedom is in the choice of assumptions. Thus we propose to *pack the hypotheses* with carefully chosen assumptions that have the power to push through the proof in a rigorous and striking manner and yet are either obvious to the student or acceptable on the basis of past experience. Existential and qualitative assumptions are more likely to meet these requirements than quantitative assumptions.

We shall illustrate this general principle of *packing the hypotheses* by an example from calculus. Perhaps our example is not well chosen, but it has the advantage of being brief. Anyone who is interested can find better examples for himself. Let us establish then the formula $d(\log_a x)/dx = c/x$. This certainly is a relevant and interesting relation for the student, since he already knows that $d(x^{n+1})/dx = (n+1)x^n$ if $n \neq -1$. The student also knows, or believes, that the graph of $\log_a x$ is continuous, monotone increasing, and has a tangent at each point. We shall use as our first hypothesis: $H_1 . d(\log_a x)/dx$ *exists for* $x>0$, *call the derivative function* $g(x)$. As our second hypothesis we shall use the one functional relation involving logarithms which the sophomore does know:

---

* From AMERICAN MATHEMATICAL MONTHLY, vol. 55 (1948), pp. 28–29.

H_2. $\log_a x + \log_a \alpha = \log_a \alpha x$ for $x > 0$ and $\alpha > 0$. Differentiating the left side of this equation, regarding α as a constant, we obtain, through H_1, $g(x)$. Differentiating the right side, with the help of the chain rule for the derivative of a function of a function, we obtain $\alpha g(\alpha x)$. Now let x have the value 1 to obtain the equation $g(1) = \alpha g(\alpha)$, or $g(\alpha) = g(1)/\alpha$, and the proof is complete. Note that the value of the proportionality constant is the slope of the graph of $\log_a x$ at the point where it crosses the x axis, an interesting fact which is often overlooked. Similar proofs can be made for the derivative formulas of the trigonometric and exponential functions.

Thus it is our conviction that by the device of *packing the hypotheses* it is possible to show our students mathematics of the highest caliber. Of course the teacher must use great care in selecting relevant topics and creating interest in them. Also he must choose existential and qualitative hypotheses which are completely acceptable to the student. If the assumptions are properly chosen, the teacher can follow through with a rigorous and striking proof which will show the student what mathematics is really like. In addition the student will soon become aware of the power of existential assumptions and, it is to be hoped that, those who go on to advanced courses will have some interest in existence proofs.

Finally, we might ask whether it is following a high or low path to *pack the hypotheses* as suggested? In reply, consider how the working mathematician grabs everything that is plausible and relevant when he sets about to verify a surmise. Afterwards he may check through the details of theorems that he uses, but initially he merely assumes that they are valid. For example, most mathematicians have used the Jordan Curve Theorem at some time or other, but few indeed have ever gone through a rigorous proof. So it appears that the principle of *packing the hypotheses* is not only a means to show our students good mathematics but it is also in keeping with the manner in which mathematicians actually do their work.

MORE STRESS ON GENERAL FORMULATIONS IN CALCULUS PROBLEMS*

LUISE LANGE, Woodrow Wilson Junior College, Chicago

In many calculus texts problems are formulated too one-sidedly in terms of particular, numerical data rather than in general terms. While pedagogically it may be wise to begin a new type of problem with some numerical examples, it is only the general formulation, and the interpretation of the answer in general terms, which can give insight into the functional relation between the given and the derived data.

We illustrate our point by some examples from the traditional "minimum fence" and "related time rate" problems. They could readily be multiplied.

1. *Numerical Formulation.* A rectangular lot of given area is to be fenced in on four sides. For what dimensions of the lot will the length of the fence be a minimum? Answer: a square. Same question, but the lot is to be fenced in on only three sides. Answer: (with one side x and two sides y) $x:y = 2:1$. Same question, but the lot is to be divided internally by three fences. Answer: (with inside fences parallel to x) $x:y = 2:5$. Same question, but the material of one fence costs twice as much as that of the three others. Answer: (with higher-cost material in x direction) $x:y = 3:2$.

General formulation. The lot is to be fenced in by n fences (inside or outside) in one direction, and m in the other. The minimum fence is that for which $x:y = m:n$, or $nx = my$; which means, if just as much fence runs in the one direction as in the other. If instead of length it is cost which is to be minimized the condition is the same: the cost is minimum if all fences in one direction cost as much as those in the other.—In fact, all these several questions are one: the minimum of the function $F = c_1 x + (c_2 A/x)$, which lies at $c_1 - c_2(A/x^2) = 0$ or $c_1 x = c_2(A/x)$. The graphical presentation of this function throws additional light on the question. F is the sum of the ordinates of the straight line $y = c_1 x$, which represent the length or cost of the fences in one direction, and of the hyperbola $y = c_2 A/x$ representing the length or cost of the fences in the other direction. The minimum of the function F lies at the intersection of the straight line with the hyperbola.

Posed in general formulation the problem of the minimum surface of a closed or open cylindrical can of given volume receives a similarly comprehensive answer. If there are n bases (e.g. inside partitions as well as bottom and top) the minimum condition is found to be $h = nr$, which, substituted into the formula for the lateral surface is seen to mean that the minimum surface is that for which the lateral surface is twice as great as that of all bases.

2. *Numerical formulation.* A street light hangs 18 ft. above the the street. A man 6 ft. tall walks along the street at a rate 3 ft/sec. At what rate is his shadow lengthening? Answer 1.5 ft/sec.

---

* From AMERICAN MATHEMATICAL MONTHLY, vol. 57 (1950), pp. 181-183.

General formulation. The light hangs H ft. high. An object h ft high moves along the street at rate v ft/sec. At what rate is its shadow lengthening? Answer:

$$ds/dt = \frac{h}{H-h} v = \frac{1}{H/h - 1} v.$$

Interpretation: (a) The rate of the shadow's lengthening is proportional to the rate at which the object moves. (b) The factor of proportionality depends only on the ratio H/h. If h gets small in comparison to H the factor approaches zero (e.g. a mouse, or even more so an ant, running along the street push their practically constant shadows ahead of them). As H/h approaches 1 the shadow lengthens ever more rapidly (e.g. a truck $17\frac{1}{2}$ ft. high under a light 18 ft. high has its shadow lengthening 35 times as fast it drives). (c) The relative rate of the shadow's lengthening, on the other hand,

$$\frac{ds/dt}{s} = \frac{v}{x},$$

does not depend on H and h, but is, for a given v, inversely proportional to the distance x from the light. (A mouse and an elephant walking side by side at the rate 2.5 ft/sec. have their respective shadows lengthening at the same rate 10 percent per second when 25 ft. away from the light, and only 1 percent per second when 250 ft. away from the light.)

We still wish to add that positively harmful, in our opinion, are the numerical specifications not only for arbitrary constants but for variables of which the required datum is a function, merely for the purpose of "coming out" with a numerical answer. Example: from a bridge 30 ft. high a boat is towed in by a rope which is pulled in at a rate 4 ft/sec. At what rate is the boat moving *when 40 ft. away from the bridge?* The last specification is pedagogically very bad, because it confuses in the student's mind the basic question as to what is constant and what variable in the problem.

THE PUPIL'S ADVOCATE*†

RALPH BEATLEY, Harvard Graduate School of Education

Once upon a time this Association appointed a committee to consider the rival claims of courses in mathematics and of professional courses in education in a fifth year program for prospective teachers of secondary mathematics. This committee reported that it naturally favored mastery of subject matter, but that it was bound to admit that the continued existence and growth of instruction in educational psychology, in the principles of teaching, and in allied subjects indicated some merit in those areas also. It recognized that instruction in "how to teach" was often feeble, and that therefore it was easy to see why such instruction was generally scorned by the academic mind. Nevertheless this committee took the position that where there was smoke there must also be fire, and decided to recommend for the fifth year an almost equal division between the claims of mathematics and professional training in education. They presumed, I suppose, that a prospective teacher whose training included both mathematics and some knowledge of the learning process would blend both ingredients in an effective teaching procedure. You can read the report of this committee in this MONTHLY, volume 42, May, 1935, page 263.

If this committee of your own Association could take such a position concerning secondary mathematics in 1935, it seems to me reasonable to expect that many members of this Association would be sympathetic to the idea that prospective teachers of freshman and sophomore mathematics in our colleges today should blend with their mastery of subject matter some knowledge of how the human mind acquires new ideas, how it schools itself in new procedures, and how the acquisition of these new ideas and procedures is helped or hurt by the attitudes and emotions that accompany them. I do not propose to tell you how you might alter the training of prospective college teachers from its present pattern so as to include some of the ideas on "how to teach" that I believe would be helpful. Three things, however, I will say: first, that the important ideas and attitudes that I have in mind can be put into such compact form that they can be picked up in relatively short time; second, that even a small amount of properly concentrated instruction in how to teach could result in great improvement in teaching in freshman and sophomore classes; and third, that it is futile to advocate the inclusion of such instruction in the training of prospective college teachers until college departments of mathematics see the worth of it and show that they want it. Once they decide that they want it, it will not be difficult to provide it. Until they decide that they want it, the obstacles to providing it will be numerous.

Instead, therefore, of wasting my time on proposals for modifying the present training program, I will endeavor to show the sorts of changes in instruction

---

* From AMERICAN MATHEMATICAL MONTHLY, vol. 57 (1950), pp. 369–381.

† Invited address before the thirty-third annual meeting of the Association, New York City' Dec. 30, 1949.

you might expect if prospective college teachers were brought to consider the human mind and how it learns. If you don't like the changes that I predict, you need only sit tight. If, in time, you come to the point where you can see advantage to the cause of college mathematics in making certain changes, you will find a way to do so.

The changes that I believe would result from giving more thought to the pupil and to the manner in which he learns would necessarily involve some changes in courses of study and in text books. These changes, as I envisage them, would be small, but their effect on the pupil would be great. The changes that interest me most all stem from one idea, namely, proper motivation of the subject-matter. This, broadly conceived, includes as a corollary the idea of withholding certain details of rigorous exposition until the pupil appreciates their import. When I say "withhold," I do not mean "suppress." However, I do mean something that you will probably not enjoy, even though you come in time to accept the idea. For it implies some disruption and some dislocation of the purely logical presentation of a subject. It implies beginning some subjects in the middle and working both ways from that first plunge. After all, has not much of mathematics been developed in this same way, beginning on the ground floor and then advancing to higher levels while new and better foundations were tucked in underneath?

The first attack on a subject, instead of being straightforward and logically ordered, would in some instances be fragmentary and messy. It would require more artful exposition than merely proceeding in logical order; for, in the interest of truth, we should wish to indicate where some of the truth is withheld, and to indicate at times also the general nature of the truth that we withhold. It implies some repetition also, in the form of orderly recapitulation; because, once the pupil appreciates the inadequacies of the initial attack, which I have described as fragmentary and messy, we do indeed want him to see the argument set down in order, with as much attention to rigor as he can accept. This explains what I had in mind by disruption and dislocation of the usual logical presentation.

The reason that I suggest a dual procedure of this sort with respect to certain topics is that the psychological approach to a subject is often quite different from the logical. I advocate the psychological approach in order to win and to hold the pupil's attention; I advocate a subsequent logical treatment because that is the goal of all our effort. I believe that if we proceed in this way the pupil will be more likely to accept and to heed the nice distinctions that interest us, and that he too will become interested in these fine points when he appreciates their significance. I think that our present approach causes us to lose many pupils, some by physical withdrawal, and others by mental and spiritual withdrawal. In one or two instances (overplaying limit of function and continuity at the start of the calculus, and overprotection against certain familiar pathological functions later on) the severely logical approach of some teachers, who seem to the pupil to personify mathematics itself, and of some authors seems to me to savor overmuch of the "Damn you, love me" technique that brings the young lover to the front pages of our newspapers, but indicates to most of us that that is no way to woo.

The reason for my prediction above, that you will not enjoy the prospect of a dual treatment of certain topics, is that what I have called "merely proceeding in logical order" already demands exposition of a high order, as you well know; the even greater demands on the teacher's powers of exposition to provide an extra higglety-pigglety treatment out of deference to the pupil's psyche will really test the depth of your devotion to the pupil and your interest in connecting him in some way with mathematics. In case these demands seem to you to be just too much, please recall that my suggestions are aimed at only a few crucial topics in mathematics and that I am counting on great changes on the part of the pupil in return for only a few changes on our part.

Please understand that I think I appreciate the conflict in interest and attitude that are inherent in the teacher's task of bringing together the severely logical subject of mathematics, that requires thinking, just thinking all unadorned, and the mind of the human animal, that finds pure thinking difficult and is easily diverted to other methods of solving problems that to greater or less degree involve the emotions. We meet this same conflict in acute form when we attempt to use congenial exploratory and inductive methods to lure secondary pupils to contemplate the less congenial deductive aspects of demonstrative geometry. The problem there is how to teach deduction inductively.

I am sure that many college teachers regard it as altogether obvious that in beginning a new subject, say the infinitesimal calculus, they should build a good foundation and that the pupil for his part should review any bits of earlier instruction that will be needed in the new. The introductory chapter of text book after text book is built on that obvious pattern. It defines important ideas such as sequence, limit of a sequence, variable, function, limit of a function, continuous function, and announces that, if two variables have limits, the limit of the sum of the two variables is the sum of their two limits. Often the author of the text book shows that he recognizes that the pupil will consider such an approach altogether dreary. He suggests that the pupil skip these introductory ideas but remember where he can find them if ever he needs them. So the pupil is led without introduction, or by a highly repellent introduction that is worse than no introduction, to contemplate the limit of the ratio of two variables each of which is tending toward zero. And now observe the trouble we are in. The pupil who has been conducted through the introductory chapter has met the necessary ideas, but has acquired a distaste for them. The pupil who has been protected against an unfortunate emotional set toward the subject by being allowed to skip the introductory chapter finds now that he must grub around in it after all. It is quite possible that a sympathetic teacher, and there are many such, will not probe too embarrassingly at the start into the pupil's ideas concerning the limit of delta-x and the limit of delta-y. The teacher may deem it better not to molest a pupil who happens to think of delta-x as assuming only rational values as it tends toward zero. But however sympathetic the teacher, he must face this fundamental conflict: that you can't begin the calculus without emphasizing the idea of limit, and that, whether or not the pupil has studied the introductory chapter, you can't count on his having an accurate notion of limit, ready for immediate use. To my mind, one of the most important outcomes of instruction in the infinitesimal calculus is the pupil's ultimate acquisition of sound ideas concerning variable and limit, and his appreciation of the vital role that these ideas play in establishing methods of calculating those quantities from geometry and physics that are the chief concern of the calculus and that the pupil recognizes as important and interesting. But the pupil will require time and patient handling if he is ultimately to acquire these ideas in the form that we desire. My plea is not for a cheapening of the ultimate goal of our instruction. I urge merely a compassionate approach that would save as many souls as possible at the beginning and along the way. At the beginning, therefore, I favor a method that relies heavily on the pupil's intuition and only gradually reveals to him that his intuitive ideas will bear examination and revision. I believe that I am correct in asserting that the usual formal introductory approach to the calculus is a direct affront to the pupil's intuition and for that reason is psychologically bad. I repeat, I think that such an approach antagonizes many pupils unnecessarily; that it loses us many pupils who could be saved to understand and enjoy mathematics; and that by saving them we could render a distinct service to the cause of mathematics.

I believe that the way to save them is easy; that it requires very little time and effort; and would necessitate very little change in courses of study or in the printed page. It may indeed require a considerable change in attitude at certain points. And this brings me back to my basic idea, the proper motivation of the subject-matter. The word motivation implies respect for the pupil's interests and attitudes; and it is good strategy for us to respect them, for interests and attitudes provide the drive to action. If you will trust my expressed intention that all the treasured ideas of limit, continuous function, and so on are eventually to be saved; and if my hope is realized that more pupils than formerly will be saved to appreciate these ideas; then why not introduce the pupil to the calculus by first showing him one or two examples of what the calculus can do and how it does it? Give him some reason for studying the calculus; appeal to his interests; stimulate in him a spirit of inquiry. For example, we can invent some plausible reason for wishing to know the maximum value of $5+6x-x^2$ as x varies—say, the number of inches $f(x)$ of growth of pole beans in terms of the number of inches x of rain in June—and show the bare technique by which this maximum value is computed. No mention of variable secant, no deltas, no limit; nothing but the rigmarole of passing from $5+6x-x^2$ to $6-2x$, of attaching to this latter expression the idea of slope of tangent, and then eliciting from the pupil the connection between relative maximum and zero slope, and allowing him to finish the computation himself then and there. The pupil's earlier instruction in algebra has probably already given him an idea or two concerning the critical point of a quadratic function against which he can check this new technique of the calculus. And then, in another few minutes, another example involving a cubic function set equal to y, where our new technique, called differentiation, produces y', the slope of the tangent. And then, with equal abandon, a third example in which the pupil employs this new technique to pass from $s = 16t^2$ to $s' = 32t$, and is given immediately a brief picture of the connection between slope of tangent and rate of change of a function in general. In this last example it is conceivable that the pupil can suggest that rate in this case can be interpreted as velocity; at least the teacher will invite him to try to make this association. And then the teacher can suggest the possibility of reversing this operation so that, given the relation between velocity and time, one can infer a relation between distance and time. There is no attempt here at explanation; the sole intent is to give a quick preview of some of the ideas that will occupy the pupil in his study of the calculus; to show him quickly and with very simple examples how the calculus is linked to geometry and physics; to contrast the two main operations of the calculus; and, above all else, to make the subject appear in prospect to be easy. The pupil is not expected to understand much of what he has been shown. The object is rather to show him that it involves concepts already familiar to him and is the sort of thing that he can understand. A further object is to establish a feeling of confidence between pupil and teacher; for at this same time the teacher must say that certain steps in reasoning that support the simple techniques just exhibited will seem to the pupil to involve nice distinctions, the need for which is not immediately apparent, but the teacher understands this and does not expect immediate and complete acceptance of every new idea on the part of the pupil; that allowance will be made for a period of incubation. I am aware that this sounds a bit childish. But if so, is it not because we commonly forget how close the adult is to the child in matters of this sort? In brief, I am suggesting that if we defer somewhat to the psychological child in the pupil before us, we increase the likelihood that he will accept the adult mathematics that we set before him. Now, just before you start to tear this suggestion to pieces, allow me to insist that I am not discussing at this point the teaching of the calculus, and that I do not want my remarks with respect to $5 + 6x - x^2$ and thereafter to be taken as a sample of how I would teach those topics in the calculus. We are not in the calculus yet; only on the doorstep. I am discussing the introduction to the calculus; and not so much how to introduce the idea of limit as how to introduce the pupil. The word calculus still means to me all that it has ever implied at the beginning level with respect to the idea of limit.

I sympathize with the desire to parade at the beginning of a book on the calculus the materials of the present introductory chapter. From the point of view of the learner, however, I incline to favor the dissemination of this material throughout the book as need for it arises. That is one of the disruptions that I mentioned earlier. Then, considering the learner and also the teacher,

I am inclined to favor a collection of all this material for ready reference in the appendix. And that is one of the dislocations that I mentioned.

If only the introduction could show the pupil why he is studying the calculus, and lead him at once to the derivative as usually explained, and to exercises and problems involving the derivative, I think the pupil would be better served and better pleased. The exposition of the derivative would suffer no change. The pupil would continue to be as naïve as formerly about variable and limit, but his morale would be better. Instead of being given the idea that he was doing less than expected and was getting off to a bad start because he had not entirely understood the necessary preliminaries about variable and limit, he would be given the idea that he was doing all that was expected at this time, that it was adequate for the present, but that he must expect very soon to acquire greater sensitivity to the techniques involved and to the ideas on which they rest. Most books require the pupil to apply the fundamental definition of the derivative to a series of ten or more expressions like $4x^2+3$ and $3x^2-5x+4$. I would allow the pupil to consider the earlier ones in this list as aimed chiefly at giving him facility in the technique of translating $f(x+\Delta x)-f(x)$ into terms appropriate to a given exercise. And then in subsequent exercises I would expect the teacher to be more demanding of the pupil with respect to the limit idea. Just how demanding depends at this stage on how much of this the pupil can accept.

It is possible, of course, to take the point of view that if we defer always to the pupil, he will stage a sit-down strike and baulk us in our attempt to reach our goal. Actually, I think the human animal reacts quite differently: that when pushed and crowded he does indeed sit-down and baulk; but when held responsible for displaying his own powers under conditions that give him a fair chance to do so, he surprises us by doing more than we expect. I believe it is all too easy to find examples of pupils who sit passively in our class-rooms and respond only briefly to our best efforts. I have an idea that if we should pay more attention to how pupils learn, we could discover ways of converting our best efforts into even better and more successful efforts.

In similar vein, would it not be worth while to devote a few minutes at the beginning of infinite series to explain why we interrupt the steady course of the calculus to include this topic? And why we make so much fuss over the matter of convergence? It is a rare book that lets the pupil in on the secret that $x \tan x$ cannot be integrated in terms of elementary functions; that admits that the strange and brief assortment of curves presented as exercises under length of arc are the best that can be assembled, again because of difficulties with integration; or that discloses why certain differential equations are offered as exercises and why others, much like them, are not offered. I believe that it would cheer the pupil immeasurably to learn that author and teacher are less than omnipotent, that there are things they cannot do. Why not tell him? In this connection I ask why we list in a table of integrals only those expressions that we can integrate in terms of elementary functions; why not include some of the closely related functions that cannot be handled so readily, with a simple notation to that effect? I believe that the pupil would take hold of infinite series much better if we told him at the outset that one reason for introducing the study of infinite series into the calculus is to circumvent this barrier of non-integrability. And then, instead of proceeding to a formal exposition of the first ideas on infinite series—handing it out cold, as the pupil would say—my own inclination would be to lead the pupil to do a bit of experimenting at this point. I suggest that he try to employ the first few terms of the binomial expansion in an effort to evaluate $(2+1)^{-2}$; $(1+2)^{-2}$; $(4+1)^{1/2}$; $(1+4)^{1/2}$; $1/(1+x)$ for $x=\frac{1}{2}$; and $1/(1+x)$ for $x=2$. Nothing would be said at this time about the validity of employing the binomial expansion in this way. Further, I would suggest that the pupil integrate from $x=0$ to $x=h$ the first six terms of the expansion of $1/(1+x)$ in order to approximate the integral $\int_0^h dx/(1+x)$ and try in this way to find $\log(1+\frac{1}{2})$ and $\log(1+2)$. And I wouldn't be above asking him to contrast the expansion of $1/(1+x)$ when x equals 1 with the value of his series for $\log(1+h)$ when h equals 1. I should expect that the pupil would see some point now in considering whether a given series converges or not.

With respect to convergence, there is some danger of piling up ingenious examples that run far beyond any use that the pupil will make of this important idea. In this respect this topic is reminiscent of factoring in algebra. With respect to both topics it can be said, and truly, that the fussy cases are introduced to test the pupil's grasp of basic principles. If we recall why we introduced series and why we decided to inquire into the matter of convergence, it will be easy to determine when the pupil has had enough. Certain series serve admirably to test the pupil's grasp of the idea of limit of a sequence and deserve consideration for that reason if for no other. By now the pupil has progressed far beyond the point where we need be solicitous about his attitude toward the idea of limit; our concern at this stage is to seek occasions that will cause him to revert to the idea and test his mastery of it.

Another opportunity to appeal to the pupil's interest is afforded by the hyperbolic functions, but I do not recall any text that seizes this opportunity. To me it seems almost unavoidable that the pupil, comparing the integrals

$$\int_0^x \sqrt{a^2 - x^2}\, dx = \frac{x}{2}\sqrt{a^2 - x^2} + \frac{a^2}{2}\sin^{-1}\frac{x}{a}$$

and

$$\int_0^x \sqrt{a^2 + x^2}\, dx = \frac{x}{2}\sqrt{a^2 + x^2} + \frac{a^2}{2}\log_e\left(\frac{x}{a} + \frac{\sqrt{a^2 + x^2}}{a}\right)$$

and their geometric counterparts right-triangle-plus-circular-sector and right-triangle-plus-hyperbolic-sector, as in Figure 1, would be compelled by the striking analogy to ruminate on the result of relabeling $\log_e (x/a + \sqrt{a^2+x^2}/a)$, which we shall call u for short, as $\sinh^{-1} x/a$. He would then be curious as to what $\sinh u$,

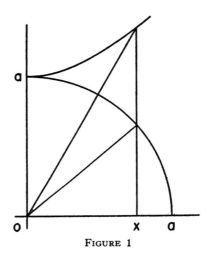

FIGURE 1

equal to x/a, might look like. He can easily satisfy his curiosity about x/a by converting

$$\log_e \left(\frac{x}{a} + \frac{\sqrt{a^2 + x^2}}{a} \right) = u$$

into

$$\frac{x}{a} + \frac{\sqrt{a^2 + x^2}}{a} = e^u,$$

which yields

$$\frac{x}{a} = \frac{e^u - e^{-u}}{2} = \sinh u.$$

Similarly, by comparing the negative integral

$$\int_a^x \sqrt{a^2 - x^2}\, dx = \frac{x}{2} \sqrt{a^2 - x^2} + \frac{a^2}{2} \sin^{-1} \frac{x}{a} - \frac{a^2}{2} \cdot \frac{\pi}{2}$$

$$= \frac{x}{2} \sqrt{a^2 - x^2} - \frac{a^2}{2} \cos^{-1} \frac{x}{a}$$

and the positive integral

$$\int_a^x \sqrt{x^2 - a^2}\, dx = \frac{x}{2}\sqrt{x^2 - a^2} - \frac{a^2}{2}\log_e\left(\frac{x}{a} + \frac{\sqrt{x^2 - a^2}}{a}\right)$$

and their geometric counterparts right-triangle-minus-(larger)circular-sector and right-triangle-minus-(smaller)hyperbolic-sector, as shown in Figure 2, the

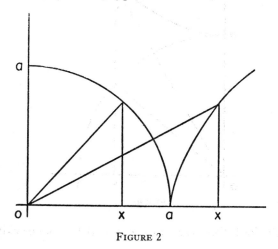

FIGURE 2

pupil would be tempted by the analogy to try the result of writing

$$\log_e\left(\frac{x}{a} + \frac{\sqrt{x^2 - a^2}}{a}\right) = v = \cosh^{-1}\frac{x}{a}.$$

From

$$\frac{x}{a} + \frac{\sqrt{x^2 - a^2}}{a} = e^v$$

he gets

$$\frac{x}{a} = \frac{e^v + e^{-v}}{2} = \cosh v.$$

Giving further rein to his fancy he invents the hyperbolic tangent, the derivatives of the hyperbolic sine, cosine, and tangent, and pokes around for the antihyperbolic tangent. He gets this by twisting

$$\tanh v = \frac{e^v - e^{-v}}{e^v + e^{-v}} = w$$

into
$$e^{2v} = \frac{1+w}{1-w},$$
from which he derives
$$v = \tanh^{-1} w = \frac{1}{2} \log \frac{1+w}{1-w}.$$
This brings the integrals
$$\int \frac{dx}{1+x^2} = \tan^{-1} x \quad \text{and} \quad \int \frac{dx}{1-x^2} = \frac{1}{2} \log \frac{1+x}{1-x}$$
much closer together; and the pupil will find further support for the results of this fanciful inquiry when he works out the series developments of these functions.

This takes a bit more time than just "handing it out cold"; but I believe that it pays dividends in the long run. My object is to arouse the pupil's interest, to insist that he sees the meaning of what he does, and to lead him to go on his own whenever possible. Neither you nor I want him to accept procedures blindly, trusting the word of teacher or of the book. But sometimes we get behind in our program, possibly just because we have listened long and patiently to pupils' difficulties, objections, doubts, and inquiries. Remembering the demands of the syllabus of the course we are teaching, we are tempted then to stifle questions and proceed with the program. Sometimes, of course, we must do, this. But the pupil knows whether our attitude in general is to prefer his interest to the demands of the syllabus, or whether we tend to give the syllabus the right of way. What good does it do if the teacher completes the course on time, but loses the pupil on the way?

I never cease to wonder how some instructors can proceed blandly on with the discussion of topics B and C when the pupils in front of them are not yet satisfied with important details of topic A. Cannot such an instructor imagine the discontent such treatment breeds in the pupil? Does he not see that the pupil who is still dissatisfied with respect to topic A is ripe to be taught; that he is emotionally set to inquire, to test one alternative after another; in short, is emotionally primed really to study? Here is an opportunity not to be missed. It does not mean necessarily more debate in class at this time; it may be better to suggest that the pupil—it is almost time to call him the "student"—follow this or that line of inquiry and report on it at the next meeting. Almost anything will serve that pays attention to the pupil at such a time. It is important that the instructor remember on his part to follow up the pupil's inquiry at their next meeting.

Ignoring a pupil who is emotionally set to learn tends to turn him against the instructor and against the course. What is the syllabus for, anyhow, that it is more important than the pupil?

What does the pupil bring to your class? Some proficiency; some misconceptions; some interest; some fear. It is the teacher's task to increase this proficiency; to replace misconception with proper understanding; to hold whatever interest the pupil may have had and to add to this interest. The teacher can do little of any of these, however, unless he dispels the pupil's fear. Well, what is the pupil afraid of? He is afraid that he will not succeed with worded problems; that he will not really understand the "why" and "what for" of the various mathematical maneuvers he is expected to perform. He inclines to seek the security of a sure rule of thumb. You know where that will lead him! Consequently the teacher's chief task is to dispel these doubts; to create an atmosphere of confidence, in which the pupil will expose his ignorance and misconceptions freely, and will even welcome the teacher's probing to uncover other weaknesses that the pupil has not confessed. In such an atmosphere doubtful points in the teacher's exposition and in the book will be cleared up. Exercises that encourage the pupil to discover mathematics himself in his own imperfect way are an aid to this end; and so are exercises that allow for plenty of learning by actual doing. Admittedly these things all take time. But it is amazing how rapidly a pupil can acquire the information we want him to acquire if only we take pains to give him the proper emotional set first of all. Teachers in general know that they ought to provide an effective motive for learning; that they ought to allow some time for the pupil to discover some of mathematics for himself; and that there is an efficacy in the pupil's own doing that outweighs that which is done for him. But there are schedules to be maintained; the class must cover the ground. In many cases, if only the teacher would begin by concentrating on the pupil and ignoring the syllabus, he would discover that the pupil finally will rescue the syllabus for himself. There is much to be said for a slow beginning, to save as many souls as possible; those who survive can tolerate cheerfully a more rapid pace at the end.

It is my belief that many teachers and authors of textbooks recognize the merit of these ideas. Why, then, are they not more generally adopted? I will hazard the guess that it is partly because of ignorance and partly because of fear. Our knowledge of mathematics is relatively sure; our knowledge of human beings is uncertain. Then, too, it is more respectable to be concerned with the subject than with the pupil. If the teacher teaches, or the author writes, with his attention chiefly on the pupil, he risks suspicion as one who does not "know his stuff." I think that there are ways by which one who is really master of his subject can protect his scholarly reputation by adroit qualifications and occasional foot-notes. Some authors have shown that they can do this. I wish that more would follow suit.

The environment in which the college teacher works is concerned chiefly with his subject. What forces act on him in behalf of the pupil? If only it could be generally understood that every author has heard of the so-called "pathological functions" and was guaranteed free of attack because he seemed to ignore them, I think the pupil would be pleased and would progress more rapidly to the point where he too could join safely in the sacred rites surrounding them. I recognize the important part that exceptional cases and famous puzzles have played in the development of mathematics. But we can never attain a completely rigorous presentation on the adult level even; we must draw the line somewhere. You can see that I think that we commonly draw it with too little regard for the pupil.

In this plea for greater consideration of the pupil's point of view I have drawn my illustrations from the calculus and have said nothing with respect to the subjects that normally precede it. You will not be surprised, however, at my opinion that, from the point of view of the subject, the best way to begin analytic geometry is in terms of directed line segments and their projections on the axes; but that, from the point of view of the pupil, such a beginning is unfortunate. My own inclination would be to rely at first on the pupil's intuition concerning the general validity of $x_2 - x_1$ as the projection on the x-axis of the directed line segment P_1P_2. I am fully aware that the generality of the algebraic processes that we employ in analytic geometry forms the very heart of the subject; but I prefer to insinuate this idea gradually as the course develops. I believe that direct frontal assault causes the loss of too many of the participants; that repeated small forays at different times and places will bring more participants to the desired goal. Furthermore, directed line segments do not quite live up to the high hopes we have of them. It is a bit of an art, indeed one of the trickiest aspects of analytic geometry for the beginnner, to know when directed line segments are to be respected and when they can be safely ignored. I might even say "profitably ignored," having in mind the possible loss of the inner loop of the limaçon $r = 3 - 4 \cos \theta$ by too stuffy insistence on positive r's only. You know when you think it well to consider length, area, and volume as unsigned quantities and when you would be relieved of this restriction in order to win some important generalization. To the pupil the rules of the game are not so clear.

Again, in trigonometry, the majority of college teachers who write texts in this subject seem to think that if they begin with general angle they can clean up on the elementary functions once and for all. In my opinion the pupil would prefer to be introduced to the elementary functions as they appear in right triangles, with some subsequent extension short of the general angle to accommodate obtuse-angled triangles. I am aware that the style has turned a bit from the solution of triangles and that analytic trigonometry is more in favor. I think I know why, and I agree. I mention, merely, that I think that the approach through triangles means more to the pupil; and I would not ignore the pupil's point of view.

In summary, I advocate giving the game back to the pupil. Let him find out a few things for himself in the trial-and-error method by which we all learn.

I would not be so solicitous for his safety in the mathematical world. A good soaking may teach him more than he would ever learn if he started out with full equipment of rubbers, raincoat, and umbrella. Actually this metaphor does not express my whole meaning, because I want him to return home after the first shower and get more equipment, once he sees what he is in for. Of course, it would make things easier at home if he would start out in the first place with all the equipment he might possibly require, even including crampons, and perhaps water and lunch, in case he is marooned on a glazed highway and cannot move. That is the way some recent textbooks impress me. They care for every emergency in advance. That way everything is neat and orderly. But most of it is done by the guide; little initiative is left to the pupil.

Changing the metaphor, I would say that on the whole our mathematical house is in good order and that we take pride in keeping it so. I am not asking for a thorough house-cleaning. Quite the contrary. The house is spick and span now and needs no house-cleaning. I think, however, that it does need a little room that the children can call all their own, where a certain amount of disorder is tolerated, provided that from time to time the children are expected to tidy up the room according to their own standards of what is neat and orderly. They will not lack examples of adult standards of neatness and order in all other parts of the house.

PARAMETRIC EQUATIONS AND MECHANICAL MANIPULATION OF MATHEMATICAL SYMBOLS*

A. D. FLESHLER, Champlain College

Many writers have pointed out the necessity of impressing upon the student the fact that mathematical symbols and formulae are not to be handled mechanically, expecting that an apparent automatic, correct operation will always yield a correct result. Examples illustrating the pitfalls involved in such a mechanical procedure are legion.† Even the better student who knows his integration and trigonometric transformations will perform the following operation:

$$\int_0^{4\pi} \sqrt{1 - \cos x}\, dx = \sqrt{2} \int_0^{4\pi} \sin \frac{x}{2}\, dx = 0(!!).$$

Most students, when the difficulty involved in the above operation is pointed out to them, still remain puzzled why an apparent correct transformation led to the wrong result.

I found that one of the most effective ways of training the student to be on his guard against wrong results arising from such automatic transformations is to ask him to plot a curve given in parametric equations: first by eliminating the parameter and plotting $y = f(x)$, and then plotting directly from the parametric equations $x = h(t)$; $y = g(t)$. The equations, of course, are chosen so that the two methods of plotting lead to two different loci. I gave the class the simple set of equations:

$$x = \sin t; \quad y = \sin t$$

and I asked them for the locus represented by these equations. The unanimous agreement was that since $x = y$ by eliminating $\sin t$, the locus was a 45 degree line through the origin. I then asked them to plot directly by making a table of values of t, x, y. They quickly realized that they could not get out of the intervals: $-1 \leq x \leq 1$; $-1 \leq y \leq 1$ and that the parametric equations represented only a finite portion of the line. I gave them innumerable variations of the above set of equations. In each case a mechanical elimination led to $x = y$; while the original set of equations represented different portions of the line.

For example: The equations $x = e^t$; $y = e^t$ represent the portion of the line in the first quadrant; while the equations: $x = \coth(t)$; $y = \coth(t)$ represent the entire line except the interval: $-1 < x < 1$; $-1 < y < 1$. Once the students grasped the significance of the above illustrations they were on their guard against such fallacies when the above method was applied to other loci.

---

* From AMERICAN MATHEMATICAL MONTHLY, vol. 58 (1951), pp. 106–107.

† See F. S. Nowlan: "Objectives in Teaching College Mathematics," this MONTHLY, vol. 57, pp. 73–82, 1950.

Time and again the student had used the following elimination: Given $x = r \cos \theta$; $y = r \sin \theta$. Squaring and adding he obtained $x^2 + y^2 = r^2$ a circle. When I gave them the following set of equations $x = a(\tanh t)$; $y = a(\text{sech } t)$ a few were tempted to say that the equations represented a full circle, since, following the above procedure we obtain:

$$x^2 + y^2 = a^2(\text{sech } t)^2 + a^2(\tanh t)^2 = a^2$$

The majority, however, realized that y must always be positive; and hence the entire locus must be above the x axis.

The advantage of this method is that once the student grasps the idea he gets real pleasure out of constructing his own illustration and is duly impressed with the importance of a proper interpretation of mathematical symbols and expressions.

PARAMETRIC EQUATIONS AND PROPER INTERPRETATION OF MATHEMATICAL SYMBOLS*

A. ERDÉLYI, California Institute of Technology

The point raised recently (this MONTHLY, vol. 58, pp. 106–107, 1951†) by A. D. Fleshler is a pertinent one and deserves further discussion. The examples given by Fleshler show very clearly that mathematical symbols have no existence of their own: they are given a lease of life by an exact definition, and cannot function except in conjunction with that definition.

The chain of operations

$$\text{(1)} \qquad \int_0^{4\pi} \sqrt{1 - \cos x}\, dx = \sqrt{2} \int_0^{4\pi} \sin \frac{x}{2}\, dx = 0$$

is reprehensible *not* because the result is wrong, but rather because the point of departure is not well defined. As a matter of fact, it is possible to advance a definition of $\sqrt{1-\cos x}$ which will make the above operation justifiable, and its result correct. However, when we write down

$$\int_0^{4\pi} \sqrt{1 - \cos x}\, dx$$

we tacitly assume that everybody knows that by $\sqrt{1-\cos x}$ we mean the non-negative square root, and with *that* definition (1) is incorrect.

In a parametric representation $x=\phi(t)$, $y=\psi(t)$ it is necessary to define unambiguously not only the functions $\phi(t)$ and $\psi(t)$, but also the region over which t varies. For example, if t is assumed to run through *all real values*, the equations

$$\text{(2)} \qquad x = e^t, \qquad y = e^t$$

represent the portion of the line $x=y$ in the first quadrant. If t varies over *positive real values* only, then our equations represent only that portion of the line which lies to the right of the point $(1, 1)$; and the equations (2) represent the entire line $x=y$ (with the exception of the origin) if t is allowed to assume complex values, and its region of variability in the complex t plane consists of two lines, the real axis and a line parallel to it at the distance π.

Often the tacit assumption will be that t runs over all real values, but this convention should be made clear to the students, and one should also explain that it is not always appropriate. Take another one of Mr. Fleshler's examples,

$$\text{(3)} \qquad x = \sin t, \qquad y = \sin t.$$

These equations can be made to represent the entire line $x=y$ by letting t vary in the complex plane over a broken line which joins the points $(-\pi/2)-i\infty$, $(-\pi/2)$, $(\pi/2)$, $(\pi/2)+i\infty$ in this order. If one lets t vary over all

---
*From AMERICAN MATHEMATICAL MONTHLY, vol. 58 (1951), pp. 629–630.
†Ed. note: Reprinted on pp. 65–66, this book.

real values, one obtains *not* so much *the segment* joining $(-1, -1)$ and $(1, 1)$, *but* rather *a curve* which covers this segment infinitely often (and has an infinite length!). If this is pointed out to a student familiar with the periodic property of trigonometric functions, he will see it, but on being challenged to specify a reasonable region of variation for t, his first guess is likely to be $0 \leq t \leq 2\pi$, which leads again not to the segment but to a closed curve covering this segment twice. Actually, one possible interval of definition is $-\pi/2 \leq t \leq \pi/2$. It is not the only one to lead to the desired result, but some such specification is necessary. In some case one will have to consider equations (3) in the interval $0 \leq t \leq \pi/2$, when they represent the segment joining the origin and the point $(1, 1)$, and so on.

By a discussion of (3) in various (real) ranges of t the student is introduced at an early stage to the necessity of a precise definition, and the ground is prepared for the demand of the more advanced theory that a correct definition of a function should always include (a) the region of variability of the independent variable, and (b) an unambiguous rule for finding the value of the function for each admissible value of the independent variable.

STUDENT'S CHOICE IN MATHEMATICS—FUNDAMENTAL REASONING OR BLIND SUBSERVIENCE TO RULES*

M. R. SPIEGEL, Rensselaer Polytechnic Institute

In a recent note (this Monthly, Vol. 58, p. 188†), F. Hawthorne proposed the following simple endpoint maxima and minima problem:

Find that point on the circle $x^2+y^2=1$ which is nearest to $(2, 0)$. As he states "—the unsuspecting student sets up this problem using x as independent variable and obtains for the square of the distance from (x, y) to $(2, 0)$" the quantity $L^2 = 5 - 4x$. Differentiating and equating the result to zero, the student obtains $-4 = 0$. He is baffled. But why should he be? The reason, of course, is obvious. He has been taught in his calculus course a rule which he must memorize and apply blindly. To find the maximum or minimum—differentiate and set equal to zero. A machine can be designed to do the same.

If the student were taught some of the fundamental concepts of the calculus instead of blind subservience to rules and formulae, he might realize that his supposedly sacred rule does not apply in this case. He might be taught after discovering his predicament, to reason, for example, as follows. $L^2 = 5 - 4x$ is a positive quantity to be made as small as possible. This can be accomplished by making x as large as possible. Since, $x^2+y^2=1$, x will be largest when $y=0$, in which case $x=1$. Thus, the required point is $(1, 0)$. He has not used calculus *per se* in the sense that he has not taken a derivative. He has, nevertheless, used calculus concepts.

Of course this fundamental reasoning cannot be acquired by the student in a day. He must receive constant practice in such approaches to problems. I believe that this blind attitude on the part of the student is but one of the sad consequences of some of our teaching methods discussed by M. Richardson in his article (this MONTHLY, vol. 58, p. 182) in which he endeavors to encourage more teaching of fundamentals in our undergraduate courses.

It may be interesting to quote a more elementary example of our student's desire to obtain results from memorized formulae rather than by fundamental reasoning. I pose the following problem:

A motorist drives from New York to Troy at an average speed of 30 miles per hour and from Troy to New York (being in a hurry) at 60 miles per hour. What is his average speed for the entire trip assuming that the distance travelled from New York to Troy is the same as from Troy to New York?

I have given this problem to many students (to say nothing of some high-powered professors—not in mathematics, of course), and all of them have given the wrong answer. What?—did you say 45 miles per hour? Perhaps then, you had better try some fundamental reasoning too.

---

*From AMERICAN MATHEMATICAL MONTHLY, vol. 59 (1952), pp. 99–100.
†*Ed. note:* Reprinted on p. 290, this book.

BRINGING CALCULUS UP-TO-DATE*

M. E. MUNROE, University of Illinois

1. Introduction. There are many conversations, committee meetings, *etc.*, today about the modernization of the undergraduate calculus course; but all too often the attack on the problem falls short of being comprehensive. Calculus has been in cold storage for over fifty years now. These have been highly productive years in mathematics, and the result is that more changes are in order than most people would like to admit.

The relevant branches of modern mathematics would seem to be real function theory and differential geometry, and a survey of these fields suggests at least the five points listed below. It is significant that point (i) can be gleaned from a study of real function theory and is already coming into vogue in calculus, while points (ii)–(v) come largely from differential geometry and are too generally overlooked. Calculus is too often erroneously classified as a part of real function theory only.

(i) It is essential to distinguish between a function f and its values $f(a)$.

(ii) The coordinate variables x and y of analytic geometry are themselves mappings; x maps points into their abscissas; y maps points into their ordinates.

(iii) There is an important difference in point of view toward coordinate variables in analytic and differential geometry. In analytic geometry x and y are defined over the entire plane, and $y=f(x)$ is a conditional equation describing a locus. In differential geometry the locus is preassigned; x and y are restricted to this locus, and $y=f(x)$ is an identity. A calculus problem starts as analytic geometry and finishes as differential geometry. Thus, somewhere in the middle the symbols change meaning. (More details on this point in Section 2.)

(iv) Modern differential geometry has finally produced a definition of the differential that is quite satisfactory for purposes of integration and differentiation theory. In this definition the concept of differential has been divorced from that of approximate increment. (Details in Section 3.)

(v) In the modern, advanced study of integration of differential forms the theory of the integral is lifted from real function theory, and the algebra of the differential forms is shown to fit into this framework. When this abstract theory is specialized to the simplified, concrete cases discussed in calculus, modern differential theory becomes an effective (and simpler) tool for certain developments in the theory of the integral. (See Sections 5 and 6.)

2. Functions and variables. To begin with, define the word *function* to mean a mapping of numbers into numbers. A later generalization will include

---

* From AMERICAN MATHEMATICAL MONTHLY, vol. 65 (1958), pp. 81–90.

mappings of n-tuples of numbers into numbers. This is the traditional use of the word function in calculus.

One also encounters spaces whose elements are called points. These take the form of lines, planes, curves, surfaces, 3-space, space-time, *etc.* An essential part of the machinery is the mapping of these points into numbers. Let these mappings be called *variables*. Linguistically incongruous as it may seem, this is traditionally the most common use of the word variable in calculus. This definition follows the principle that language is established by usage, not by logic. It retains the word and gives a more enlightened description of what it has been used to mean. It should be noted that probability theory has already established this usage with the phrase *random variable*.

The coordinate variables x and y are variables in the sense just defined. In plane analytic geometry each of these symbols stands for a mapping of the entire plane onto the real number system. If x has this meaning and f is a function, then $f(x)$ is used to denote a composite mapping:

$$\text{point} \xrightarrow{x} \text{number} \xrightarrow{f} \text{another number.}$$

The form

$$y = f(x) \tag{1}$$

is now a *conditional equation*. Here there are two symbols, y and $f(x)$, each standing for a mapping; but in analytic geometry (1) is not regarded as an assertion that these are two names for the same mapping. Rather, (1) appears (or should appear) only as a noun clause in the phrase, "the locus of $y=f(x)$." This locus is, of course, the set of all points p in the common domain of x and y for which $y(p)=f[x(p)]$. Note that this last "$=$" means what it should.

Suppose, now, that (1) has been given in proper context and its locus has been found and named C. The next step in the logical analysis of a calculus problem is to restrict the mappings x and y to C. Note that this could not be done before because the superstructure described above was used to define C. It would help if the restricted mappings were given new names; say, start with X and Y and boil them down to x and y. However, this is probably asking for too much of a change in set habits. In any case, with x and y restricted to C, (1) becomes an *identity*—an assertion that two mappings are the same. At this stage one has a special case of the type structure studied in modern differential geometry and so can turn to the literature of that discipline for further enlightenment. What follows is an informal summary of the specialization of this work to calculus. For a more complete discussion of the general theory see, for example, Chevalley, *Theory of Lie Groups*, Princeton, 1946.

3. Differentials. A *one-dimensional manifold* is a structure consisting of a point set C and a set of *coordinate variables* defined thereon. Precise postulates can be given, but a very loose description will suffice for the present discussion. The set C is a smooth curve, and each of the coordinate variables is continuous and locally one-to-one. Since each is locally one-to-one, each pair u, v is related at least locally by an identity of the form $v = f(u)$, and it is convenient to assume that each such connecting function f has three derivatives. Note that a locus in the plane with the rectangular coordinate variables restricted to it will generally form such a structure.

Consider three classes of variables on C.

X: all variables on C,
Y: those related to a coordinate by a differentiable function,
Z: those related to a coordinate by a thrice-differentiable function.

Let D be a mapping of Y into X, and let Du denote the map of u by D. If p is a point on the curve, denote the value of Du at p by $(Du)_p$. Such a mapping D is called a *derivative operator* at p provided

(a) $[D(au + bv)]_p = a(Du)_p + b(Dv)_p$,
(b) $[D(uv)]_p = u(p)(Dv)_p + v(p)(Du)_p$.

Here u and v are variables and a and b real numbers.

Regarding a constant a as a variable, observe that

(2) $$(Da)_p = 0,$$

because by (a), $[D(au)]_p = a(Du)_p$, and by (b), $[D(au)]_p = a(Du)_p + u(p)(Da)_p$; and if u is chosen so that $u(p) \neq 0$, then (2) follows. Observe also that setting $u = v$ in (b) yields

(3) $$[D(u^2)]_p = 2u(p)(Du)_p.$$

Now, let u be a coordinate, and let $v = f(u)$. Assume f has three derivatives; then there is a differentiable function g such that

(4) $$v = v(p) + f'[u(p)][u - u(p)] + g(u)[u - u(p)]^2.$$

Recalling (2) and (3), operate on (4) with D at p to obtain

(5) $$(Dv)_p = f'[u(p)](Du)_p.$$

Each of the other terms on the right vanishes either because of (2) or because it contains the factor $u(p) - u(p)$. Note that in this step D must be defined on Y because it must operate on $g(u)$. However, (5) is proved only for $v \in Z$ because this is required in order to set up (4).

Let D_u be the operator such that $D_u u = 1$ at each point on the curve. By (5) this determines D_u on v where $v = f(u)$, and indeed

(6) $$D_u v = f'(u).$$

The variable $D_u v$ is called the *derivative of v with respect to u*. The notation f' denotes a purely function theoretic concept; f' is defined from f in terms of limits of difference quotients. On the other hand, D_u is an operator defined by purely algebraic conditions. Thus, (6) is not a definition; it is a theorem relating these two basically different notions.

Consider now D_1 and D_2, two derivative operators at p. If for some variable u, $(D_1u)_p = a(D_2u)_p$, then for $v = f(u)$,

$$(D_1v)_p = f'[u(p)](D_1u)_p = af'[u(p)](D_2u)_p = a(D_2v)_p.$$

That is, at p any two derivative operators are constant multiples one of the other. So, the set of derivative operators at p has the algebraic structure of a straight line. Identify it with T_p, the tangent line to the curve at p, and define the *tangent bundle* for the given curve as the set of all ordered pairs (p, D) where p is a point of the curve and D is a derivative operator at p.

Finally, the *differential* is defined as follows. For u a variable on the curve, du is a variable on the tangent bundle. Its value at (p, D) is denoted by $du_p(D)$, and du is defined by

(7) $$du_p(D) = (Du)_p.$$

Substitute (6) into (5) to obtain

(8) $$(Dv)_p = (D_uv)_p(Du)_p.$$

By (7), $(Dv)_p$ and $(Du)_p$ are values of appropriate differentials; so (8) may be written

$$dv_p(D) = (D_uv)_p du_p(D).$$

This is for an arbitrary p on the curve and D on T_p; so

(9) $$dv = D_uv\,du$$

over the tangent bundle. An appropriate name for (9) is "Fundamental Theorem on Differentials." At any rate, in the modern theory it is a theorem, not a revolving definition that changes meaning with every change of "independent" variable.

Multidimensional cases follow the same pattern. Let C be an n-dimensional manifold—intuitively, a uniformly n-dimensional, smooth set with variables attached. The basic form relating variables will be $v = f(u_1, \cdots, u_n)$. Derivative operators are defined by the same postulates as before, and by the n-dimensional generalization of (4) it is shown that they form a family of n-dimensional vector spaces. A derivative operator is determined if specified on n variables, and the operators $\partial/\partial u_i$ defined by

$$\frac{\partial u_j}{\partial u_i} = \begin{cases} 1 & \text{for } j = i, \\ 0 & \text{for } j \neq i \end{cases}$$

form a basis in the tangent bundle. Differentials are defined again by (7), and the fundamental theorem reads

(10) $$dv = \sum_{i=1}^{n} \frac{\partial v}{\partial u_i} du_i.$$

4. Some immediate advantages. The ultimate advantage of saying things correctly rather than incorrectly need not be discussed. Let it be understood, then, that the following list—which is quite probably nonrepresentative—merely points out a few of the common difficulties that can be cleared up by the modern approach.

(a) *Differentials identified with variables.* A variable v on C generates a differential dv on the tangent bundle, and this differential retains its identity whether $v = f(u)$ or $v = g(w)$. Note that df and dg are not defined. Understanding of this will help to clear up many notation obscurities.

(b) *Second differentials.* No one seems to know exactly what these ought to be. Function theorists sometimes want $d^2v = D_u^2 v du^2$; at other times (when changes of variable are in order) they want $d^2v = D_u^2 v du^2 + D_u v d^2u$. In working, say, with parametric equations the student is apt to use the first of these forms as though it had the invariance properties of the second. It seems advisable to leave these out of calculus completely, and the development outlined above does this very nicely. There the differential of dv is not defined because the domain of dv is the tangent bundle while the differential is defined only for a variable whose domain is C.

(c) *Chain rules.* Given $f(x, y, u, v) = g(x, y, u, v) = 0$, it is common practice to find, say, $\partial u/\partial x$ in terms of other partial derivatives by writing out various expressions modeled after (10), making substitutions and equating coefficients of an appropriate differential. This is an effective technique for deriving chain rules, and in the modern theory it is quite acceptable. However, if (10) is a definition rather than a theorem, it must be proved invariant under coordinate changes. One must often know that the differential form is invariant under the very chain rule he is seeking to find!

(d) *Increments and approximations.* The undergraduate is afraid to use the differential—in such manipulations as (c), for example—because of the approximation bugaboo introduced with the now outmoded definition. In the modern theory this fear is never introduced, and the approximate increment problem is properly classified—under Taylor's theorem.

5. Line and surface integrals. Let u and v be variables on a curve C. Partition C by points $p_0, p_1, \cdots, p_n$. For each i, let q_i be the point on the tangent line T_{p_i} whose distance from p_i is the length of the arc $p_i p_{i+1}$ on C. Intuitively, "unroll" each increment of arc onto a tangent line. As noted in Section 3, the set of derivative operators at p_i forms a "line," to be identified with the tangent line T_{p_i}. To obtain the usual geometric interpretation, specify that if s is the arc length variable on C, then D_s is identified with the unit tangent vector. In any case, each point q_i constructed above is identified with a derivative operator at p_i; so the notation $dv_{p_i}(q_i)$ has an obvious meaning. Now, form sums

(11) $$\sum_{i=0}^{n-1} u(p_i) dv_{p_i}(q_i).$$

Take an appropriate limit of these, and call the result

(12) $$\int_C u\,dv.$$

The important feature to this development of the line integral is that the points q_i on the tangent lines are determined by an intrinsic notion (arc length) on C and have no connection with the variables u and v. Thus, the substitution theorem, $\int_C u\,dv = \int_C uD_w v\,dw$, is an immediate consequence of the fundamental theorem on differentials (9). The approximating sums (11) are invariant under the substitution; hence so is the integral. The familiar $\int_a^b f(x)dx$ is now a special case of (12)—in which C is an interval on the x-axis. In this development the differential is clearly the same thing in differential and integral calculus.

In a double integral—carefully to be distinguished from an iterated integral—*exterior multiplication* of differentials must be introduced. This is an operation denoted by $*$ and characterized by the rule

(13) $$du * dv = - dv * du.$$

This stems from the fact that $du * dv$ is to be a signed measure on oriented plane regions. Setting $u = v$ in (13) yields the corollary $du * du = 0$. Using these rules of multiplication (plus a distributive law) and substituting from the two-dimensional case of (10), one has

(14) $$\begin{aligned} du * dv &= \left(\frac{\partial u}{\partial x}dx + \frac{\partial u}{\partial y}dy\right) * \left(\frac{\partial v}{\partial x}dx + \frac{\partial v}{\partial y}dy\right) \\ &= \frac{\partial u}{\partial x}\frac{\partial v}{\partial x}dx*dx + \frac{\partial u}{\partial x}\frac{\partial v}{\partial y}dx*dy \\ &\quad + \frac{\partial u}{\partial y}\frac{\partial v}{\partial x}dy*dx + \frac{\partial u}{\partial y}\frac{\partial v}{\partial y}dy*dy \\ &= \left(\frac{\partial u}{\partial x}\frac{\partial v}{\partial y} - \frac{\partial u}{\partial y}\frac{\partial v}{\partial x}\right)dx*dy \\ &= \frac{\partial(u,v)}{\partial(x,y)}dx*dy. \end{aligned}$$

In general, a pair (du_p, dv_p) forms a set of affine coordinates in the tangent plane T_p. If the pair is orthogonal (and some pairs always are), then multipliers a and b can be chosen so that (adu_p, bdv_p) forms a set of rectangular coordinates in T_p. In general, a and b depend on p; for example, on the tangent planes to the unit sphere $(d\phi, \sin \phi \, d\theta)$ generates rectangular coordinates. In any case, if (adu_p, bdv_p) is a rectangular set on T_p, and if A is a region in T_p positively oriented with respect to this coordinate system, then define $ab(du * dv)_p(A)$ to be the area of A. Since the Jacobian of a rotation is unity, (14) shows that this is consistent for all such rectangular sets. Values of other exterior product operators are then determined by (14).

Now, the substitution rule for multiple integrals consists merely of making the substitution (14) behind the integral sign. A definition of the surface integral patterned after that given above for the line integral makes this remarkably easy to justify.

Take a smooth surface S and partition it in an arbitrary fashion—not necessarily following coordinate lines—into subsets S_i. At some point p_i in S_i take a tangent plane. Here a minor complication appears; one cannot "unroll" the S_i as he did the arcs and preserve measure. However, define surface area—there are many tricks for accomplishing this—and construct on each tangent plane a set having the same area as S_i. Call this plane set A_i. The shape of A_i is immaterial because (see definition above) $(du * dv)_{p_i}(A_i)$ depends only on the area and orientation of A_i, not on its shape. Now, form sums

$$(15) \qquad \sum_{i=1}^{n} w(p_i)(du * dv)_{p_i}(A_i).$$

Take the usual limit, and call it

$$\iint_S w \, du * dv.$$

Since the A_i depend in no way on u and v, the sums (15) are invariant under the substitution (14); and there emerges the familiar result

$$\iint_S w \, du * dv = \iint_S w \, \frac{\partial(u, v)}{\partial(x, y)} dx * dy.$$

6. Theorems of Green, Gauss, and Stokes. This view of the multiple integral brings an elegant unification to the theory relating an integral over a manifold to one over the boundary. There is a master formula:

$$(16) \qquad \int \cdots \int_B u \, dv_1 * \cdots * dv_k = \iint \cdots \int_M du * dv_1 * \cdots * dv_k$$

where M is an oriented $(k+1)$-dimensional manifold and B is its boundary.

This is proved in the usual way by reducing the integral on the right to an iterated integral and performing one integration.

Now, if A is a region in the plane and C its boundary, then (16) yields

$$\int_C (udx + vdy) = \iint_A (du * dx + dv * dy).$$

However,

$$du * dx = \left[\frac{\partial u}{\partial x} dx + \frac{\partial u}{\partial y} dy\right] * dx = -\frac{\partial u}{\partial y} dx * dy,$$

$$dv * dy = \left[\frac{dv}{\partial x} dx + \frac{\partial v}{\partial y} dy\right] * dy = \frac{\partial v}{\partial x} dx * dy;$$

so the result is Green's theorem. A similar maneuver with the form

$$\int_C (udx + vdy + wdz)$$

where C is an appropriate space curve yields the classical theorem of Stokes. The divergence theorem is also a special case of (16) as is easily shown by direct computation. Minus signs do not appear in the divergence theorem because in all this work the permutations of differentials are cyclic permutations, and a cyclic permutation of three things is an even permutation.

When presented in this light, all these theorems appear as generalizations of the fundamental theorem of calculus. For, this latter may be written $u(q) - u(p) = \int_p^q du$, and this is a specialization of (16) in which the difference on the left is regarded as an integral over the zero-dimensional boundary consisting of two points. Subtraction results because the boundary is considered oriented.

7. Notation. Roughly, the structure studied in calculus seems to consist of mappings (variables) "up" from point sets into the real number system together with other mappings (functions) "across the top." It is well to preserve this stratification in the notation, and the usual use of f and g for functions and the last letters of the alphabet for variables is all to the good. However, there are other distinctions that common notational practice fails to make.

The derivative of a function is another function whose values are appropriate limits of difference quotients. Note that a function is not differentiated "with respect to" anything in particular. Thus, an appropriate notation for the derivative of f is f', and it would be well if the prime symbol were reserved for functions only.

Derivatives of variables are generated by the derivative operators introduced above, and the basic form is $D_x y$. The differential form dy/dx is not merely an alternative to this. It means $dy \div dx$, but it follows at once from the fundamental theorem on differentials that $dy/dx = D_x y$.

Hybrid notation such as y' and df/dx abounds in the literature, but it is not really well-defined. One step in the modernization of calculus is to break some well-formed (too well-formed) habits in this respect. Frequently, the confusion stems from writing $y = f(x)$ and then confusing the symbols y and f. As for habits, however, it is sad to relate that the author of a recent textbook made essentially the comment just made here about confusing y and f and then proceeded to do exactly that three pages later.

Similar distinctions need to be made in the multidimensional cases. If f maps n-tuples of numbers into numbers, its arguments are distinguished by position only. For example, $f(a, b) = a - b$ and $f(b, a) = b - a$ define the same function f. Thus, f_i is appropriate notation for the partial derivative of f with respect to its ith argument. As noted above, $\partial/\partial x$ is a derivative operator in manifold theory and should be reserved for application to variables.

8. Existence theorems. Little has been said so far about the role of real function theory in a program of modernizing calculus. In a sense this is a question quite independent of the ones raised so far here, but certainly the following question is pertinent to the present discussion. The algebraic theory of derivative operators and differentials outlined above yields formulas and techniques in a very elegant fashion, but it certainly is not self contained. Is any additional real function theory required as a background to tighten up the logic in this development?

In integral calculus one must develop the notion of integral of a function alongside that of integral of a variable (outlined above). The two are fairly easily related, and the usual existence proof appears in the function theory.

In differential calculus the existence questions are answered *a priori* in the postulates for a manifold. That is, differentiability conditions are imposed by fiat on the connecting functions in such a way that the algebraically defined derivative operators do represent differentiations in the function theoretic sense. However, to keep from operating in a vacuum, one must show that certain things are manifolds.

For purposes of calculus one gets an adequate supply of manifolds by taking the rectangular coordinates in $(n+k)$-space and considering the locus of n equations in them. The implicit function theorem (a standard item in advanced calculus) gives conditions under which k of these variables become local coordinates on the locus, and the Brouwer theorem on invariance of domain (now appearing in the better advanced calculus texts) guarantees that the locus has the proper dimension.* To verify the hypotheses of these theorems, one must know the differentiability properties of the elementary functions. These may be established in the usual way through theorems on differentiability of sums, products, composites, *etc.*

---

*Added in proof: A recent note by Yamabe, this MONTHLY, vol. 64, 1957, p. 725, gives precisely the result needed. [*Ed. note:* Reprinted on pp. 210–211, this book.]

It is probably fair to say, then, that the introduction of the modern theory of the differential changes very little the number or the nature of the real function theory existence proofs required for a logically sound development of calculus. This, of course, is subject to the stipulation that calculus is restricted to the well-behaved cases. A general attack on the topological problems of manifold theory is much more difficult.

9. Two facets to the problem. Most programs for the improvement of calculus amount to the injection into the course of more and better real function theory. It is the purpose of the present paper not to discourage this, but to suggest that a more imperative project is the injection of better—if not more—differential geometry. This latter is more imperative because the garden variety calculus contains more downright errors in differential geometry than in real function theory. On the other hand, it is more difficult for at least two reasons. First, to bring the differential geometry up to date will change the appearance of a calculus text; it will modify definitions, terminology, notation, procedures. Second, while most calculus teachers are well grounded in real function theory, for many of them graduate study came before some of the important developments in differential geometry.

Despite these difficulties an effort should be made. In failing to bring calculus up to date we are transmitting to the next generation not the information available to our contemporaries, but that available to our grandfathers.

CONCERNING DOMAINS OF REAL FUNCTIONS*

HUGH A. THURSTON, University of British Columbia

By the domain of a function ϕ, I mean the set of all x for which $\phi(x)$ exists.

In modern textbooks of calculus the concept of domain is emphasized when real functions are defined, but is apt to take a back seat thereafter. For example, in Begle's [1] a theorem

$$\{\phi + \psi\}' = \{\phi' + \psi'\}$$

is stated (the prime denotes differentiation). This is not quite true: the most that is proved is $\{\phi+\psi\}' \supseteq \{\phi'+\psi'\}$. The subject is evidently a tricky one.

The usual instances of functions with restricted domains are too trivial to be effective. If ϕ is defined by

$$\phi(x) = 2x + 5 \quad \text{whenever} \quad 0 \leq x \leq 2,$$

so having for domain the closed interval $[0, 2]$, the restriction appears artificial. If a student is told that the domain of the function ψ defined by

$$\psi(x) = x^{-1} \text{ whenever the right-hand side exists}$$

is the set of all *nonzero* numbers, he merely thinks "Of *course* it is."

It is therefore interesting that a paradox in maxima-and-minima depends on the domain of a function not being properly determined. The paradox is Example 2 on page 517 of Thomas [3]: To find the least value of $x^2+y^2+z^2$ when $x^2-z^2-1=0$.

We argue that the least value will be a local minimum; so we eliminate one variable, say z, from the minimand via the given equation, and look for local minima of the result; *i.e.*, of $2x^2+y^2-1$. Its partial derivatives are $4x$ and $2y$; these are simultaneously zero only when $x=0$ and $y=0$, ... and then z is undefined. Geometrical intuition suggests that there *is* a minimum, so we try again, this time eliminating x instead of z, and find minima at $(1, 0, 0)$ and $(-1, 0, 0)$. But this does not save the paradox; there is no logical reason why the second try should necessarily avoid the snag.

We must define "local minimum" carefully. There are two reasonably common definitions:

1. ϕ has a local minimum at a if there is an open interval N containing a such that $\phi(a) \leq \phi(x)$ whenever x is in N.

2. ϕ has a local minimum at a if there is an open interval N containing a such that $\phi(a) \leq \phi(x)$ whenever x is in N and *also in the domain of ϕ*.

---

* From AMERICAN MATHEMATICAL MONTHLY, vol. 66 (1959), pp. 900–902.

The first definition is equivalent to that in Courant [2]; the second to that in Begle [1].

The difference shows most often at the end of an interval: if

$$\alpha(x) = +\sqrt{(1-x^2)} \text{ with domain } [-1, 1],$$

then α has local minima at 1 and -1 under Definition 2 but not under Definition 1.

It is now obvious how to define a local minimum of a function under a given condition. *E.g.*:

1*. ϕ has a *local minimum* at (a, b) under the condition $\mathfrak{P}$ if $\mathfrak{P}(a, b)$ is true and there is an open interval N containing (a, b) such that $\phi(a, b) \leq \phi(x, y)$ whenever (i) (x, y) is in N and (ii) $\mathfrak{P}(x, y)$ is true.

2*. is the same but for the addition of "and (iii) (x, y) is in the domain of ϕ."

It is also clear how to define a minimum under a condition:

ϕ has a *minimum* at (a, b) under $\mathfrak{P}$ if $\mathfrak{P}(a, b)$ is true and $\phi(a, b) \leq \phi(x, y)$ whenever (i) (x, y) is in the domain of ϕ, and (ii) $\mathfrak{P}(x, y)$ is true.

The highbrow method of finding minima when the condition is expressible by equations is by Lagrange's multipliers. I am interested in lowbrow methods; I do not mean nonrigorous, but simple and obvious. The usual lowbrow method is to eliminate variables, so getting a function of fewer variables, and to find the minima of this. In our example, this function was χ, where

(1) $$\chi(x, y) = 2x^2 + y^2 - 1.$$

Let us consider the general case: the minima of $\phi(x_1, \cdots, x_{m+n})$ under n conditions $\psi_i(x_1, \cdots, x_{m+n}) = 0$. We can eliminate the last n variables if the equations are equivalent to $x_i = \xi_i(\mathbf{x})$ for i from $m+1$ to $m+n$ where $\mathbf{x}$ is short for $x_1, \cdots, x_m$. Let us just consider this case; there is not much loss of generality. If the function got by elimination is χ, then

$$\chi(\mathbf{x}) = \phi(\mathbf{x}, \xi_{m+1}(\mathbf{x}), \cdots, \xi_{m+n}(\mathbf{x}))$$

and its domain is the set of all $\mathbf{x}$ for which $(\mathbf{x}, \xi_{m+1}(\mathbf{x}), \cdots, \xi_{m+n}(\mathbf{x}))$ is in the domain of ϕ. With this definition of χ it is easy to prove that ϕ has a minimum at $(a_1, \cdots, a_{m+n})$ under the given conditions if and only if χ has a minimum at $\mathbf{a}$ and $a_i = \xi_i(\mathbf{a})$ for i from $m+1$ to $m+n$.

In our example, we now see that formula (1) for χ is incomplete. It needs the legend "for every x and y for which there is a z such that $x^2 - z^2 - 1 = 0$ and $x^2 + y^2 + z^2$ is defined." This is clearly equivalent to "for every x and y for which $|x| \geq 1$." Minima are to be suspected (i) where the partial derivatives are all zero, (ii) where they do not all exist, and (iii) on the boundary of the domain of

χ. Here (i) and (ii) yield nothing; (iii) yields $x=1$ or -1, $z=0$, y arbitrary. Investigating our suspects, we find that $y=0$ does give minima.

Thus, although minima of χ correspond to those of ϕ under $\mathfrak{P}$ if Definition 1 is used, local minima do not necessarily correspond to local minima. In fact, whether a minimum of ϕ under $\mathfrak{P}$ corresponds to a local minimum of the function we get by elimination may depend on which variables we eliminate.

The lowbrow method works if students are taught to look for all three types of suspected extrema. Either definition of local extremum, properly presented, will lead them to do so, *provided* that the domain of χ is not neglected.

References

1. E. G. Begle, Introductory Calculus, New York, 1954.
2. R. Courant, Differential and Integral Calculus, Glasgow, 1937.
3. G. B. Thomas, Calculus and Analytic Geometry, Cambridge, Massachusetts, 1953.

THE AREA OF THE ELLIPSE*

ROGER BURR KIRCHNER, Harvard University

Consider the ellipse E whose equation is $x^2/a^2 + y^2/b^2 = 1$, where $a, b > 0$. Suppose that we make the transformation $(x, y) \rightarrow (ax, by)$. The x- and y-axes are stretched (or compressed) into new axes which we call the x' and y' axes. If the x-y coordinates of a point P are (x, y) and its x'-y' coordinates are (x', y'), then clearly

(1) $$x = ax', \quad y = by'.$$

Thus, in the x'-y' system, E has the equation $x'^2 + y'^2 = 1$.

Let Δ be a triangle with one side parallel to the x-axis. Suppose that the x-y coordinates of its vertices are (x_1, y_1), (x_2, y_1), and (x_3, y_3), and the corresponding x'-y' coordinates are (x_1', y_1'), (x_2', y_1'), and (x_3', y_3'). If the area of Δ in the x-y system is denoted by A, and in the x'-y' system by A', then

(2) $$A = \tfrac{1}{2} |x_2 - x_1| \cdot |y_3 - y_1|, \quad A' = \tfrac{1}{2} |x_2' - x_1'| \cdot |y_3' - y_1'|.$$

Hence, from the equations (1), we find that $A = abA'$.

Since any region bounded by a polygonal curve can be broken up into triangles of the type Δ, we see that its area in the x-y system is just ab times its area in the x'-y' system. The same must then be true for regions whose boundaries are limits of polygonal curves. Such a region is the ellipse E. Since E is just the unit circle in the x'-y' system, its x'-y' area is π. Hence the area of the ellipse with equation $x^2/a^2 + y^2/b^2 = 1$ is πab.

---

* From AMERICAN MATHEMATICAL MONTHLY, vol. 68 (1961), p. 653.

CALCULUS—A NEW LOOK*

D. E. RICHMOND, Williams College

1. Introduction. It is something of an historical accident that analytic geometry was discovered before calculus, so that the differential calculus was formulated in terms of the tangent problem and the integral calculus in terms of the area problem. It is now possible to see that these formulations are geometrical applications of more fundamental ideas: (1) The idea of a local multiplier from the domain space of a function to its range space; (2) The idea of the average value of a function over an interval. These ideas simplify the understanding of calculus to a surprising extent. It is the purpose of this paper to explain and justify this statement. We shall confine ourselves, for simplicity, to functions of one variable.

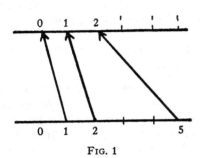

Fig. 1

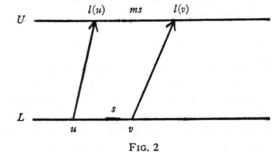

Fig. 2

2. Mappings. Constant and linear functions. Many recent elementary texts introduce functions as mappings from a set D, the *domain*, onto a set R, the *range*. For our purposes D and R shall be subsets of the real numbers. They may be represented as point-sets, usually intervals, on two parallel lines, provided with identical scales.

Thus the function

$$f: x \to \sqrt{(x-1)}$$

maps the reals $x \geq 1$ onto the nonnegative reals (Fig. 1). This function is *increasing*, since an increase in x corresponds to an increase in $f(x) = \sqrt{(x-1)}$.

The simplest functions are the *constant functions*,

$$c: x \to c$$

(the function c maps all reals onto the constant c) and the *linear functions*

$$l: x \to mx + b,$$

(m and b constant, $m \neq 0$). Clearly l maps 0 into b. To interpret m, note that $l(v) - l(u) = m \cdot (v - u)$. Hence the step $s = v - u$ on the lower line L corresponds

---

* From AMERICAN MATHEMATICAL MONTHLY, vol. 70 (1963), pp. 415–423.

to the step ms on the upper line U. Thus m is the constant *multiplier* in passing from the domain space to the range space. (l increases if $m>0$, decreases if $m<0$).

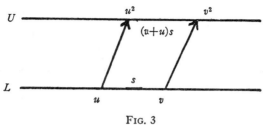

Fig. 3

3. The derivative. We turn now to functions that are not linear, but which are *locally linear* at points of their domain.

For example, if $f: x \to x^2$,

$$f(v) - f(u) = v^2 - u^2 = (v+u)\cdot(v-u) = (v+u)s.$$

The multiplier $v+u$ is no longer constant but depends on the coordinates of both end points. If we let u and v approach x, we obtain the limiting value $2x$, the *local multiplier* at x. For X near x, and x fixed, f is approximated by the function $l: X \to x^2+2x(X-x)$ (Fig. 4), which, as we observe, is *linear* in X.

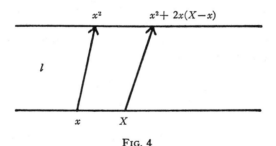

Fig. 4

Generally, if for a given function f and given x,

$$f(v) - f(u) = m(u, v)\cdot(v - u),$$

where the multiplier $m(u, v)$ approaches a limit $f'(x)$ as u and v approach x, f is said to be *differentiable* at x and $f'(x)$, the local multiplier at x, is called the *derivative* of $f(x)$. (We assume that $u \leq x \leq v$ and, if necessary, that $v-u$ be small.)

It is easy to see that if f is differentiable at x, it is continuous there. The proofs of the formulas for the derivatives of sums, products, powers, and quotients of differentiable functions require no discussion here. However, the chain rule becomes particularly intuitive in terms of the mapping picture.

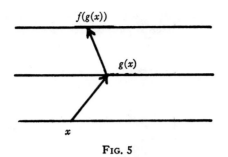

Fig. 5

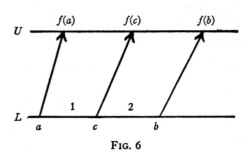

Fig. 6

Let x be mapped into $g(x)$ and $g(x)$ into $f(g(x))$ (Fig. 5). The local multiplier of g at x is $g'(x)$. At $g(x)$ the local multiplier of f is $f'(g(x))$. The over-all local multiplier at x is the product, $f'(g(x)) \cdot g'(x)$. This intuitive argument is easily turned into a simple rigorous proof. The interpretation of the chain rule in terms of the graphs of $u = g(x)$, $y = f(u)$, and $y = f(g(x))$ introduces needless complications.

For a one-one function f, we may represent the inverse function by reversing the direction of the arrows. The formula for the derivative of the inverse is immediate at points x at which $f'(x) \neq 0$.

4. Averages. Let f be differentiable on the closed interval $[a, b]$, $a < b$. To obtain $f(b) - f(a)$ on U from $b - a$ on L, we multiply by $\{f(b) - f(a)\}/(b - a)$ which may be called the average multiplier over the interval $[a, b]$. In fact, we shall take this ratio to be the *mean* or *average* value of the local multiplier $f'(x)$ over $[a, b]$ and accordingly write

(1) $$M(f') = \frac{f(b) - f(a)}{b - a}.$$

This terminology is justified by the following considerations.

Let c be any point of division. The suggested averages over the subintervals are

$$M_1(f') = \frac{f(c) - f(a)}{c - a}$$

and

$$M_2(f') = \frac{f(b) - f(c)}{b - c}.$$

Since $(c - a) M_1(f') + (b - c) M_2(f') = f(b) - f(a)$,

I. $$\frac{c - a}{b - a} M_1(f') + \frac{b - c}{b - a} M_2(f') = M(f').$$

Thus each subinterval contributes to the over-all average an amount which corresponds to the fraction of the whole interval which it represents. This is surely a property which we should want an average to have.

Moreover, our suggested average has the property that as the interval over which it is taken shrinks to a point, the average approaches the value of the function f' at that point. In fact, if we replace a by u and b by v

II. $$M(f') \text{ over } [u, v] \to f'(x)$$

as $u \to x$ and $v \to x$.

The average $M(f')$ given by (1) is the only average of f' which has the properties I and II. For let $A(f')$ be a second average with these two properties. Adjust the notation, if necessary, so that $[a, b]$ denotes a closed subinterval of the original interval for which $A(f')$ and $M(f')$ have distinct values. Assume for definiteness that $A(f') > M(f')$ (the other case is handled in a similar fashion), so that

$$A(f') = M(f') + d \qquad (d > 0).$$

Bisect $[a, b]$ and average over each half. Then either

(2) $$A_1(f') \geq M_1(f') + d \text{ or } A_2(f') \geq M_2(f') + d.$$

Otherwise by (I)

$$A(f') = \tfrac{1}{2}[A_1(f') + A_2(f')] < \tfrac{1}{2}[M_1(f') + M_2(f')] + d$$
$$= M(f') + d = A(f'),$$

a manifest contradiction.

By choosing a subinterval for which (2) holds and continuing the bisecting process, we obtain a sequence of intervals which shrink to a point x on $[a, b]$. Since for each of these intervals the A average exceeds the M average by at least d, condition II is contradicted.

It is apparent that the use of (1) enables us to compute the averages of many functions with great ease. Thus to find the mean value of $f' : x \to x^2$ over the interval $[1, 3]$, we set $x^2 = f'(x)$ and $x^3/3 = f(x)$. Then

$$M(f') = \frac{f(3) - f(1)}{3 - 1} = \frac{9 - \tfrac{1}{3}}{2} = \frac{13}{3}.$$

(The uniqueness proof shows that another choice of "antiderivative" for $f(x)$ can give no different answer.)

We may also use (1) to obtain the average of f' over an *open* interval (a, b), provided that f is continuous on the *closed* interval $[a, b]$. This result is obtained if we define $M(f')$ over (a, b) to be the limit of $M(f')$ over $[u, v]$ as $u \to a$ and $v \to b$.

For example, if $f(x) = 2\sqrt{x}$, $f'(x) = 1/\sqrt{x}$ and $M(f')$ over $(0, b)$ is equal to $(2\sqrt{b} - 0)/(b - 0) = 2/\sqrt{b}$.

Thus, to average a given function g over $[a, b]$ or (a, b), it suffices to discover a function f, continuous on $[a, b]$, such that $f'(x) = g(x)$ on (a, b) and then apply (1).

It is easy to extend this method of averaging to functions for which such antidifferentiation may be performed piecewise. It is sufficient to select antiderivatives on the subintervals so that the resultant function f is continuous at each function. This device is useful in the case of step-functions (Section 6).

It would be inappropriate in this place to discuss the well-known techniques of antidifferentiation.

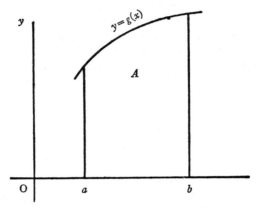

Fig. 7

5. Applications of averages. Area, volume, work. The standard applications of integration to the determination of areas, volumes, work, and so on, easily reduce to the problem of finding the average of an appropriate function. This obviates the necessity of introducing "elements" of a sum in the customary way.

For example, if $g(x) > 0$ on (a, b), we define the area A bounded by the graph of $y = g(x)$, the x-axis and the ordinates at a and b, to be the average value of the ordinate times the length of the base $(b-a)$. Hence, if $f'(x) = g(x)$,

$$A = (b - a)M(g) = f(b) - f(a).$$

The modifications to be made if $g(x) < 0$ on (a, b) or if $g(x)$ changes sign are immediate and intuitively clear.

Similarly, the area between $y = g_1(x)$, $y = g_2(x)$, $(g_1(x) < g_2(x))$, $x = a$ and $x = b$, is defined to be $(b-a)M(g_2 - g_1)$ in a very natural manner.

In polar coordinates, the area bounded by $r = g(\theta)$, $\theta = \alpha_1$ and $\theta = \alpha_2$ is defined to be

$$(\alpha_2 - \alpha_1) M\left(\frac{g^2}{2}\right)$$

as is immediately suggested by the equivalent circular sector.

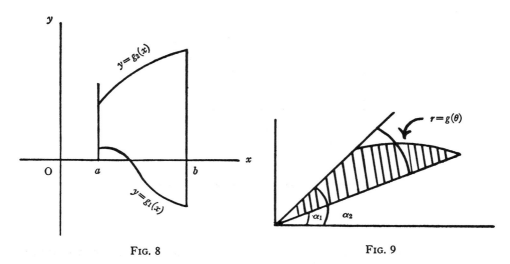

Fig. 8   Fig. 9

We turn next to *volumes*. Let $A(x)$ be the area of the cross section obtained by cutting a solid by a plane perpendicular to the x-axis at abscissa x. Then V, the volume between the planes $x=a$ and $x=b$, may be defined to be the average cross-sectional area, $M(A)$, times the altitude $(b-a)$. For a solid of revolution we obtain

$$(b - a)M(\pi g^2) = \pi[f(b) - f(a)],$$

where $f'(x) = g^2(x)$.

The volumes customarily obtained by the use of cylindrical shells may be found by averaging the areas of cylindrical surfaces. If this is not sufficiently intuitive, the correct formula is easily derived from the previously discussed case by integration by parts. This involves no appeal to the theorems of Duhamel or Bliss and no sloppy neglect of pieces.

The reader will easily be able to see how work and pressure problems, centroids and the like can be handled in a similar way. He might, however, wonder how arc lengths fit into this scheme.

Let $x=f(t)$ and $y=g(t)$, $a \leq t \leq b$, be parametric equations of a plane curve where f and g are functions with continuous derivatives.

Imagine a particle moving along the curve with horizontal and vertical components of velocity given by $f'(t)$ and $g'(t)$. Then at time t the speed is

$$v = \sqrt{(f'(t)^2 + g'(t)^2)}.$$

It is then natural to define the arc length to be $(b-a)M(v)$, where the average is taken over the time interval $[a, b]$. The specialization to the case $f(t) = t$ gives the familiar result.

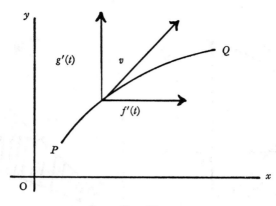

Fig. 10

These examples show that by focusing on the idea of average, we obtain a unifying concept which simplifies the application of calculus to geometry and the natural sciences. We shall not insist upon the obvious importance of this concept for probability and statistics.

To some readers this formulation will no doubt seem subversive because it short-circuits the important idea of the limit of a sum. The answer is twofold. First, if it is possible to make calculus a less formidable subject than it has been and currently is, we shall make this powerful tool accessible to a large number of people who could profitably use it. Secondly, this approach can be made every bit as rigorous as the conventional one. We have proved the *uniqueness* of the averages obtained. We have not, of course, proved the *existence* of an average for an arbitrary continuous g. This is not too difficult to do (see Section 6). But quite apart from such a proof, the student rightly feels that if he can actually find an answer (using antidifferentiation) in a given case, there is no *need* to prove existence.

6. Approximations. If we are unable to antidifferentiate a given $g(x)$ which we wish to average over an interval, it is natural to substitute an approximation, $h(x)$, which we *can* antidifferentiate.

If we can choose an $h(x)$ so that

$$-\epsilon \leq g(x) - h(x) \leq \epsilon$$

throughout $[a, b]$, then

(3) $$-\epsilon \leq M(g) - M(h) \leq \epsilon.$$

This result rests upon the following theorems.

1. *If $f(x) \geq 0$ on $[a, b]$, then $M(f) \geq 0$.*

The assumption that $M(f) < 0$ leads to a contradiction by repeated bisection as in Section 4.

2. *M is a linear functional.* That is, if $M(g)$ and $M(h)$ exist over $[a, b]$, and C_1 and C_2 are constants, then

$$M(C_1 g + C_2 h) = C_1 M(g) + C_2 M(h).$$

This follows from the uniqueness theorem, since the suggested average, $C_1 M(g) + C_2 M(h)$, satisfies I and II.

3. $M(c) = c$, *where c is the function which has the value c over* $[a, b]$ *or* (a, b).

Since $\epsilon + h(x) - g(x) \geq 0$ on $[a, b]$, then $M(\epsilon + h - g) \geq 0$ by 1. Also $M(\epsilon) + M(h) - M(g) \geq 0$ by 2 (twice), and $M(g) - M(h) \leq \epsilon$, since $M(\epsilon) = \epsilon$ by 3.

The other half of the inequality follows similarly.

Now, if we can choose ϵ in (3) arbitrarily small, we can approximate the desired $M(g)$ as closely as we please.

We consider two especially important choices of $h(x)$.

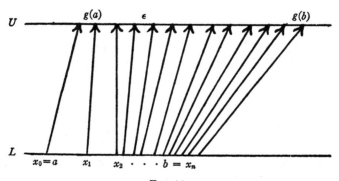

FIG. 11

a) *Step Functions.* Let g, a continuous function, increase from $g(a)$ to $g(b)$ as x increases from a to b. Subdivide this interval on U into n equal parts of length $\epsilon = (g(b) - g(a))/n$. Let the points of subdivision correspond to the x-values, $x_1, x_2, \cdots, x_{n-1}$ and let $x_0 = a$, $x_n = b$. On each half-open interval $[x_k, x_{k+1})$ replace $g(x)$ by its minimum value $g(x_k)$ on $[x_k, x_{k+1}]$. Let $h(x)$ be the "step function" so constructed. On each subinterval $[x_k, x_{k+1}]$, the average of h is $g(x_k)$. By repeated use of I, the average $M(h)$ over $[a, b]$ is seen to be

$$M(h) = \frac{1}{b-a} \sum_{0}^{n-1} g(x_k)(x_{k+1} - x_k).$$

By replacing $g(x)$ on each subinterval by its *maximum* value, $g(x_{k+1})$, we obtain an approximation $h_1(x)$ from above, whose average is

$$M(h_1) = \frac{1}{b-a} \sum_{0}^{n-1} g(x_{k+1})(x_{k+1} - x_k).$$

It is easy to show that $M(h_1) - M(h) = \epsilon$.

The required $M(g)$ must lie between $M(h)$ and $M(h_1)$. Thus $M(g)$ is the one and only one number which corresponds to the intersection of all intervals $[M(h), M(h_1)]$. The argument is easily extended to establish the existence of $M(g)$ if g is continuous and consists of a succession of a finite number of increasing and decreasing "pieces," and indeed, if g is an *arbitrary* continuous function. However, it will be a long time before the student will encounter functions for which this degree of generality is necessary.

The use of upper and lower sums to evaluate $M(g)$ with any desired accuracy gives concrete meaning to the usual existence proofs. Also, of course, the conventional notation for the definite integral, $\int_a^b g(x)dx$, is easily motivated by these considerations.

b) *Polynomials.* Let a given $g(x)$ have successive derivatives $g'(x)$, $g''(x), \cdots, g^{(n+1)}(x)$ with

(4) $$-K \leq g^{(n+1)}(x) \leq K \text{ on } [a, b].$$

If we antidifferentiate each term of (4) from a to x, $n+1$ times, we obtain

$$\frac{-K(x-a)^{n+1}}{(n+1)!} \leq g(x) - h(x) \leq \frac{K(x-a)^{n+1}}{(n+1)!},$$

where

$$h(x) = g(a) + g'(a)(x-a) + \cdots + \frac{g^{(n)}(a)(x-a)^n}{n!}.$$

For sufficiently large n, $h(x)$ gives an arbitrarily good approximation to $g(x)$ over the whole interval, provided that (replacing K by K_{n+1})

$$K_{n+1} \frac{(b-a)^{n+1}}{(n+1)!} \to 0$$

as n becomes infinite. The generalization to a Taylor series about an interior point is obvious.

It will be seen that this treatment of series fits naturally into the general pattern of thought which we have been developing.

AREAS AND VOLUMES WITHOUT LIMIT PROCESSES*

D. E. RICHMOND, Williams College

1. Introduction. It is universally believed that areas under polynomial graphs can be found only by some sort of limit process. This is not the case!

We shall assume as usual that the area of a plane region bounded by a simple closed curve is a positive number, that congruent regions have the same area, and that the area of the union of two nonoverlapping regions is the sum of their areas. Finally, we assume that the area of a rectangle is the product of the length of its base by its altitude.

We begin with a simple case.

2. The area under a parabola. Let $F(x)$ be the area under the parabola $y = x^2$ above the interval $[0, x]$ and $F(x')$ the area above $[0, x']$ ($x' > x$). Then the area above $[x, x']$ is $F(x') - F(x)$ (additivity). This area is greater than that of the rectangle of base $x' - x$ and altitude x^2, and less than the area of the rectangle of base $x' - x$ and altitude x'^2 (additivity and positivity). Hence,

$$(1) \qquad (x' - x)x^2 < F(x') - F(x) < (x' - x)x'^2.$$

This must be true for all $0 \leq x < x'$.

It is very easy to find a function F such that the double inequality (1) is satisfied for all $0 \leq x < x'$. $F(x)$ is clearly less than half the area of the rectangle of base x and altitude x^2, so that $F(x) = x^3/3$ (and $F(x') = x'^3/3$) is a good guess.

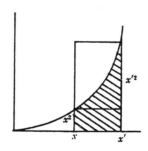

Fig. 1

In fact,

$$\frac{x'^3}{3} - \frac{x^3}{3} = (x' - x)\left(\frac{x^2 + xx' + x'^2}{3}\right)$$

is less than $(x' - x)x'^2$ and greater than $(x' - x)x^2$.

If desired, one can start with the easily derived inequality

$$(x' - x)3x^2 < x'^3 - x^3 < (x' - x)3x'^2$$

and divide by 3.

---

* From AMERICAN MATHEMATICAL MONTHLY, vol. 73 (1966), pp. 477–483.

$F(x) = x^3/3$ is of course the correct solution. To justify it we must show that if there exists an area function F for which $F(0) = 0$ and

$$(x' - x)x^2 < F(x') - F(x) < (x' - x)x'^2$$

for all $0 \leq x < x'$, then this function is unique. We need not prove the *existence* of such a function, since a bird in the hand exists.

The essence of the method is to note that for a function f which increases on an interval $[a, b]$, an area function $F(x)$ must satisfy the double inequality

(2) $$(x' - x)f(x) < F(x') - F(x) < (x' - x)f(x')$$

for all $x < x'$ on the interval. For elementary functions, F can be found easily without dividing by $x' - x$ and passing to the limit as $x' \to x$ as is customary. The exhibition of such a function F makes it unnecessary to prove its existence. If $F(a) = 0$, which is easy to arrange, the theorem to be proved in the next section shows that $F(x)$ is the only possible answer for the required area above $[a, x]$.

Since we can treat decreasing functions by reversing the inequality signs in (2), this simple method is sufficient to handle elementary functions which are piecewise monotone.

3. The uniqueness proof.

THEOREM 1. *Let f be a nonnegative strictly increasing function on the interval $a \leq x \leq b$. Let $F(x)$ be the area below the graph of $y = f(x)$ and above the interval $[a, x]$. If $F(a) = 0$ and*

(2) $$(x' - x)f(x) < F(x') - F(x) < (x' - x)f(x')$$

for all $a \leq x < x' \leq b$, F is uniquely defined.

Proof. Assume that there exists a different function $\overline{F}$ satisfying (2) with $\overline{F}(a) = 0$ so that for some $c (a < c \leq b)$, $\overline{F}(c) \neq F(c)$.

Let

$$x_k = a + \frac{k(c - a)}{n},$$

n a positive integer and $k = 0, 1, 2, \cdots, n$. Then

$$\frac{c - a}{n} f(x_{k-1}) < F(x_k) - F(x_{k-1}) < \frac{c - a}{n} f(x_k) \qquad (k = 1, 2, \cdots, n).$$

Summing from $k = 1$ to $k = n$,

$$\frac{c - a}{n} \sum_0^{n-1} f(x_k) < F(c) < \frac{c - a}{n} \sum_1^n f(x_k).$$

Similarly,

$$\frac{c-a}{n}\sum_0^{n-1} f(x_k) < \overline{F}(c) < \frac{c-a}{n}\sum_1^{n} f(x_k).$$

Hence, $|\overline{F}(c) - F(c)| < [(c-a)/n][f(c) - f(a)]$ and

$$n < \frac{(c-a)[f(c) - f(a)]}{|\overline{F}(c) - F(c)|}.$$

Since the Archimedean axiom assures us that there exists a positive integer n for which

$$n |\overline{F}(c) - F(c)| > (c-a)[f(c) - f(a)],$$

the assumption that $\overline{F}(c) \neq F(c)$ leads to a contradiction and must be rejected.

The proof is easily modified to treat decreasing functions where the inequality signs in (2) are reversed.

It will be observed that this proof does not require that f be continuous. (I am indebted to Professor Robert H. Breusch for suggesting an improvement of a previous proof which did assume the continuity of f. See also Apostol, *Calculus*, Vol. I, pp. 7–8.) It therefore appears that no use has been made of the limit concept. Nor has the Cantor continuity postulate for the reals been used.

4. Areas under polynomial graphs. The double inequality

(2) $\quad (x' - x)f(x) < F(x') - F(x) < (x' - x)f(x') \qquad (0 \leq x < x')$

is easily solved if $f(x) = cx^n$, $c > 0$ and n a positive integer. Since

$$x'^{n+1} - x^{n+1} = (x' - x)(x^n + x^{n-1}x' + \cdots + x'^n)$$

is between $(x'-x)(n+1)x^n$ and $(x'-x)(n+1)x'^n$ we readily find

$$F(x) = \frac{cx^{n+1}}{n+1}.$$

Of course, $F(0) = 0$.

To handle polynomial functions, we need two simple theorems.

THEOREM 2. *If f and g are nonnegative strictly increasing functions on $x \geq 0$ and F and G are their respective area functions, then $F+G$ is the area function for $f+g$.*

Proof. We add

$$(x' - x)f(x) < F(x') - F(x) < (x' - x)f(x')$$

and

$$(x' - x)g(x) < G(x') - G(x) < (x' - x)g(x')$$

and obtain

$$(x'-x)[f(x)+g(x)] < [F(x')+G(x')] - [F(x)+G(x)] < (x'-x)[f(x')+g(x')].$$

Note that $(F+G)(0) = F(0) + G(0) = 0$.

Repeated use of this theorem gives the areas under polynomial graphs over any interval $[a, b]$, $a \geq 0$, provided that all the coefficients are positive.

For polynomials some of whose coefficients are negative, we may write the polynomial function as the difference of two increasing polynomial functions, p and q. ($f = p - q$.) We assume that we are dealing with an interval for which $p(x) > q(x)$. The graph of $f(x)$ consists of a finite number of pieces over which f increases or decreases. For definiteness, assume that f increases on the interval $[a, b]$.

THEOREM 3. *Let p and q be nonnegative, strictly increasing functions on $[a, b]$ and let P and Q be the corresponding area functions. If $f = p - q$ is positive and increasing (decreasing) on $[a, b]$, then $F = P - Q$ is the area function for f.*

Geometric Proof. The area above $[x, x']$ ($a \leq x < x' \leq b$) and below $y = p(u)$ ($x \leq u \leq x'$) is $P(x') - P(x)$. The corresponding area below the horizontal line $y = p(x)$ is $(x'-x)p(x)$. Hence the area R (see Figure 2) is

$$P(x') - P(x) - (x'-x)p(x).$$

Similarly, the area S is $Q(x') - Q(x) - (x'-x)q(x)$.

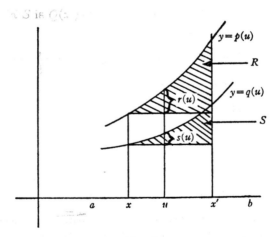

FIG. 2

Since $p - q$ increases on $[x, x']$,

$$p(u) - q(u) > p(x) - q(x) \qquad (x < u)$$

or $r(u) = p(u) - p(x) > q(u) - q(x) = s(u)$. Hence, $R > S$, that is,

$$P(x') - P(x) - (x' - x)p(x) > Q(x') - Q(x) - (x' - x)q(x)$$

and

$$[P(x') - Q(x')] - [P(x) - Q(x)] > (x' - x)[p(x) - q(x)].$$

Thus

$$F(x') - F(x) > (x' - x)f(x).$$

Similarly from $p(u) - q(u) < p(x') - q(x')$, $(u < x')$, we find that

$$F(x') - F(x) < (x' - x)f(x').$$

Since $F(0) = P(0) - Q(0) = 0$, $F = P - Q$, is the required area function.

Analytic Proof. To make the previous proof analytic we need to prove two results without an appeal to the figure.

1. For *fixed* x, $P(u) - P(x) - (u - x)p(x)$ is the area function for $p(u) - p(x)$ which vanishes at $u = x$. The analogous statement for Q and q immediately follows.

Proof. For $x \leq u < u' \leq x'$, $p(u)(u' - u) < P(u') - P(u) < p(u')(u' - u)$. Add $-p(x)(u' - u)$ throughout. Then

$$[p(u) - p(x)](u' - u) < [P(u') - p(x)u'] - [P(u) - p(x)u]$$
$$< [p(u') - p(x)](u' - u).$$

If we add $-P(x) + xp(x)$ within each of the central brackets, we have the required result.

2. Let $r(u)$, $s(u)$ and $r(u) - s(u)$ be nonnegative and increasing on $x \leq u \leq x'$ and $r(x) = s(x) = 0$, and let $R(u)$ and $S(u)$ be the area functions of $r(u)$ and $s(u)$ which vanish at $u = x$. Then

$$R = R(x') > S = S(x').$$

Proof. Subdivide $[x, x']$ into $2n$ intervals each of length $\Delta u = (x' - x)/2n$. On each subinterval $[u_{k-1}, u_k]$

(3) $$R(u_k) - R(u_{k-1}) > r(u_{k-1})\Delta u$$

and

(4) $$S(u_k) - S(u_{k-1}) < s(u_k)\Delta u.$$

Change the signs in (4) and add. Then

$$(R - S)(u_k) - (R - S)(u_{k-1}) > [r(u_{k-1}) - s(u_k)]\Delta u.$$

Summing from $k = 1$ to $k = 2n$,

$$(R - S)(x') = R - S > \left[\sum_{1}^{2n-1} r(u_k) - \sum_{1}^{2n} s(u_k)\right]\Delta u$$

$$= \sum_{1}^{2n} [r(u_k) - s(u_k)]\Delta u - r(x')\Delta u$$

$$> \sum_{n}^{2n} [r(u_k) - s(u_k)]\Delta u - r(x')\Delta u$$

$$> [r(u_n) - s(u_n)] \frac{x' - x}{2} - r(x')\Delta u.$$

Let

$$r(u_n) - s(u_n) \left[\text{or } r\left(\frac{x + x'}{2}\right) - s\left(\frac{x + x'}{2}\right) \right] = d,$$

and take $\Delta u < (x'-x)d/2r(x')$. Then $R - S > 0$ and $R > S$.

5. Volumes. Let a solid be cut by a plane perpendicular to the x-axis at abscissa $x(\geq 0)$. Let its cross-sectional area $S(x)$ increase with x. If $F(x)$ is the volume between the planes at 0 and x, and $F(x')$ the volume between the planes at 0 and x', we have the double inequality

(5) $\qquad (x' - x)S(x) < F(x') - F(x) < (x' - x)S(x'), \qquad x' > x.$

For example, if the cross-section is a square of area x^2

(6) $\qquad (x' - x)x^2 < F(x') - F(x) < (x' - x)x'^2.$

This is the same inequality (1) that arose in connection with the area below the parabola. But in (6), $F(x)$ is the volume of a square pyramid. Hence $F(x) = \frac{1}{3}x^2 x = x^3/3$ and $F(x') = x'^3/3$. Therefore, (1) could have been solved by referring to the square pyramid. The area under a parabola is just as elementary as the volume of a pyramid!

It should be clear that problems which involve the distance for a given velocity, the work against a variable force and the like, are equally easy to set up and solve by a simple use of inequalities.

6. Remarks. It is immediately clear that the ordinate function f is the derivative of what we have called the area function F. The method of this paper therefore gives a way of finding derivatives without the use of a limit process.

When f is increasing on $[a, b]$, the graph of $y = F(x)$ is convex and the tangent is characterized by the fact that the curve lies above it except at one point. In fact, for $a < c < x \leq b$, $(x-c)f(c) < F(x) - F(c)$, that is,

$$F(x) > F(c) + f(c)(x - c).$$

For $a \leq x < c$, $F(c) - F(x) < (c-x)f(c)$ and $F(x) > F(c) + f(c)(x-c)$. Thus $F(x)$ is above $y = F(c) + f(c)(x-c)$ for all $x \neq c$ on $[a, b]$.

Similarly, if f decreases, $F(x) < F(c) + f(c)(x-c)$ for $x \neq c$.

It is easy to show that if $F(x)$ has a derivative on (a, b) its graph has no finite jumps. It follows that if f itself has a derivative, then f is unique. It is also the case that this method enables one to obtain results which closely parallel those of elementary calculus.

It is frequently emphasized that inequalities should be studied in secondary school, but the applications customarily given do not convince most students of their importance. The fact that results traditionally found by calculus methods can be obtained so easily from the algebra of inequalities immediately opens up significant applications of this algebra. Moreover, it enables the student to handle these applications without the subtleties of limit theory. This is a possible answer to the problem of teaching calculus in secondary school.

It is also our conviction that elementary calculus is now being taught in an oversophisticated way and that we are discouraging many talented students by refined theoretical considerations which are actually unnecessary. If mathematics is as important as we say it is, it is also important that students with modest theoretical appetite should understand it much better than they obviously do. It is therefore incumbent upon us to make mathematics as simple as it can be made without sacrifice of accuracy and power. It is hoped that this paper will contribute to this end.

BIBLIOGRAPHIC ENTRIES: PEDAGOGY

Except for the entry labeled MATHEMATICS MAGAZINE, the references below are to the AMERICAN MATHEMATICAL MONTHLY.

1. Moore, E. H., A note on mean values, vol. 2, p. 303.

    Points out the need for a precise definition of the concept of the mean value of a function defined on an infinite set.

2. Ford, W. B., The teaching of the calculus, vol. 17, p. 77.
3. Graham, B., Some calculus suggestions by a student, vol. 24, p. 265.

    Describes difficulties of students in learning about integrals first as antiderivatives and then as limits of sums.

4. Rietz, H. L., On the teaching of the first course in calculus, vol. 26, p. 341.
5. Ballantine, J. P., Note on the introduction of integral calculus into a college course in solid geometry, vol. 32, p. 252.
6. Brown, A. B., To text-book writers—and readers, vol. 47, p. 375.

    Criticisms of arguments found in many calculus texts of 1940.

7. Moulton, E. J., Comments on the preceding paper, vol. 47, p. 381.

    A rebuttal of a small sample of Brown's paper (see entry 6).

8. Ransom, W. R., Introducing $e=2.718+$, vol. 55, p. 572.
9. Menger, K., Are variables necessary in calculus?, vol. 56, p. 609.
10. Nowlan, F. S., Objectives in the teaching of college mathematics, vol. 57, p. 73.
11. Richardson, M., Fundamentals in the teaching of mathematics, vol. 58, p. 182.
12. Prenowitz, W., Insight and understanding in the calculus, vol. 60, p. 32.
13. James, Glenn, The gist of the calculus, MATHEMATICS MAGAZINE, vol. 22, p. 29.

3

FUNCTIONS

(a)

CONCEPTS AND NOTATION

A KIND OF PROBLEM THAT EFFECTIVELY TESTS FAMILIARITY WITH FUNCTIONAL RELATIONS*

K. O. MAY, Carleton College

The kind of problem described here probes the student's understanding of the functional concept and notation, tests his familiarity with the functions being studied, and is easy to run off on a duplicator and to correct.

Present the student with the graph of a rather simple function, perhaps piecewise constant or at least piecewise linear. Label this function $f(x)$ and then require the student to graph $-f(x), f(-x), |f(x)|, f(|x|), e^{f(x)}, \sin f(x)$ or whatever functions involve the ideas whose understanding you wish to test. The correct graphs are often decorative and always easily recognized. The student's deviations indicate clearly in many cases the nature of his difficulties. While the resulting graphs are easy to check, it is not hard to construct examples that require considerable thought and challenge the best students. On the other hand, the simpler examples, such as $-f(x)$ or the inverse function of $f(x)$, appeal to the manipulative students. Obvious modifications, such as the use of two functions, will occur to any teacher.

As an example suppose that $y = f(x)$ has the graph:

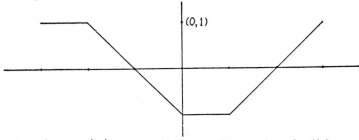

Sketch: $y = |f(x)|$, $y = f(|x|)$, $y = -f(x)$, $y = e^{f(x)}$, $y = \text{Arc sin } f(x)$.

---

* From AMERICAN MATHEMATICAL MONTHLY, vol. 60 (1953), p. 624.

A GEOMETRIC CONSTRUCTION OF COMPOSITE FUNCTIONS*

KURT KREITH, University of California, Davis

Let f and g be real functions of a real variable and suppose that the domain of g contains the range of f. Then the composite function $F(x) = g(f(x))$ is defined for all x in the domain of f. The purpose of this note is to give a simple geometrical description of the graph of the function $F(x)$ which adds to the student's insight into the operation of composition.

In a 3-dimensional coordinate system we begin by graphing $y = f(x)$ and $z = g(y)$ in the planes $z = 0$ and $x = 0$, respectively. The problem is to describe the graph of $z = F(x)$ in the plane $y = 0$.

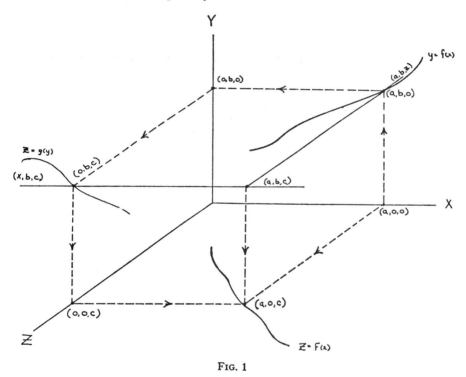

Fig. 1

Beginning at $x = a$ we follow the mapping $b = f(a)$, $c = g(b)$, and arrive at the conclusion that $(a, 0, c)$ belongs to the graph of $z = F(x)$.

The point $(a, 0, c)$ could also have been reached as follows: Construct the lines (a, b, z) and (x, b, c). These intersect at (a, b, c), and the projection of (a, b, c) onto the plane $y = 0$ yields $(a, 0, c)$.

---

* From AMERICAN MATHEMATICAL MONTHLY, vol. 69 (1962), pp. 293–294.

Now to find *all* the points in the graph $z = F(x)$ we consider the cylinders

$$S_1 = (x, f(x), z),$$
$$S_2 = (x, y, g(y)).$$

The intersection of these cylinders is a space curve C, and the projection of C onto the plane $y = 0$ clearly yields the graph of $z = F(x)$.

This construction of the composite function in a 3-dimensional space also facilitates a graphical explanation of certain theorems on composite functions (e.g. the composition of continuous functions is continuous).

BIBLIOGRAPHIC ENTRY: CONCEPTS AND NOTATION OF FUNCTIONS

Nyberg, J. A., The presentation of the notion of function, AMERICAN MATHEMATICAL MONTHLY, vol. 24, p. 309.

(b)
TRIGONOMETRIC FUNCTONS

POLAR COORDINATE PROOFS OF TRIGONOMETRIC FORMULAS*

OSWALD VEBLEN, University of Chicago

1. Graphical, that is to say analytic geometrical methods, seem at present to be on the gain in the teaching of Trigonometry. Particularly true is this in courses conducted by the "Laboratory Method." This fall, I have obtained rather pleasing results by adopting a suggestion of Professor Moore to use polar coordinates. The geometric simplicity of these graphs, the sine and cosine being represented by circles and the secant and cosecant by straight lines (see Fig. 1), not only makes them attractive to the student but, unlike the Cartesian graphs, makes them useful in proving theorems.

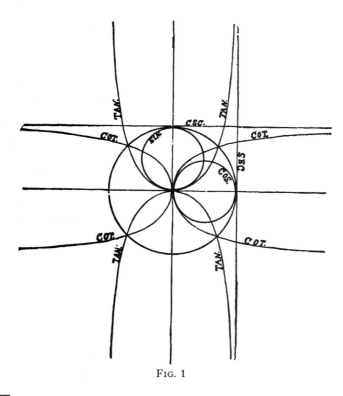

Fig. 1

---

* From AMERICAN MATHEMATICAL MONTHLY, vol. 11 (1904). pp. 6–12.

The proofs* given below, it is hoped, will demonstrate this latter point. That they contain some elements of simplicity I am convinced by the fact that several of my students worked out the proof of the formula for sin $(x+y)$ with no other help than the mere suggestion to use polar coordinates. §§6, 7, 8 are based on memoranda given me by Professor Moore of work intended for his elementary calculus course. The proofs are made only for positive angles less than $\frac{1}{2}\pi$.

2. *If an angle is inscribed in a circle of unit diameter its sine is the chord of the arc subtended.*

If one side OA of the angle AOB is a diameter of the circle (see Fig. 2), then since OBA is a right angle, $AB/OA = AB/1$ is the sine of AOB. If the angle is inscribed in any other way, by a familiar theorem, it subtends the same chord as AOB.

The theorem is also true in the limiting case where one side of the angle is tangent to the circle. This is the polar coordinate case and thus, in Fig. 3, $OA = \sin AOX$. We may note also that in a circle of unit diameter the length of the arc subtended by an inscribed angle is the measure of that angle in radians.

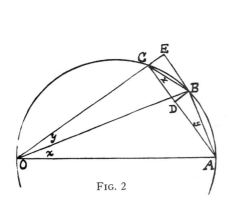

Fig. 2

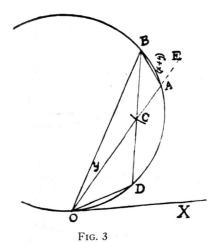

Fig. 3

3. Proof of the formula sin $(x+y) = \sin x \cos y + \cos x \sin y$.
In a circle ABO (Fig. 2)† of unit diameter, let

$$\angle AOB = x, \qquad \therefore AB = \sin x.$$
$$\angle BOC = y, \qquad \therefore BC = \sin y.$$
$$\therefore \angle AOC = x + y, \qquad \therefore AC = \sin (x + y).$$

---

* While these proofs are probably to be found somewhere in the literature, I have not been able to find them.

† Of course, in this proof OA need not be a diameter.

Let BD be perpendicular to AC. Then

$$\angle BAC = y \text{ (subtending same arc as } \angle BOC),$$
$$\angle BCA = x \text{ (subtending same arc as } \angle AOB).$$
$$\therefore AD = AB \cos y = \sin x \cos y,$$
$$DC = BC \cos x = \sin y \cos x.$$

Since $AC = AD + DC$, $\sin(x+y) = \sin x \cos y + \cos x \sin y$.

This proof applies directly to the polar graph (see Fig. 4) if OA is taken tangent to the circle (so that $A = 0$). The present form is intended to suggest its use to those who do not care to introduce polar coordinates in a beginning course.

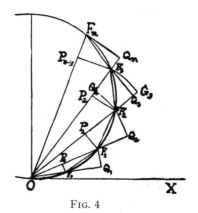

FIG. 4

4. Proof that $\cos(x+y) = \cos x \cos y - \sin x \sin y$. Let OAB (Fig. 2) be the polar cosine curve, i.e., OA is the diameter $(=1)$ of the circle OAB.

$$OB = \cos x, \quad OC = \cos(x+y), \quad BC = \sin y.$$

Let BE be perpendicular to OC.

$$\therefore OE/OB = \cos y. \quad \therefore OE = \cos x \cos y.$$

EB and CA are parallel, both being perpendicular to OC.

$$\therefore \angle CBE = \angle BCA = x. \quad \therefore CE/CB = \sin x. \quad \therefore CE = \sin x \sin y.$$

Since $OC = OE - CE$, $\cos(x+y) = \cos x \cos y - \sin x \sin y$.

5. Proof that

$$\sin x - \sin y = 2 \cos \frac{x+y}{2} \sin \frac{x-y}{2}.$$

In the polar sine circle, Fig. 3, $OA = \sin x$, $AB = \sin y$. On OE, lay off $AC = AB$ and let D be the point in which BC meets the circle.

$$\therefore OC = \sin x - \sin y.$$

By elementary geometry, $\angle BAE = \angle AOB + \angle ABO = x+y$.

$$\therefore \angle CBA = \angle BCA = \frac{x+y}{2}. \qquad \therefore \angle OCD = \frac{x+y}{2}.$$

Now $\angle COD$ is measured by the arc AD and hence

$$\angle COD = \angle ABD = \frac{x+y}{2}$$

Hence OCD is an isosceles triangle and

$$OC = 2 \cdot OD \cos \frac{x+y}{2}.$$

But

$$OD = \sin(\angle OBD) = \sin\left(x - \frac{x+y}{2}\right) = \sin \frac{x-y}{2}.$$

$$\therefore \sin x - \sin y = OC = 2 \sin \frac{x-y}{2} \cos \frac{x+y}{2}.$$

6. First proof of the formulas

$$D_x \sin x = \cos x, \tag{1}$$

$$\int_{x_0}^{X} \cos x \, dx = \sin X - \sin x_0. \tag{2}$$

The proofs in this section make use of Fig. 4, in which $\angle XOF_0 = x_0$; $\angle F_0OF_1 = \Delta x = \angle F_{i-1}OF_i (i=1\ldots n)$; or, if one prefers to speak of the arcs,

$$\Delta x = \text{arc } F_0F_1 = \text{arc } F_1F_2 = \cdots = \text{arc } F_{n-1}F_n.$$

$F_{i-1}P_{i-1}$ is perpendicular to OF_i and F_iQ_i is perpendicular to OF_{i-1}. To prove (1), we make use of only one of the Δx portions of the figure, for example, the third. OG_2 is taken equal to OF_2 and $OG_3 = OF_3$. Then by elementary geometry

$$P_2F_3 > G_2F_3 = G_3F_2 > F_2Q_3 \tag{3}$$

But if we call $\angle XOF_2 = x$, $\angle F_2F_3O = x$, and $\angle F_3F_2Q_3 = x+\Delta x$,

$$F_2G_3 = OF_3 - OF_2 = \sin(x + \Delta x) - \sin x,$$

$$P_2F_3 = F_2F_3 \cos x = \sin \Delta x \cos x,$$

$$F_2Q_3 = F_2F_3 \cos(x + \Delta x) = \sin \Delta x \cos(x + \Delta x).$$

Hence (3) says that

$$\sin \Delta x \cos x > \sin(x + \Delta x) - \sin x > \sin \Delta x \cos(x + \Delta x). \tag{4}$$

$$\therefore \frac{\sin \Delta x}{\Delta x} \cos x > \frac{\sin(x + \Delta x) - \sin x}{\Delta x} > \frac{\sin \Delta x}{\Delta x} \cos(x + \Delta x).$$

Since

$$L_{h=0} \frac{\sin h}{h} = 1$$

and $\cos x$ is continuous, both extremes of this double inequality approach $\cos x$ as Δx approaches zero.

Therefore the middle term approaches $\cos x$ and we have

$$D_x \sin x = L_{\Delta x = 0} \frac{\sin(x + \Delta x) - \sin x}{\Delta x} = \cos x.$$

In this, according to our figure, Δx was always positive. But if XOF_3 had been taken as x the same figure with similar reasoning would prove (4) for that case also.

Of course the theorem that for continuous functions integration is the inverse of differentiation shows that (2) is a corollary of (1). But for some purposes of instruction it is worth while to compute (2) directly from the definition of an integral as the limit of a sum. Assuming the existence of a definite integral for $\cos x$ we have

$$\int_{x_0}^{X} \cos x \, dx = L_{n = \infty} \sum_{k=1}^{n} \cos(x_0 + k \Delta x) \Delta x \tag{5}$$

where $\Delta x = (X - x_0)/n$, and also*

$$\int_{x_0}^{X} \cos x \, dx = L_{n = \infty} \sum_{k=0}^{n-1} \cos(x_0 + k \Delta x) \Delta x \tag{6}$$

Since

$$L_{\Delta x = 0} \frac{\sin \Delta x}{\Delta x} = 1,$$

---

* In the familiar Cartesian figure, (5) corresponds to the inner set of rectangles and (6) to the outer.

(5) and (6) can be replaced by (7) and (8)*:

$$\int_{x_0}^{X} \cos x \, dx = L \sum_{n=\infty}^{n} \sum_{k=1}^{n} \cos(x_0 + k\Delta x) \sin \Delta x = L_{n=\infty} S_n, \qquad (7)$$

$$\int_{x_0}^{X} \cos x \, dx = L \sum_{n=\infty}^{n-1} \sum_{k=0}^{n-1} \cos(x_0 + k\Delta x) \sin \Delta x = L_{n=\infty} S'_n, \qquad (8)$$

and (2) will be proved if we show that $S_n < \sin X - \sin x < S'_n$.

From the quadrilateral $F_0 P_0 F_1 Q_1$, we obtain, as we obtained (4),

$$\sin \Delta x \cos x_0 > \sin(x_0 + \Delta x) - \sin x_0 > \sin \Delta x \cos(x_0 + \Delta x).$$

From the second quadrilateral $F_1 P_1 F_2 Q_2$ we similarly get

$$\sin \Delta x \cos(x_0 + \Delta x) > \sin(x_0 + 2\Delta x) - \sin(x_0 + \Delta x) > \sin \Delta x \cos(x_0 + 2\Delta x),$$

and so on. From the last quadrilateral we obtain, calling $X \angle XOF_n = x_0 + n\Delta x$,

$$\sin \Delta x \cos[x_0 + (n-1)\Delta x] > \sin X - \sin[x_0 + (n-1)\Delta x] > \sin \Delta x \cos(x_0 + n\Delta x).$$

Adding together these inequalities we see that the sum of the first terms is S'_n, and of the last terms is S_n. In the middle terms everything else cancels, leaving only $\sin X - \sin x_0$. So we have as we desired

$$S'_n > \sin X - \sin x_0 > S_n.$$

This result can be seen still more directly by noting that

$$S'_n = P_0 F_1 + P_1 F_2 + \cdots + P_{n-1} F_n,$$
$$S_n = F_0 Q_1 + F_1 Q_2 + \cdots + F_{n-1} Q_n,$$
$$OF_0 = \sin x_0, \qquad OF_n = \sin X.$$

If the rays centering at O be imagined to fold together like a fan from OF_0 to OF_n it is evident that S_n is less and S'_n greater than $OF_n - OF_0$.

7. Second Proof of (1) and (2).

In some quarters there is a tendency to reverse the old order and present the integral calculus before the differential. The definitions of the two operations of differentiation and integration are certainly independent of each other; and whatever order may be preferred for pedagogical reasons, it is not amiss to see that in either case precisely similar methods can be used in deriving the formulas for the usual functions. That such is the case depends on the following theorem. †

---

* *Ed. note:* To justify this step, use the relation $\Delta x = \sin \Delta x + o(\Delta x)$ as $\Delta x \to 0$.

† The proof of the first part of this theorem is made possible by the fact that any monotonic function is integrable,—a monotonic function being such that if $a < b$ either always $f(a) >$ or $= f(b)$ or always $f(a) <$ or $= f(b)$. Just as we did for the special case of $\sin x$ in the last section we can let $b - a = \Delta x$ and add up n inequalities like (9) and thus have $S_n > F(x) - F(x) > S'_n$. To prove the second part divide by $b - a$ and pass to the limit as b approaches a.

If on an interval $x_0, \ldots, X$, two functions $f(x)$ and $F(x)$ have the property that for every two values of x, a, and b ($x_0 \leq a < b \leq X$),

$$f(a)(b-a) > F(b) - F(a) > f(b)(b-a); \tag{9}$$

then, first,

$$\int_{x_0}^{X} f(x)\, dx = F(X) - F(x_0);$$

second, if $f(x)$ is continuous, $D_x F(x) = f(x)$.

In view of this theorem, to prove (1) and (2) we need only to prove the inequality

$$\cos x_0 \cdot (x - x_0) > \sin x - \sin x_0 > \cos x \cdot (x - x_0), \quad 0 \leq x_0 < x \leq \pi/2. \tag{10}$$

To this end we make use of the inner part of Fig. 5 in which
$$\angle XOS_0 = x_0, \qquad \angle XOS = x.$$

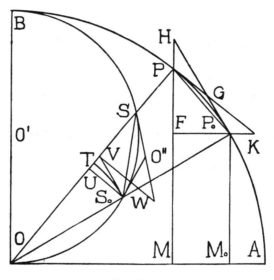

Fig. 5

$S_0 U$ is perpendicular to OS, $S_0 V$ to OS_0, and SO'' and $S_0 O''$ are tangents to the sine curve $OS_0 B$. About O'' a circle is described with radius $O''S = O''S_0$ and meeting $O''S$ in W. Since

$$\angle S_0 VS = \tfrac{1}{2}\pi + (x - x_0) = \pi - [\tfrac{1}{2}\pi - (x - x_0)],$$

and

$$\angle S_0 O''S = \pi - 2(x - x_0) = 2[\tfrac{1}{2}\pi - (x - x_0)],$$

the circle about O'' must pass through V. Hence $SVW = \frac{1}{2}\pi$. From these considerations it follows that

$$\sin x - \sin x_0 = TS > VS = \cos VSW \cdot WS$$
$$= \cos x \cdot (S_0 O'' + O''S) > \cos x \cdot \text{arc } S_0 S = \cos x \cdot (x - x_0),$$

and

$$TS < US = \cos USS_0 \cdot S_0 S < \cos x_0 \cdot \text{arc } S_0 S < \cos x_0 \cdot (x - x_0),$$

which proves (10).

8. Second Proof of (1) and (2).

Without going into details I will add the outline of a second proof of Professor Moore's for the inequality (10) and hence for (1) and (2). This, unlike the others, is not a polar coordinate proof, but uses the unit circle. In the outer part of Fig. 5,

$$\sin x - \sin x_0 = FP = \cos FPK \cdot PK > \cos x \cdot (PG + GP_0) > \cos x \cdot \text{arc } P_0 P$$
$$= \cos x \cdot (x - x_0),$$

$$FP = \cos FPP_0 \cdot P_0 P = \cos \frac{x + x_0}{2} \cdot P_0 P < \cos x_0 \cdot P_0 P$$

$$= \cos x \cdot (x - x_0).$$

$$\therefore \cos x \cdot (x - x_0) < \sin x - \sin x_0 < \cos x_0 \cdot (x - x_0),$$

which is (10).

The outer part of Fig. 5 can also be used to prove that

$$\sin x - \sin x_0 = 2 \cos \frac{x + x_0}{2} \sin \frac{x - x_0}{2}.$$

RELATING TO FINDING DERIVATIVES OF TRIGONOMETRICAL FUNCTIONS*

T. H. HILDEBRANDT, University of Michigan

In most textbooks on the elementary calculus the derivatives of the trigonometric functions are based on the derivative of the sine function, which, in turn, is derived from the definition of derivative. The proofs dealing with the value of this derivative seem to have something indirect about them. All goes well until the point is reached where the expression

$$\lim_{\Delta x \doteq 0} \frac{\sin(x + \Delta x) - \sin x}{\Delta x}$$

is to be evaluated, and then one of two methods is used. Either $\sin(x+\Delta x)$ is expanded by the formula for the sine of the sum of two angles and the formula for $1 - \cos x$ in terms of half angles is used, or the formula for the difference of two sines is used. Both of these latter formulæ have long since escaped the memory of the average sophomore student—if they ever had lodging there—and he practically accepts this part of the derivation on faith.

While it must be admitted that the most natural beginning for a chapter on the derivatives of trigonometrical functions is a paragraph devoted to finding the derivative of the sine, this advantage is more than counterbalanced by the simplicity with which it is possible to obtain the derivative of the tangent function directly from the definition of derivative—a fact which seems almost to have escaped the attention of writers of textbooks on calculus. For this purpose it is possible to proceed in either of two ways, both of which are elegant and altogether natural and direct. If we take $\tan x = \sin x/\cos x$, then

$$\tan(x + \Delta x) - \tan x = \frac{\sin(x + \Delta x)}{\cos(x + \Delta x)} - \frac{\sin x}{\cos x}$$

$$= \frac{\sin(x + \Delta x)\cos x - \cos(x + \Delta x)\sin x}{\cos(x + \Delta x)\cos x}$$

$$= \frac{\sin \Delta x}{\cos(x + \Delta x)\cos x}.$$

If we divide now by Δx and take the limit as Δx approaches zero, then $\lim_{\alpha \doteq 0} (\sin \alpha)/\alpha = 1$ is applicable for evaluating the derivative. Or, proceeding directly, we have

$$\tan(x + \Delta x) - \tan x = \frac{\tan x + \tan \Delta x}{1 - \tan x \tan \Delta x} - \tan x$$

$$= \frac{(1 + \tan^2 x)\tan \Delta x}{1 - \tan x \tan \Delta x}.$$

---

* From AMERICAN MATHEMATICAL MONTHLY, vol. 25 (1918), pp. 125–126.

Then by using the fact that $\lim_{\alpha \to 0} (\tan \alpha)/\alpha = 1$, we again get at once the value of the derivative of the tangent. Either of these methods yields this derivative without any troublesome trigonometrical transformations.

On the basis of the derivative of the tangent, the remaining derivatives are easily obtained in the order sec x, cos x, and sin x: the secant by taking derivatives in the relation

$$\tan^2 x + 1 = \sec^2 x,$$

the cosine through the fact that it is the reciprocal of the secant, and the sine from one of the relations

$$\sin x = \cos (\pi/2 - x), \qquad \sin x = \cos x \tan x.$$

We get, then, in this way the derivatives which are used most, and we employ only trigonometrical relations which are familiar to the average sophomore student.

AN ELEMENTARY LIMIT*

M. S. KNEBELMAN, State College of Washington

The proof of the fundamental theorem

$$\lim_{\theta \to 0} \frac{\sin \theta}{\theta} = 1,$$

as ordinarily given in elementary books, usually depends on two unproved theorems. The following proof is at least simpler, if not more rigorous.

If P is the perimeter of a regular n-gon inscribed in a circle of radius r, then $P = 2nr \sin \pi/n$ and we know from plane geometry that $\lim_{n \to \infty} P = 2\pi r$. Hence $\lim_{n \to \infty} (n/\pi) \sin \pi/n = 1$ and if we let $\pi/n = \theta$ then $\theta \to 0$ as $n \to \infty$ and conversely. Hence $\lim_{\theta \to 0} \sin \theta/\theta = 1$.

---

* From AMERICAN MATHEMATICAL MONTHLY, vol. 50 (1943), p. 507.

TRIGONOMETRY FROM DIFFERENTIAL EQUATIONS*

D. E. RICHMOND, Williams College

1. Introduction. This note shows how analytic trigonometry may be developed in an elementary manner (with no use of infinite series) from the differential equation

(I) $$\frac{d^2y}{dt^2} + y = 0$$

to which one is naturally led through the study of simple harmonic motion.

First write (I) as

(1) $$\frac{dx}{dt} = -y$$

by setting

(2) $$\frac{dy}{dt} = x.$$

Multiplying (1) by x and (2) by y and adding,

$$x\frac{dx}{dt} + y\frac{dy}{dt} = 0.$$

Hence $x^2+y^2=r^2$ where r is a constant. For given t, x and y are the coordinates of a point on a circle, omitting the trivial case $r=0$.

The radius of this circle becomes 1 if we set $x=rX$, $y=rY$. Then

(3) $$\frac{dX}{dt} = -Y$$

(4) $$\frac{dY}{dt} = X$$

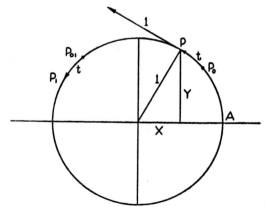

---

* From AMERICAN MATHEMATICAL MONTHLY, vol. 61 (1954), pp. 337–340.

where $X^2+Y^2=1$. Solutions $Y(t)$ so obtained will be said to be *normalized*.

Clearly (3) and (4) give the components of a motion with the uniform velocity 1 in a counterclockwise direction along the unit circle C. If arc length s is measured along C in this direction, $ds/dt = 1$ and $s = t+c$. In fact, $s = t$ if s is measured from P_0, the position at $t=0$.

Two normalized solutions of (I), $Y(t)$ and $Y_1(t)$, can differ only in the positions of their initial points, P_0 and P_{01} respectively. But then $Y_1(t) = Y(t+a)$ where a is the arc length from P_0 to P_{01}. Moreover, given one normalized solution of (I), $Y(t)$, then all functions $Y(t+a)$ with constant a are normalized solutions of (I). The general solution of (I) is therefore

$$y = rY(t+a)$$

with arbitrary constants r and a.

Let us define sin t to be that solution $Y(t)$ for which $Y=0$ and $X(=dY/dt)=1$ at $t=0$, that is, that solution for which $P_0 = A$ (see Figure). The corresponding $X(t) = dY(t)/dt$ will be defined to be cos t. From (3) and (4)

$$\frac{d \sin t}{dt} = \cos t,$$

$$\frac{d \cos t}{dt} = -\sin t.$$

It follows at once that cos t (as well as sin t) is a solution of (I).

It is geometrically obvious that $\sin(t+2\pi) = \sin t$ and $\cos(t+2\pi) = \cos t$ for all t, and also that

$$\sin(-t) = -\sin t, \qquad \cos(-t) = \cos t.$$

The radian (or circular) measure of angles becomes extremely natural.

2. The addition formulas. To derive the addition formulas, we verify by differentiation that

$$c_1 \sin t + c_2 \cos t$$

is a solution of (I) for arbitrary constants c_1 and c_2. Then for some r and some a,

(5) $$c_1 \sin t + c_2 \cos t = r \sin(t+a).$$

Differentiating,

(6) $$c_1 \cos t - c_2 \sin t = r \cos(t+a).$$

Substituting $t=0$ in (5) and (6),

$$c_1 = r \cos a, \qquad c_2 = r \sin a.$$

Inserting these values in (5) and (6) and cancelling the factor r,

(7) $$\sin(t+a) = \sin t \cos a + \cos t \sin a,$$

(8) $$\cos(t+a) = \cos t \cos a - \sin t \sin a.$$

The remaining formulas of analytic trigonometry follow without difficulty.

3. Calculation of sin t and cos t. If it is desired to *calculate* $\sin t$ and $\cos t$, it suffices to consider the case $t>0$ and integrate both members of $\cos t \leq 1$ from 0 to t, obtaining $\sin t < t$. Continuing,

$$1 - \cos t < \frac{t^2}{2}$$

or

$$\cos t > 1 - \frac{t^2}{2};$$

$$\sin t > t - \frac{t^3}{3!};$$

$$\cos t < 1 - \frac{t^2}{2!} + \frac{t^4}{4!};$$

$$\cdots\cdots\cdots\cdots.$$

The theorem used here is the intuitively obvious one that if $f(t)$ and $g(t)$ are two different integrable functions such that $f(t) \leq g(t)$, $(t>0)$, then

$$\int_0^t f(t)\,dt \leq \int_0^t g(t)\,dt, \qquad (t>0).$$

It is hoped that this note will nourish the suspicion that the conventional semester course in trigonometry involves considerable educational waste.

ON THE $\lim_{\theta \to 0} \cos \theta$*

M. J. PASCUAL, Siena College

In most texts on calculus, the proof of the fundamental limit equation $\lim_{\theta \to 0} \sin \theta/\theta = 1$ involves the assumption that the limit equation $\lim_{\theta \to 0} \cos \theta = 1$ is evident. The apparent reason for this supposedly obvious conclusion must have its basis in the fact that $\cos \theta$ is continuous at $\theta = 0$. But the notion of limit is a more primitive one than that of continuity. Hence to avoid the possible confusion which may easily arise in the mind of a beginner in calculus, the teacher should at least assert that the $\lim_{\theta \to 0} \cos \theta$ could be proved to be 1 by direct appeal to the definition of limit. Indeed, we could give the following proof, which would simultaneously introduce the student to the analytical method of proof and also prepare an example for the lesson on continuity, in which the instructor shows that $\cos \theta$ is actually continuous at $\theta = 0$. Perhaps it could be left to the students to show it is continuous at any value of θ.

Since this proof would be given at an early stage of a calculus course, prior to which inverse trigonometric functions may not have been covered, it will be desirous to avoid such functions, so we shall define $\cos \theta$ by the conventional method employing rectangular coordinates. This will also prove helpful in establishing the continuity at $\theta = 0$, as the old triangle definition will not suffice, there being no triangle.

Placing the vertex of θ at the origin of the rectangular system, with its initial side along the positive x-axis, and measuring θ as usual counterclockwise, we define $\cos \theta$ by choosing an arbitrary point other than $(0, 0)$ on the terminal side of θ having coordinates (x, y) and let $\cos \theta = x/\sqrt{x^2+y^2}$. Since $y \to 0$ as $\theta \to 0$, we wish to establish the equation $\lim_{y \to 0} x/\sqrt{x^2+y^2} = 1$. We shall restrict θ to the first or fourth quadrants and for convenience we may choose $x = 1$.

To fulfill the definition of limit, for any $\epsilon > 0$, we must produce a δ_ϵ such that

$$\left| \frac{1}{\sqrt{1+y^2}} - 1 \right| < \epsilon \quad \text{for} \quad |y| < \delta_\epsilon.$$

We analyze as follows: for any $y \neq 0$

$$0 < \frac{1}{\sqrt{1+y^2}} < 1 \quad \text{so that} \quad \left| \frac{1}{\sqrt{1+y^2}} - 1 \right| = 1 - \frac{1}{\sqrt{1+y^2}}.$$

Hence we wish to have

$$1 - \frac{1}{\sqrt{1+y^2}} < \epsilon$$

---

* From AMERICAN MATHEMATICAL MONTHLY, vol. 62 (1955), pp. 252–253.

and for $0 < \epsilon < 1$ (for $\epsilon \geq 1$ this inequality obviously holds)

$$1 - \epsilon < \frac{1}{\sqrt{1 + y^2}}$$
$$1 + y^2 < (1 - \epsilon)^{-2}$$
$$|y| < \sqrt{(1 - \epsilon)^{-2} - 1}$$

giving the desired δ, as

$$\sqrt{(1 - \epsilon)^{-2} - 1}.$$

ON THE DERIVATIVES OF TRIGONOMETRIC FUNCTIONS

M. R. SPIEGEL, Rensselaer Polytechnic Institute

In the usual course in calculus it is the custom to introduce the trigonometric functions and their derivatives before considering the derivatives of the inverse trigonometric functions. In the so-called unified courses in calculus where some of the elementary concepts of integration are taken up early, there seems to be some advantage in introducing these derivatives in the reverse manner.

To illustrate the possibilities in this direction we show how to find the derivative of arcsin x. Referring to the figure, ACB represents one quarter of

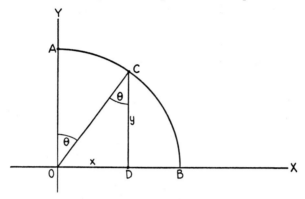

the circumference of a circle with center at the origin and having unit radius. The area of $OACD$ given by

$$\int_0^x \sqrt{1-x^2}\, dx$$

is also equal to the sum of the areas of OCD and OAC. This sum is given by

$$\frac{x}{2}\sqrt{1-x^2} + \frac{1}{2}\text{arc sin } x$$

from elementary formulas for the area of a triangle and sector of a circle. Hence

(1) $$\int_0^x \sqrt{1-x^2}\, dx = \frac{x}{2}\sqrt{1-x^2} + \frac{1}{2}\text{arc sin } x.$$

By differentiation it follows that

$$\sqrt{1-x^2} = \frac{d}{dx}\left(\frac{x}{2}\sqrt{1-x^2}\right) + \frac{1}{2}\frac{d}{dx}(\text{arc sin } x)$$

which yields easily enough

---

* From AMERICAN MATHEMATICAL MONTHLY, vol. 63 (1956), pp. 118–120.

(2) $$\frac{d}{dx}(\text{arc sin } x) = \frac{1}{\sqrt{1-x^2}}$$

or, if one likes,

(3) $$\frac{d}{dx}(\text{arc sin } u) = \frac{1}{\sqrt{1-u^2}}\frac{du}{dx}$$

where u depends on x.

From this result it is easy to obtain the derivatives of the other inverse trigonometric functions. Thus

$$\frac{d}{dx}(\text{arc cos } x) = \frac{d}{dx}\left(\frac{\pi}{2} - \text{arc sin } x\right) = \frac{-1}{\sqrt{1-x^2}}$$

$$\frac{d}{dx}(\text{arc tan } x) = \frac{d}{dx}\left(\text{arc sin }\frac{x}{\sqrt{1-x^2}}\right) = \frac{1}{1+x^2}$$

$$\frac{d}{dx}(\text{arc cot } x) = \frac{d}{dx}\left(\frac{\pi}{2} - \text{arc tan } x\right) = \frac{-1}{1+x^2}$$

$$\frac{d}{dx}(\text{arc csc } x) = \frac{d}{dx}\left(\text{arc sin }\frac{1}{x}\right) = \frac{-1}{x\sqrt{x^2-1}}$$

$$\frac{d}{dx}(\text{arc sec } x) = \frac{d}{dx}\left(\frac{\pi}{2} - \text{arc csc } x\right) = \frac{1}{x\sqrt{x^2-1}}.$$

We may now obtain the derivatives of the trigonometric functions. The general procedure is illustrated by determining $d(\sin x)/dx$. Let

$$y = \sin x$$

so that

$$x = \text{arc sin } y.$$

Then it follows at once that

$$\frac{dx}{dy} = \frac{1}{\sqrt{1-y^2}}$$

or

(4) $$\frac{dy}{dx} = \sqrt{1-y^2} = \cos x.$$

Derivatives of the other trigonometric functions can be obtained in a similar way.

The writer believes that the mode of presentation of this paper is more easily understood by the student than the conventional procedure for several reasons. First, there is no need for the investigation of $\lim_{\theta \to 0} (\sin \theta/\theta)$ which is involved in the conventional procedure. Second, there is no need to manipulate a difference quotient into a form suitable for limiting processes. Third, it is tied in with an interesting geometric interpretation. Incidentally the integration formula (1) is a useful by-product which normally would involve for its evaluation the techniques of integration by parts or trigonometric substitution.

It should be remarked that in obtaining all of the above results, angles were restricted to those lying in the first quadrant. It is not at all difficult to consider other angles, the procedure being entirely analogous to that used in the conventional approach.

PRINCIPAL VALUES OF CERTAIN INVERSE TRIGONOMETRIC FUNCTIONS*

C. B. READ AND FERNA WRESTLER, University of Wichita

Students often feel the definitions of the principal values of the inverse trigonometric functions are purely arbitrary. Texts in trigonometry may fail to point out that alternative definitions exist; they sometimes justify the definitions given on the basis of later use in the calculus.

The calculus student in turn will be fortunate if his text explains why the particular definitions used were selected. He may for example wonder why $-\pi/2 \leq \text{Arc sin } x \leq \pi/2$, rather than $\pi/2 \leq \text{Arc sin } x \leq 3\pi/2$. The first choice results in a positive slope throughout the interval, in agreement with the usual value of the derivative. This explanation may not be a satisfactory reason why, for $x \leq -1$, we define $-\pi \leq \text{Arc sec } x < -\pi/2$ rather than $\pi/2 < \text{Arc sec } x \leq \pi$ since the second choice would give a positive slope throughout while the first does not. Some calculus texts give the second definition; to many students this seems more consistent since it agrees with the definition for Arc cos x.

Two suggestions are offered in justification of the choice, when $x \leq -1$, $-\pi \leq \text{Arc sec } x < -\pi/2$. If in the usual derivation of the derivative of Arc sec x we let $x = \sec y$, then $dx/dy = \sec y \tan y = x \tan y$. If $x \geq 1$ and $0 \leq y < \pi/2$, then $\tan y = \sqrt{x^2-1}$; if $x \leq -1$ and $\pi/2 < y \leq \pi$, then $\tan y$ is $-\sqrt{x^2-1}$, hence $dy/dx = \pm 1/x\sqrt{x^2-1}$, requiring the ambiguous sign. However, if when $x \leq -1$ we define $-\pi \leq y < -\pi/2$, $\tan y$ is $\sqrt{x^2-1}$ and for all cases $dy/dx = +1/x\sqrt{x^2-1}$ (dy/dx is not defined at $x = \pm 1$).

As the second suggestion, consider the area bounded by $y = 1/(x\sqrt{x^2-1})$ between $x = -2$ and $x = -\sqrt{2}$. Clearly the area is negative but if the principal value of Arc sec x is defined as $0 \leq \text{Arc sec } x \leq \pi$, a positive result is obtained, suggesting reasons for another definition.

These explanations may still leave the student (or the teacher) puzzled as to why the "natural" definition Arc sec $x = \text{Arc cos } (1/x)$ should not be adopted. The ambiguity of signs discussed in the first suggestion just presented can be readily removed by defining $0 \leq \text{Arc sec } x = \text{Arc cos } (1/x) \leq \pi$ and using absolute values in the formula

$$\frac{d}{dx} \text{Arc sec } x = \frac{1}{|x|\sqrt{x^2-1}}.$$

In the case of integration, we may write, for negative x,

$$\int \frac{dx}{x\sqrt{x^2-1}} = \int \frac{d(-x)}{(-x)\sqrt{x^2-1}} = \text{Arc sec } (-x) + C$$

$$= \int \frac{d|x|}{|x|\sqrt{x^2-1}} = \text{Arc sec } |x| + C$$

$$x \leq -1$$

hence for x either positive or negative $\int dx/x\sqrt{x^2-1} = \text{Arc sec } |x| + C (|x| \geq 1)$.

---

* From AMERICAN MATHEMATICAL MONTHLY, vol. 63 (1956), pp. 184–185.

Use of this formula (analogous to $\int dx/x = \ln |x| + C$) and the definition $0 \leq \text{Arc sec } x \leq \pi$ gives, for the integration example presented, a negative result. It would appear, then, that there is really no necessity for an apparently artificial definition.

It would not be out of order to suggest that authors might well devote a little more space to explaining why the particular choice of definition was selected.

Note: The authors wish to thank the referee for a suggestion for expanding and improving this note.

A CLASSROOM PROOF OF $\lim_{t\to 0}(\sin t)/t = 1$*

STEPHEN HOFFMAN, Trinity College, Hartford, Connecticut

The usual proof of $\lim_{t\to 0} (\sin t)/t = 1$ requires that the inequalities: $\sin t < t < \tan t$ be established for sufficiently small (positive) values of t. The latter part of this string of inequalities is usually determined by comparing the area of a unit-circular sector of angle t with the area of a triangle. If the finding of the area of a circular sector exhausts the students' reservoir of thinking, then an understanding of the remainder of the proof is lost. The following method of proof avoids the use of area.

Let it already have been established (either by definition or by some kind of geometric-trigonometric methods) that the point with coordinates $(\cos t, \sin t)$ is at a distance, measured appropriately along the circumference of the unit circle, of $|t|$ units from the point $(1, 0)$.

THEOREM 1.
$$\lim_{t\to 0} (1 - \cos t)/t = 0.$$

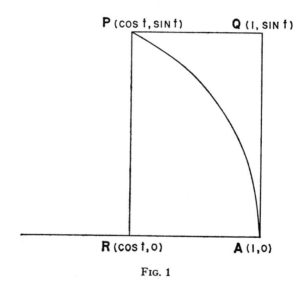

FIG. 1

Proof. Since the straight-line distance between $(\cos t, \sin t)$ and $(1, 0)$ does not exceed the arc length along the circle between the same two points, we have

$$0 \leq \sqrt{\{(\cos t - 1)^2 + (\sin t - 0)^2\}} \leq |t|.$$

---

* From AMERICAN MATHEMATICAL MONTHLY, vol. 67 (1960), pp. 671–672.

Squaring, simplifying, and dividing by $2t$ yields

$$0 \leq (1 - \cos t)/t \leq \tfrac{1}{2}t, \quad t > 0; \quad \tfrac{1}{2}t \leq (1 - \cos t)/t \leq 0, \quad t < 0.$$

By the "pinching theorem" for limits, the theorem is proved.

THEOREM 2.
$$\lim_{t \to 0} (\sin t)/t = 1.$$

Proof.* If $0 < t < \tfrac{1}{2}\pi$, let P be the point $(\cos t, \sin t)$, Q, the point $(1, \sin t)$, A, the point $(1, 0)$, and R, the point $(\cos t, 0)$. (Fig. 1) Then $\overline{PR} \leq$ arc $PA \leq \overline{PQ} + \overline{QA}$ and therefore

$$\sin t \leq t \leq 1 - \cos t + \sin t.$$

But $\sin t < t$ implies $1 - \cos t + \sin t < 1 - \cos t + t$ so that

$$t \leq 1 - \cos t + \sin t \leq 1 - \cos t + t.$$

Hence, dividing by t,

$$1 \leq \frac{1 - \cos t}{t} + \frac{\sin t}{t} \leq \frac{1 - \cos t}{t} + 1.$$

By Theorem 1 and the "pinching theorem" for limits, the theorem is proved.

---

* For $t \to 0+$, the proof for $t \to 0-$ being similar.

DERIVATIVES OF sin θ AND cos θ*

C. S. OGILVY, Hamilton College

The derivatives of the sine and cosine functions may be presented handily with the aid of vectors, following a suggestion of R. T. Coffman of Richland, Washington.

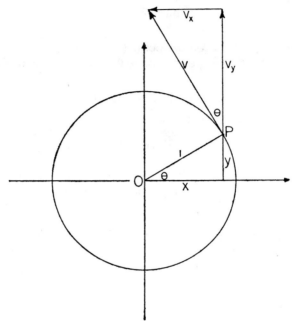

FIG. 1

Consider the point $P(x, y)$ moving around the unit circle (Fig. 1) in the direction of increasing θ; then $x = \cos \theta$ and $y = \sin \theta$. We note that $ds/d\theta = 1$ if θ is measured in radians. The direction of motion of P is tangential to the circle; hence the vector v is perpendicular to OP at P. From the vector triangle,

$$\sin \theta = \frac{|v_x|}{|v|} = \left|\frac{dx}{dt} \frac{dt}{ds} \frac{ds}{d\theta}\right| = \left|\frac{dx}{d\theta}\right|,$$

and we must make it minus because it is in the direction of decreasing x. Exactly similarly, $dy/d\theta = \cos \theta$.

---

* From AMERICAN MATHEMATICAL MONTHLY, vol. 67 (1960), p. 673.

ALTERNATE CLASSROOM PROOF THAT $(\sin t)/t \to 1$ AS $t \to 0$*

GARY PERRY, Trinity College, Hartford, Connecticut

Hoffman [this MONTHLY, 67 (1960) 671–672†] offers a method of proving $\lim_{t\to 0} (\sin t)/t = 1$ which avoids the use of area. The following proof also avoids the use of area and makes use of several important concepts usually encountered in a first year analytic geometry and calculus course.

THEOREM. $\lim_{t\to 0} (\sin t)/t = 1$.

Proof. We assume for simplicity that $t > 0$. Let $P(x, y)$ be a point in the first quadrant on $x = (1 - y^2)^{1/2}$, and let t denote the length of arc along the curve from the x-axis to P. The arc length formula yields

$$t = \int_0^y \frac{du}{(1-u^2)^{1/2}}, \qquad (0 \leq y < 1).$$

Since $\sin t = y$,

$$\frac{t}{\sin t} = \frac{1}{y} \int_0^y \frac{du}{(1-u^2)^{1/2}}, \qquad (0 < y < 1).$$

Clearly $y \to 0^+$ as $t \to 0^+$, and we write

$$\lim_{t\to 0^+} \frac{t}{\sin t} = \lim_{y\to 0^+} \frac{1}{y} \int_0^y \frac{du}{(1-u^2)^{1/2}}.$$

If F is the function defined by

$$F(y) = \int_0^y \frac{dw}{(1-w^2)^{1/2}}, \qquad (0 \leq y < 1),$$

then $F(0) = 0$, so that

$$\lim_{y\to 0^+} \frac{1}{y} \int_0^y \frac{du}{(1-u^2)^{1/2}} = \lim_{y\to 0^+} \frac{F(y) - F(0)}{y} = F'(0)$$

by the definition of the right-hand derivative. Finally, by the fundamental theorem of calculus,

$$F'(y) = \frac{1}{(1-y^2)^{1/2}}, \qquad (0 \leq y < 1).$$

Then $F'(0) = 1$ and the theorem is proved.

---

* From AMERICAN MATHEMATICAL MONTHLY, vol. 70 (1963), pp. 426–427.
† Ed. note: Reprinted on pp. 124–125, this book.

A SIMPLE WAY OF DIFFERENTIATING TRIGONOMETRIC FUNCTIONS AND THEIR INVERSES IN AN ELEMENTARY CALCULUS COURSE*

H. A. THURSTON, University of British Columbia

We assume that the student knows (i) how to differentiate inverse functions, and (ii) that areas under graphs can be obtained by integration. The converse of (ii) is also true; if the area is known, it can be used to find an integral. This is

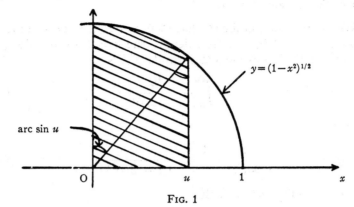

FIG. 1

the quickest way of finding $\int_0^u (1-x^2)^{1/2} dx$. From the diagram it follows that this integral is

$$\tfrac{1}{2} \text{arc sin } u + \tfrac{1}{2} u \cdot (1 - u^2)^{1/2}.$$

From the fundamental theorem we now have

$$D_x[\text{arc sin } x + x \cdot (1 - x^2)^{1/2}] = 2(1 - x^2)^{1/2}$$

which quickly yields the usual formula for the derivative of arc sin. Because sin is an inverse of arc sin, its derivative on the interval $[-\pi/2, \pi/2]$ is soon found, and the relation $\sin(x+\pi) = -\sin x$ gives the derivative in general. The rest of the trigonometrical results follow as usual, and finally the limit as $h \to 0$ of $h^{-1} \cdot \sin h$ can be found (if wanted) by L'Hôpital's rule.

BIBLIOGRAPHIC ENTRIES: TRIGONOMETRIC FUNCTIONS

Except for the entry labeled MATHEMATICS MAGAZINE, the references below are to the AMERICAN MATHEMATICAL MONTHLY.

1. Whitaker, H. C., An elementary derivation of the series of sin x and cos x, vol. 7, p. 99.

> The author assumes the existence of power series expansions for sin x and cos x and uses the addition formulas for sin $(x+y)$ and cos $(x+y)$ to determine the coefficients.

---

* From AMERICAN MATHEMATICAL MONTHLY, vol. 70 (1963), p. 424.

2. Porter, M. B., Relating to infinite products for sin z and cos z, vol. 24, p. 246.

> Uniform convergence of series used to derive the infinite products for the sine and cosine.

3. Frumveller, A. F., A theory and generalization of the circular and hyperbolic functions, vol. 26, p. 280.

> A unified treatment of circular and hyperbolic functions obtained by parametrizing a central conic $q_1^2 x^2 - q_2^2 y^2 = 1$, where q_1, q_2 can be complex.

4. Uspensky, J. V., A curious case of the use of mathematical induction in geometry, vol. 34, p. 247.

> The author proves that $(\sin x)/x$ is a decreasing function in $(0, \pi/2)$ by first using induction and a limiting argument to prove the following purely geometric theorem: If α and β denote the angles at A and B, respectively, in $\triangle ABC$, then $CA/BA > \beta/(\alpha+\beta)$.

5. Ward, L. E., Some functions analogous to trigonometric functions, vol. 34, p. 301.

> Three complex-valued solutions of $y''' + y = 0$ are chosen and are shown to have properties not unlike the sine and cosine.

6. Lubin, C. I., Differentiation of the trigonometric functions, vol. 54, p. 465.
7. Anderson, A. G., The derivative of cos x, vol. 60, p. 255.
8. Pease, D. K., Limits used in the derivative of sin x, vol. 60 p. 477.
9. Vaughan, H. E., Characterization of the sine and cosine, vol 62, p. 707.
10. Kearns, D. A., An analytic approach to trigonometric functions, vol. 65, p. 616.
11. Mancill, J. D., The sine and cosine functions, MATHEMATICS MAGAZINE, vol. 36, p. 302.

(c)
LOGARITHMIC FUNCTIONS

AN EXISTENCE PROOF FOR LOGARITHMS*

F. E. HOHN, University of Arizona

The purpose of this article is to give a simple proof of the following theorem, customarily assumed in college algebra courses:

THEOREM. *Given a real number $b>1$ and any positive number N, then there always exists a real number q, called the logarithm to the base b of N, such that $b^q = N$.*

The method of proof is to *construct the number q*, and to this end we prove first the following preliminary result:

LEMMA. *Given a number $b>1$ and any positive number N, we can construct a sequence of decimal fractions of the form $q_0 = p_0$, $q_1 = p_0 + p_1/10$, $q_2 = p_0 + p_1/10 + p_2/100$, $\cdots$, such that b^{q_n} approaches N as n increases.*

Proof: Since $N>0$ there is a unique integer p_0, not necessarily positive, such that
$$b^{p_0} \leq N < b^{p_0+1}.$$
Then
$$1 \leq N/b^{p_0} < b,$$
and hence
$$1 \leq N^{10}/b^{10 p_0} < b^{10}.$$
There is then a unique integer p_1, $0 \leq p_1 \leq 9$, such that
$$b^{p_1} \leq N^{10}/b^{10 p_0} < b^{p_1+1}.$$
Then
$$1 \leq N^{10}/b^{10 p_0 + p_1} = N^{10}/b^{10 q_1} < b,$$
so that
$$1 \leq N^{100}/b^{100 q_1} < b^{10}.$$
There is then a unique integer p_2, $0 \leq p_2 \leq 9$, such that
$$b^{p_2} \leq N^{100}/b^{100 q_1} < b^{p_2+1}.$$
Then
$$1 \leq N^{100}/b^{100 q_1 + p_2} = N^{100}/b^{100 q_2} < b.$$

---

* From AMERICAN MATHEMATICAL MONTHLY, vol. 50(1943), pp. 115–116.

This process can evidently be continued to obtain

$$1 \leq N^{10^n}/b^{10^n q_n} < b,$$

which gives

$$1 \leq N/b^{q_n} < b^{10^{-n}},$$
$$0 \leq (N/b^{q_n}) - 1 < b^{10^{-n}} - 1,$$
(1) $$0 \leq N - b^{q_n} < b^{q_n}(b^{10^{-n}} - 1) \leq N(b^{10^{-n}} - 1).$$

The lemma follows at once if we assume that $b^{10^{-n}}$ approaches 1 as the integer n increases. This can be considered a consequence of the fact that $\lim_{x \to 0} b^x = b^0 = 1$, which most students will be willing to grant.

If we now define q and b^q by

$$q = \lim_{n \to \infty} q_n, \qquad b^q = \lim_{n \to \infty} b^{q_n},$$

then $N = b^q$, or $q = \log_b N$.

It is worth noting that as far as most applications of logarithms are concerned we never need to use anything beyond relation (1). For instance, a table accurate to 5 places can be constructed from (1) with $n = 6$.

Although this method of computing logarithms is not of much practical value, it can be carried out without too much labor. With the help of a table of 10th powers, the value of log 3 can be computed to 5 places in about ten minutes.

Note by the Editor. The proof that $\lim_{n \to \infty} b^{10^{-n}} = 1$, though probably not suitable for an elementary class, is quite simple. In the first place, $b^{10^{-n}} > 1$ for all positive n, for if $b^{10^{-n}} \leq 1$, then $b = (b^{10^{-n}})^{10^n} \leq 1$ contrary to assumption. Hence if $\lim_{n \to \infty} b^{10^{-n}} \neq 1$, we must have $b^{10^{-n}} > 1 + \epsilon$ for some $\epsilon > 0$ and arbitrarily large n. Then $b = (b^{10^{-n}})^{10^n} > (1+\epsilon)^{10^n} > 1 + 10^n \epsilon$ by the binomial theorem, and this cannot be true for arbitrarily large n. R.J.W. [*Ed. note:* The editor of Discussions and Notes for this issue is Marie J. Weiss.]

A PROPERTY OF THE LOGARITHM*

D. S. GREENSTEIN, Northwestern University

One important limit involving the logarithm is

$$\lim_{x \to \infty} \frac{\log x}{x} = 0.$$

Since it is becoming customary in elementary calculus courses to use the definition

$$\log x = \int_1^x t^{-1}\, dt$$

to deduce properties of the logarithm (including the logarithm of a product equalling the sum of the logarithms of its factors), the following simple demonstration of the above limit should be of interest.

From the defining integral, it is easily shown (or intuitively demonstrated by means of areas) that for $x > 1$,

$$0 < \log x < x.$$

Since $\log x = 2 \log x^{1/2}$, it necessarily follows that

$$0 < \frac{\log x}{x} < \frac{2x^{1/2}}{x} = 2x^{-1/2},$$

from which the limit is easily deduced. Finally, it should be noted that the substitution $x = t^{-1}$ also yields $\lim_{t \to 0} t \log t = 0$.

BIBLIOGRAPHIC ENTRIES: LOGARITHMIC FUNCTIONS

Except for entries labeled MATHEMATICS MAGAZINE, the references below are to the AMERICAN MATHEMATICAL MONTHLY.

1. Nicholson, J. W., A simple deduction of the differential of log x, vol. 4, p. 306.

    Based on the functional equation $f(xy) = f(x) + f(y)$.

2. Porter, M. B., The derivative of the logarithm, vol. 23, p. 204.
3. Kennedy, E. C., A note on logarithms, vol. 48, p. 465.

    Polynomial approximations for log x obtained from consideration of area.

4. Yates, R. C., Differentiating the logarithm, vol. 61, p. 120.
5. Wilansky, A., Remarks on Yates' note "Differentiating the logarithm," vol. 61, p. 634.
6. Seebeck, C. L., Jr., and Jewett, J. W., A development of logarithms using the function concept, vol. 64, p. 667.
7. Leader, S., On the definition of ln a, vol. 65, p. 622.

---

* From AMERICAN MATHEMATICAL MONTHLY, vol. 72 (1965), p. 767.

8. Seebeck, C. L., Jr., and Miller, H. C., Jr., More about logarithms, vol. 65, p. 697.

9. Rouse, H. R., The natural logarithm is transcendental, vol. 73, p. 187.

10. Milkman, J., The logarithmic function is unique, MATHEMATICS MAGAZINE, vol. 24, p. 11.

The only continuous solution of $f(xy)=f(x)+f(y)$ for $x>0$, $y>0$ is $f(x)=c \log x$.

11. Lightstone, A. H., On exp and log in elementary calculus, MATHEMATICS MAGAZINE, vol. 36, p. 17.

12. Matlak, R. F., A proof of the formula representing the logarithm as the limit of a sequence, MATHEMATICS MAGAZINE, vol. 39, p. 64.

An elementary proof that $\lim_{n\to\infty} n(x^{1/n}-1) = \log x$ if $x>0$.

(d)

EXPONENTIAL AND HYPERBOLIC FUNCTIONS

RELATING TO THE EXPONENTIAL FUNCTION*

OTTO DUNKEL, Washington University

An interesting treatment of the elementary transcendental functions was given by A. Hurwitz in the *Mathematische Annalen*, Vol. 70, 1911, entitled "Über die Einführung der elementaren transzendenten Funktionen in der algebraischen Analysis." The method used by Hurwitz was somewhat similar to the one used by Professor Huntington in his article "An Elementary Theory of the Exponential and Logarithmic Functions" in the September, 1916, number of the MONTHLY, pp. 241–246, except that Hurwitz discussed first the function log x and then derived sequences for the definition of e^x of the type used by Huntington. About the time of the appearance of Hurwitz's article the writer developed a treatment of the exponential function similar to the one given by Huntington but somewhat simpler in the fact that the inequality $(1+d)^m > 1+md$ was not required. This makes the proof more elementary, as the proof of this inequality in Huntington's article is made to depend partly upon the binomial theorem and partly upon an additional proof for a remaining case. This advantage is gained by using sequences in which each exponent is double the preceding, for then it is a very easy matter to prove the increasing and decreasing character of the two sequences, and having done this the sequences themselves supply the place of the above mentioned inequality in the subsequent reasonings. A brief sketch of how this can be effected is given below.

Given the sequences

$$(A) \quad 1+x, \left(1+\frac{x}{2}\right)^2, \left(1+\frac{x}{4}\right)^4, \left(1+\frac{x}{8}\right)^8, \cdots,$$

$$\left(1+\frac{x}{m}\right)^m, \left(1+\frac{x}{2m}\right)^{2m}, \cdots$$

$$(B) \quad (1-x)^{-1}, \left(1-\frac{x}{2}\right)^{-2}, \left(1-\frac{x}{4}\right)^{-4}, \left(1-\frac{x}{8}\right)^{-8}, \cdots,$$

$$\left(1-\frac{x}{m}\right)^{-m}, \left(1-\frac{x}{2m}\right)^{-2m}, \cdots.$$

where x is any real number, and considering only the terms for which $m > |x|$,

---

* From AMERICAN MATHEMATICAL MONTHLY, vol. 24 (1917), pp. 244–246.

we shall prove the following: (a) Sequence (A) increases; (b) Sequence (B) decreases; (c) Each A is less than the corresponding B and the difference approaches zero as m increases.

We have for any real value of x

$$\left(1 + \frac{x}{2m}\right)^2 = 1 + \frac{x}{m} + \frac{x^2}{4m^2} > 1 + \frac{x}{m},$$

and, since $1+x/m$ is positive, we have also

$$\left(1 + \frac{x}{2m}\right)^{2m} > \left(1 + \frac{x}{m}\right)^m.$$

This proves (a), and (b) follows at once from the above inequality by replacing x by $-x$ and reversing the resulting inequality.

To prove (c) we have

$$\left(1 - \frac{x}{m}\right)^{-m} - \left(1 + \frac{x}{m}\right)^m = \left(1 - \frac{x}{m}\right)^{-m}\left[1 - \left(1 - \frac{x^2}{m^2}\right)^m\right]$$

$$< \left(1 - \frac{x}{m}\right)^{-m} \frac{x^2}{m} \quad \text{when } m > x^2.$$

It is easily seen that each of the two factors on the right in the equality is positive and hence the first part of (c) is true. The inequality is seen to be true by replacing x in (A) by $-x^2/m$ and using (a). Since the factor $(1-x/m)^{-m}$ decreases and x^2/m approaches zero as m increases the second part of (c) is true.

If we call the common limit of the sequences (A) and (B) exp x, we have the important inequalities

$$1 + h < \left(1 + \frac{h}{m}\right)^m < \exp h < \left(1 - \frac{h}{m}\right)^{-m} < (1-h)^{-1}, \quad |h| < 1, \; m = 2^t,$$

from which follow the remaining properties of the function. For example, since

$$\frac{\left(1 + \frac{x}{m}\right)^m \left(1 + \frac{y}{m}\right)^m}{\left(1 + \frac{x+y}{m}\right)^m} = \left[1 + \frac{xy}{m(m+x+y)}\right]^m,$$

if we take m so large that $xy/(m+x+y) = h$ is less than unity in absolute value, the above inequalities show that

$$\frac{\exp x \, \exp y}{\exp(x+y)} = 1.$$

The derivative of exp x follows at once from the same inequalities. There is therefore an advantage in beginning the sequences with $1+x$ and $(1-x)^{-1}$ respectively. The sequences given here contain, of course, those in Huntington's paper.

EDITORIAL—THE PROOF OF EULER'S EQUATION*

C. B. ALLENDOERFER, Haverford College

Judging from the volume of mail addressed to "Classroom Notes"† it appears that one of the greatest mysteries of undergraduate mathematics is the equation of Euler: $e^{ix} = \cos x + i \sin x$. The objective of these correspondents is to develop this formula without the use of infinite series; and judging from the desperate devices employed by these writers in seeking to attain this end, it is highly desirable that a rigorous simple proof of this formula be available. The present note is therefore dedicated to those authors whose papers have been rejected, but whose ideas have helped me to prepare the following presentation.

The first point to be emphasized is that the expression e^{ix} has to be *defined*, and that certain properties must be ascribed to it. Otherwise any proof falls to the ground. Rigorous treatments of this appear in the classical literature; for example, see G. H. Hardy, *Pure Mathematics*, p. 409 (fifth edition), or E. T. Whittaker and G. N. Watson, *Modern Analysis*, p. 581 (fourth edition). Since these have more general objectives in view, it may be complained that they are too complicated for the purpose of defining the relatively simple expression e^{ix}. If, on the other hand, one wishes simplicity, he may straight off define e^{ix} to be the expression $\cos x + i \sin x$. This would settle the whole matter, but such a definition is unsatisfactory on intuitive grounds and appears to be drawn out of the air. It is hoped that the following definition is satisfactory on all three grounds: rigor, simplicity, and intuition.

DEFINITION. e^{ix} *is a complex valued function of the real variable* x *having the properties*:

(1) $$e^{i0} = 1;$$

(2) $$de^{ix}/dx = ie^{ix}.$$

THEOREM. $e^{ix} = \cos x + i \sin x$.

Proof: Let $e^{ix} = \rho(\cos \theta + i \sin \theta)$, $0 \leq \theta < 2\pi$; $\rho \geq 0$. From property (1) of the definition:

when $x = 0$; then $\rho = 1$ and $\theta = 0$.

From property (2) of the definition:

$$\frac{d\rho}{dx}(\cos \theta + i \sin \theta) + \rho(-\sin \theta + i \cos \theta)\frac{d\theta}{dx} = i\rho(\cos \theta + i \sin \theta).$$

---

* From AMERICAN MATHEMATICAL MONTHLY, vol. 55 (1948), pp. 94–95.

† *Ed. note:* C. B. Allendoerfer was, at the time this paper was published, editor of the department entitled "Classroom Notes" in the AMERICAN MATHEMATICAL MONTHLY.

Equating real and imaginary parts we have:

$$\frac{d\rho}{dx}\cos\theta - \rho\sin\theta\,\frac{d\theta}{dx} = -\rho\sin\theta$$

$$\frac{d\rho}{dx}\sin\theta + \rho\cos\theta\,\frac{d\theta}{dx} = \rho\cos\theta.$$

Solving, we obtain

$$\frac{d\rho}{dx} = 0; \quad \rho\,\frac{d\theta}{dx} = \rho.$$

Hence: $\rho = 1$ and $\theta = x$.

THE ELEMENTARY TRANSCENDENTAL FUNCTIONS*

W. F. EBERLEIN, University of Wisconsin

It is difficult to differentiate the undefined.

1. Introduction. We outline a unified approach to the circular and hyperbolic functions motivated by the traditional geometric considerations (Figs. 1a and 1b), yet leading immediately to a sound analytic definition. The treatment in whole or part seems adaptable to modern calculus courses in which integration appears early and some knowledge of trigonometry is presupposed.

2. Geometric approach to the circular and hyperbolic functions. Consider the unit circle (right branch of the equilateral hyperbola)

(1a) $\quad x^2 + y^2 = 1 \quad\quad$ (1b) $\quad x^2 - y^2 = 1$

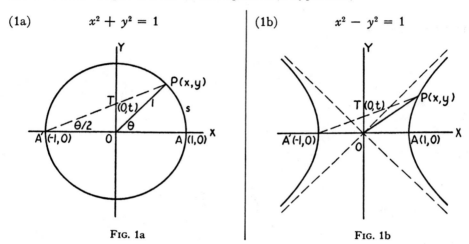

FIG. 1a | FIG. 1b

Letting θ denote twice the signed area of the sector OAP, regard the coordinates (x, y) of the variable point P as functions of θ:

(2a) $\quad \begin{aligned} x &= \cos\theta, \\ y &= \sin\theta. \end{aligned} \quad\quad$ (2b) $\quad \begin{aligned} x &= \cosh\theta, \\ y &= \sinh\theta. \end{aligned}$

[$\theta = s = \widehat{AP}$ is then the radian measure of $\angle AOP$.]

In particular

(3a) $\quad \begin{aligned} \cos 0 &= 1, \\ \sin 0 &= 0. \end{aligned} \quad\quad$ (3b) $\quad \begin{aligned} \cosh 0 &= 1, \\ \sinh 0 &= 0. \end{aligned}$

---

* From AMERICAN MATHEMATICAL MONTHLY, vol. 61 (1954), pp. 386–392.

It follows from Figure 1 that in both cases

(4) $$\theta = xy - 2\int_1^x y\,dx.$$

Taking differentials in equations (1) and (4) yields

(5a) $\quad 0 = x\,dx + y\,dy,\quad$ (5b) $\quad 0 = x\,dx - y\,dy.$

(6) $\quad d\theta = x\,dy + y\,dx - 2y\,dx = x\,dy - y\,dx.$

Now solve equations (5) and (6) for the unknowns dx and dy:

(7a) $\quad\begin{array}{l} dx = -y\,d\theta, \\ dy = x\,d\theta. \end{array}\quad$ (7b) $\quad\begin{array}{l} dx = y\,d\theta, \\ dy = x\,d\theta. \end{array}$

Hence

(8a) $\quad\begin{array}{l} D_\theta \cos\theta = -\sin\theta, \\ D_\theta \sin\theta = \cos\theta. \end{array}\quad$ (8b) $\quad\begin{array}{l} D_\theta \cosh\theta = \sinh\theta, \\ D_\theta \sinh\theta = \cosh\theta. \end{array}$

To justify the formal differentiation consider the line $A'P$ joining $A':(-1, 0)$ and the point $P:(x, y)$, where $P \neq A'$. Let $A'P$ intersect the y axis in the point $T:(0, t)$. It follows from the figure that

(9a) $\quad t = \tan\dfrac{\theta}{2},\quad$?

(10) $\quad t = y/(1 + x).$

Solving for the intersections of the line $A'T$ (10) and the curves (1) yields for the coordinates of P:

(11a) $\quad\begin{array}{l} x = \dfrac{1 - t^2}{1 + t^2}, \\ y = \dfrac{2t}{1 + t^2} \end{array}\ (-\infty < t < \infty)\quad$ (11b) $\quad\begin{array}{l} x = \dfrac{1 + t^2}{1 - t^2}, \\ y = \dfrac{2t}{1 - t^2}, \end{array}\ -1 < t < 1).$

Equations (11) define a one-to-one differentiable parametric mapping of the open interval $-\infty < t < \infty$ ($-1 < t < 1$) onto the unit circle minus the point A' (the right branch of the rectangular hyperbola):

(12a) $\quad\begin{array}{l} dx = -\dfrac{4t}{(1 + t^2)^2}\,dt, \\ dy = \dfrac{2(1 - t^2)}{(1 + t^2)^2}\,dt. \end{array}\quad$ (12b) $\quad\begin{array}{l} dx = \dfrac{4t}{(1 - t^2)^2}\,dt, \\ dy = \dfrac{2(1 + t^2)}{(1 - t^2)^2}\,dt. \end{array}$

Since (4) now defines θ as a differentiable function of t, equations (5) and (6) are justified. Substitute the values (11) and (12) into (6) to obtain*

(13a) $\qquad d\theta = \dfrac{2dt}{1+t^2},$ \qquad (13b) $\qquad d\theta = \dfrac{2dt}{1-t^2}.$

Note that equations (7) and (8) follow directly from (11), (12), and (13). Since $\theta=0$ when $t=0$,

(14a) $\qquad \theta = 2\displaystyle\int_0^t \dfrac{du}{1+u^2} \quad (-\infty < t < \infty),$ \qquad (14b) $\qquad \theta = 2\displaystyle\int_0^t \dfrac{du}{1-u^2} \quad (-1 < t < 1).$

The equations (9a), (11a), and (13a) are the familiar substitutions used to integrate rational functions of $\sin x$, $\cos x$. Moreover, since it follows from Figure 1a that $\theta = \tfrac{1}{2}\pi$ when $t=1$ and $\theta \to \pi$ when $t \to \infty$,

(15)
$$\dfrac{\pi}{4} = \int_0^1 \dfrac{du}{1+u^2},$$
$$\dfrac{\pi}{2} = \int_0^\infty \dfrac{du}{1+u^2}.$$

To obtain an analytic theory of the circular and hyperbolic functions we need only retrace our steps carefully.

3. Analytic theory of the circular functions. Define π by the first of equations (15). The second equation then follows on making the change of variable $u = v^{-1}$:

$$\int_1^\infty \dfrac{du}{1+u^2} = \lim_{t \to \infty} \int_1^t \dfrac{du}{1+u^2} = \lim_{t \to \infty} \int_{t^{-1}}^1 \dfrac{dv}{1+v^2} = \int_0^1 \dfrac{dv}{1+v^2}.$$

Equation (14a) defines θ as an odd increasing function of $t(-\infty < t < \infty)$ with positive derivative (13a). Invert (14a) to obtain t as an odd increasing differentiable function of $\theta(-\pi < \theta < \pi)$. Equations (11a), (2a) now define $\cos\theta$, $\sin\theta$ as functions of θ over the open interval $-\pi < \theta < \pi$.† $\cos\theta$, $\sin\theta$ extend to continuous functions over the closed interval $-\pi \leq \theta \leq \pi$ on setting

$$\cos(\pm\pi) = \lim_{\theta \to \pm\pi} \cos\theta = \lim_{t \to \pm\infty} \dfrac{1-t^2}{1+t^2} = -1,$$

$$\sin(\pm\pi) = \lim_{\theta \to \pm\pi} \sin\theta = \lim_{t \to \pm\infty} \dfrac{2t}{1+t^2} = 0.$$

---

* If θ is defined by the arc length s, (13a) appears as a consequence of the relation $ds^2 = dx^2 + dy^2$.

† $t=1$ or $\theta=\pi/2$ is then the smallest positive zero of $x(t) = \cos\theta$. This result is usually taken as the definition of π in the power series approach to the circular functions.

It is then legitimate to extend $\cos \theta$, $\sin \theta$ to arbitrary real values of θ by the requirement

(16a)
$$\sin (\theta + 2\pi) = \sin \theta,$$
$$\cos (\theta + 2\pi) = \cos \theta.$$

The relations $\cos(-\theta) = \cos \theta$, $\sin(-\theta) = -\sin \theta (-\pi < \theta < \pi)$ resulting from (11a) are clearly preserved under extension.

Consider now the differentiation formulae (8a). For the interval $-\pi < \theta < \pi$ they follow again from (11a), (12a), and (13a). Their validity for $\theta = \pm \pi$ results from a familiar application of the law of the mean: For example, $\cos \pi - \cos \theta = -(\pi - \theta) \sin \xi$ $(\theta < \xi < \pi)$; whence the left derivative

$$\cos'(\pi) = \lim_{\theta \to \pi} (\cos \pi - \cos \theta)/(\pi - \theta) = -\lim_{\xi \to \pi} \sin \xi = 0$$

exists and has the correct value. The validity of (8a) for unrestricted values of θ then follows from the periodicity condition (16a).

Consider the remaining equations. (3a) and (10) follow directly from (11a); (1a) requires an additional continuity and periodicity argument. The equations (5a), (6a), (7a), (12a), (13a) leading from (4) to (14a) were previously justified, whence (4) and (14a) define the same quantity θ.* Note finally that the relation (9a) follows from (10) via the double angle corollaries of the addition formulae:

LEMMA:

(19a)
$$u(\theta) \equiv \sin (\theta + \theta_0) - \sin \theta \cos \theta_0 - \cos \theta \sin \theta_0 = 0;$$
$$v(\theta) \equiv \cos (\theta + \theta_0) - \cos \theta \cos \theta_0 + \sin \theta \sin \theta_0 = 0.$$

Proof: (8a) implies that $u'(\theta) = v(\theta)$, $v'(\theta) = -u(\theta)$, whence $D_\theta(u^2 + v^2) = 2uv - 2uv = 0$. Hence $u^2 + v^2 \equiv [u(0)^2 + v(0)^2] = 0$ by (3).

4. Analytic theory of the hyperbolic and exponential functions. The development of the hyperbolic functions is even simpler in that no continuity or periodicity arguments are required. Equation (14b) defines θ as an odd increasing function of $t(-1 < t < 1)$ with positive derivative (13b). That $\theta \uparrow \infty$ when $t \uparrow 1$ follows readily from inequalities of the type

$$\int_{1-1/n}^{1-1/(2n)} \frac{du}{(1+u)(1-u)} > \frac{1}{2} \int_{1-1/n}^{1-1/(2n)} \frac{du}{1-u} \geq \frac{1}{2} \int_{1-1/n}^{1-1/(2n)} \frac{du}{n^{-1}} = \frac{1}{4}.$$

Invert (14b) to obtain t as an odd increasing differentiable function of θ ($-\infty < \theta < \infty$). Equations (11b), (2b) then define $\cosh \theta$, $\sinh \theta$ for *all* real values of θ, and $\cosh(-\theta) = \cosh \theta$, $\sinh(-\theta) = -\sinh \theta$. Equations (8b) and the remaining equations now follow even more readily than before.

---

*Changing to the variable t in (4) leads to an integral which is difficult to transform directly into (14).

To obtain the missing equation (9b) introduce the function

(17b) $$E(\theta) = \cosh \theta + \sinh \theta = x + y = \frac{1+t}{1-t} > 0 \qquad (-1 < t < 1),$$

and observe that

(18b) $$E'(\theta) = E(\theta),$$
$$E(0) = 1.$$

Moreover, $E(\theta) \uparrow \infty$ when $\theta \uparrow \infty$ $(t \uparrow 1)$, while $E(\theta) \downarrow 0$ when $\theta \downarrow -\infty$ $(t \downarrow -1)$.

Now set $e = E(1)$ and identify $E(\theta)$ with e^θ in standard fashion via the addition formula:

(19b) $$E(\theta_1 + \theta_2) = E(\theta_1)E(\theta_2).$$

LEMMA: *Let $E(\theta)$ be any differentiable function of θ satisfying the equations $E'(\theta) = aE(\theta) (-\infty < \theta < \infty)$ for some constant a, and $E(0) = 1$. Then $E(\theta)$ satisfies (19b).*

Proof: Set $u(\theta) = E(\theta+\theta_0)E(-\theta)$. Then $u'(\theta) = aE(\theta+\theta_0)E(-\theta) - aE(\theta+\theta_0) \cdot E(-\theta) = 0$. Hence $u(\theta) \equiv u(0)$, or $E(\theta+\theta_0)E(-\theta) = E(\theta_0)$. The symmetric form (19b) results on writing $\theta_0 = \theta_1+\theta_2$, $\theta = -\theta_2$.

Noting finally that (17b) implies the familiar formulae

(20b) $$\cosh \theta = \tfrac{1}{2}(e^\theta + e^{-\theta}),$$
$$\sinh \theta = \tfrac{1}{2}(e^\theta - e^{-\theta}),$$

solve (17b) for t to obtain

(9b) $$t = \frac{e^\theta - 1}{e^\theta + 1} = \frac{\tfrac{1}{2}(e^{\theta/2} - e^{-\theta/2})}{\tfrac{1}{2}(e^{\theta/2} + e^{-\theta/2})} \equiv \tanh \frac{\theta}{2}.$$

The function $w = \ln v (v > 0)$ may be defined in elementary fashion by the equation $v = e^w \cdots$

5. Euler's formula and the Fourier constant. The fundamental relations between the circular and hyperbolic functions are just the missing analogues (16b), (17a), (18a), (20a). Under the formal changes of variable $t = -is$, $u = -iv$, where $i = \sqrt{-1}$, (14a) and (11a) become:

$$i\theta = 2\int_0^s \frac{dv}{1-v^2},$$

$$\cos \theta = \frac{1+s^2}{1-s^2} = \cosh i\theta,$$

$$\sin \theta = -i\frac{2s}{1-s^2} = -i \sinh i\theta.$$

The relation $e^{i\theta} = \cosh i\theta + \sinh i\theta$ then yields Euler's formula:

(17a) $$e^{i\theta} = \cos\theta + i\sin\theta = \frac{1+it}{1-it}.$$

Now take (17a) as a definition and observe that

(16b) $$e^{i(\theta+2\pi)} = e^{i\theta};$$

(18a) $$D_\theta e^{i\theta} = ie^{i\theta},$$
$$e^{i0} = 1;$$

(20a) $$\cos\theta = \frac{1}{2}(e^{i\theta} + e^{-i\theta}).$$
$$\sin\theta = \frac{1}{2i}(e^{i\theta} - e^{-i\theta}).$$

Since the lemma of section 4 remains valid for complex-valued functions of a real variable, (18a) implies

(19a') $$e^{i(\theta_1+\theta_2)} = e^{i\theta_1} \cdot e^{i\theta_2},$$

which reduces to the addition formula (19a) on separating the real and imaginary parts. Moreover, (9a) can now be derived by solving (17a) for t.

Finally, the significance of the isolated definition (15) of π lies in its connection with the Fourier inversion formula:

$$F(u) = \int_{-\infty}^{\infty} e^{iux} f(x) dx,$$

$$f(x) = \frac{1}{C} \int_{-\infty}^{\infty} e^{-iux} F(u) du.$$

[The first equation defines $F(u)$ as a continuous function of u if $f \in L'(-\infty, \infty)$, in particular. The additional assumption that $F \in L'(-\infty, \infty)$ implies that the second equation holds at all points of continuity of $f(x)$.] In the classical theory the constant C is defined by the integral

$$C = 2 \int_{-\infty}^{\infty} \frac{\sin x}{x} dx.$$

In the modern theory C appears as a normalizing factor for the dual group measure on the (self-dual) additive group of the reals.* In either case the problem is to prove that $C = 2\pi$ by real variable arguments.

---

* For an introduction to the theory of harmonic analysis on commutative groups *cf.* the author's article Spectral theory and harmonic analysis, Proceedings of the Symposium on Spectral Theory and Differential Problems, Stillwater, Oklahoma, 1951, pp. 209–219.

Set $f(x) = e^{-|x|}$. Equation (17a) or elementary integrations by parts yield

$$F(u) = 2\int_0^\infty e^{-x} \cos ux\, dx = \frac{2}{1+u^2}.$$

Then

$$1 = f(0) = \frac{1}{C}\int_{-\infty}^\infty F(u)du = \frac{2}{C}\int_{-\infty}^\infty \frac{du}{1+u^2}.$$

Hence

$$C = 2\int_{-\infty}^\infty \frac{du}{1+u^2} = 2\pi$$

by (15).

BIBLIOGRAPHIC ENTRIES: EXPONENTIAL AND HYPERBOLIC FUNCTIONS

The references below are to the AMERICAN MATHEMATICAL MONTHLY.

1. Smith, W. B., The exponential development for real exponents, vol. 3, p. 163.
2. Llano, A., Properties of the function $(1+a)^x$, vol. 10, p. 244.
3. Roe, E.D., Jr., On the extension of the exponential theorem. vol. 16, p. 101.
4. Roe, E. D., Jr., Note on the extension of the exponential theorem, vol. 16, p. 159.
5. Elder, F. S., Note on a memory device for hyperbolic functions, vol. 21, p. 51.
6. Huntington, E. V., An elementary theory of the exponential and logarithmic functions, vol. 23, p. 241.

> The results of this paper were later simplified in a paper by Dunkel, vol. 24, p. 244, reproduced on pp. 123–124 in this collection.

7. Nyberg, J., The exponential and logarithmic functions, vol. 25, p. 337.

> Suggests some problems from physics or chemistry to motivate the study of exponentials and logarithms.

8. Bennett, A. A., An inequality in connection with logarithms, vol. 31, p. 86.
9. Hutchinson, C. A., Line representation of the hyperbolic functions, vol. 40, p. 413.
10. Price, G. B., On the definition of $e^{i\theta}$, vol. 43, p. 632.
11. MacDonald, J. K. L., Elementary rigorous treatment of the exponential limit, vol. 47, p. 157.
12. Pall, G., Limits by "consecutive rationals," vol. 56, p. 682.
13. Klamkin, M. S., On a chainomatic analytical balance, vol. 62, p. 118.
14. Kemeny, John G., The exponential function, vol. 64, p. 158.
15. Ghurye, S. G., A characterization of the exponential function, vol. 64, p. 255.

(e)
FUNCTIONAL EQUATIONS AND INEQUALITIES

THE LINEAR FUNCTIONAL EQUATION*

G. S. YOUNG, University of Michigan

In many undergraduate courses such as advanced calculus, it is proved that the only solutions of the functional equation

(1) $$f(x + y) = f(x) + f(y)$$

which are continuous, or continuous at a point, are the functions $f(x) = mx$, m a constant. To me, the usual proof, of showing that a solution is linear for x rational and then extending by continuity of the irrationals, lacks elegance, and repeats much of what was done—or should have been done—in setting up the real number system. There are more general hypotheses that imply the same conclusion, for example that f is bounded on some interval, or bounded on a set of positive measure, or measurable. In this note I prove the first of these generalizations by an argument suitable for undergraduates and which is perhaps shorter than the usual one for the continuous case. The proof is at least not well-known, though I can hardly believe it has escaped notice. Green and Gustin have given an excellent discussion of the history of this equation in [1].

THEOREM. *If $F(x)$ is a solution to* (1) *which is bounded over an interval* $[a, b]$, *then it is of the form $F(x) = mx$ for some real number m.*

Proof. We show first that $F(x)$ is bounded over $[0, b-a]$. Suppose that for all y in $[a, b]$, $|F(y)| < M$. If x is in $[0, b-a]$, then $F(x+a)$ is in $[a, b]$, so that from

$$F(x) = F(x + a) - F(a),$$

we get

$$|F(x)| < M + F(a).$$

Accordingly, if $b-a=c$, $F(x)$ is bounded in $[0, c]$. Let $m = f(c)/c$, and let $\phi(x) = F(x) - mx$. Then $\phi(x)$ also satisfies (1). We have $\phi(c) = F(c) - mc = 0$. It follows that $\phi(x)$ is periodic, with period c, for

$$\phi(x + c) = \phi(x) + \phi(c) = \phi(x).$$

Further, as the difference of two functions bounded over $[0, c]$, $\phi(x)$ is bounded

---

* From AMERICAN MATHEMATICAL MONTHLY, vol. 65 (1958), pp. 37–38.

over $[0, c]$, and from the periodicity, $\phi(x)$ is therefore bounded over the entire x-axis.

Suppose x_0 is a number such that $\phi(x_0) \neq 0$. By an easy induction, we have $\phi(nx_0) = n\phi(x_0)$. We can make $|n\phi(x_0)|$ as large as we please by increasing n, which would contradict the boundedness of $\phi(x)$. Therefore $\phi(x) \equiv 0$, or $F(x) \equiv mx$.

The existence of pathological solutions is something in the reach of quite superior students in advanced calculus, and I have several times had students who were able to report (to me) on Jones' example [2] of a pathological solution to (1) that has a connected graph.

It is easy to modify the argument to prove that the only solutions of

$$f(xy) = f(x) + f(y)$$

bounded on an interval $[1, a]$ are of the form $f(x) = k \log x$. Let $\phi(x) = f(x) - f(a)(\log x)/(\log a)$. Instead of periodicity, show that $\phi(ax) = \phi(x)$, which implies that $\phi(x)$ has the same bound in each interval $[1, a]$, $[a, a^2]$, $\cdots$. But $\phi(x^n) = n\phi(x)$.

References

1. J. W. Green and W. Gustin, Quasi-convex sets, Canad. J. Math., vol. 2, 1950, pp. 489–507.
2. F. B. Jones, Connected and disconnected plane sets and the functional equation $f(x) + f(y) = f(x+y)$, Bull. Amer. Math. Soc., vol. 48, 1942, pp. 115–120.

INEQUALITIES

Except for the entry labeled MATHEMATICS MAGAZINE, the references below are to the AMERICAN MATHEMATICAL MONTHLY.

1. Moulton, E. J., The real function defined by $x^y = y^x$, vol. 23, p. 233.

    An attempt to characterize the set of all (x, y) satisfying $x^y = y^x$.

2. Franklin, P., Relating to the real locus defined by the equation $x^y = y^x$, vol. 24, p. 137.

    Parametric equations are given for the continuous portion of the graph of $x^y = y^x$.

3. Slobin, H. L., The solutions of $x^y = y^x$, $x > 0$, $x \neq y$, and their graphical representation, vol. 38, p. 444.
4. Thielman, H. P., On a pair of functional equations, vol. 57, p. 544.

    Continuous solutions of the pair of equations

    $$f(xy) = g(x)^{p(y)} h(y)^{q(x)} \quad \text{and} \quad r(xy) = p(x)q(y) \quad \text{for} \quad x > 0.$$

5. Wright, E. M., An inequality for convex functions, vol. 61, p. 620.
6. Klee, V. L., Jr., Solution of a problem of E. M. Wright on convex functions, vol. 63, p. 106.

7. Kenyon, Hewitt, Note on convex functions, vol. 63, p. 107.
8. Flett, T. M., Continuous solutions of the functional equation $f(x+y)+f(x-y)=2f(x)f(y)$, vol. 70, p. 392.
9. Wetzel, J. E., On the functional inequality $f(x+y) \geq f(x)f(y)$, vol. 74, p. 1065.
10. Goldberg, R. R., Pseudo-multiplicative functions, MATHEMATICS MAGAZINE, vol. 30, p. 145.

Nondecreasing nonnegative functions satisfying $f(xy) \geq f(x)f(y)$.

4

CONTINUITY, $\in$ AND δ, DISCONTINUITIES

A TYPE OF FUNCTION WITH k DISCONTINUITIES*

RAYMOND GARVER, University of Rochester

Professor Pierpont, in his *Theory of Functions of Real Variables*, gives an interesting expression† for a function which has the following properties: (I) It is defined for $x \geq 0$. (II) It represents one arbitrary function in $0 \leq x < 1$, and a second arbitrary function when $x > 1$. (III) It is, in general, discontinuous at $x = 1$. (It is the third property which I am particularly interested in here, though Pierpont stresses the second and does not mention the third.) The expression is

$$y = \lim_{n \to \infty} \frac{x^n f(x) + g(x)}{x^n + 1},$$

where $f(x)$ and $g(x)$ are supposed to be defined in the necessary intervals. We clearly have $y = g(x)$ when $x < 1$, $y = \frac{1}{2}[f(x) + g(x)]$ when $x = 1$, and $y = f(x)$ when $x > 1$. Property (III) follows, and of course there may be other discontinuities arising from the $f(x)$ and $g(x)$.

We can extend the above to define a type of function possessing the first property as above, and two others which are extensions of (II) and (III). The function is, in general, discontinuous at $x = 1, 2, \cdots, k$, (k any positive integer) and this seems to me to be its most interesting property. Further, it represents $(k+1)$ arbitrary functions in different parts of its interval of definition. The expression for the function is

$$F(x) \equiv \lim_{n \to \infty} \frac{x^n f_1(x) + \left(\frac{x^2}{2!}\right)^n f_2(x) + \left(\frac{x^3}{3!}\right)^n f_3(x) + \cdots + \left(\frac{x^k}{k!}\right)^n f_k(x) + f_0(x)}{x^n + \left(\frac{x^2}{2!}\right)^n + \left(\frac{x^3}{3!}\right)^n + \cdots + \left(\frac{x^k}{k!}\right)^n + 1},$$

where it would be more compact, but perhaps less clear, to employ summations.

---

* From AMERICAN MATHEMATICAL MONTHLY, vol. 34 (1927), pp. 362–363.
† Volume 1 (1905), p. 204.

We now have at once
$$F(x) = f_0(x),\ 0 \leq x < 1\ ;\ \text{and}\ F(1) = \tfrac{1}{2}[f_0(1) + f_1(1)].$$

Dividing numerator and denominator by x^n, and letting n approach ∞, we obtain
$$F(x) = f_1(x),\ 1 < x < 2\ ;\ \text{and}\ F(2) = \tfrac{1}{2}[f_1(2) + f_2(2)].$$

Multiplying numerator and denominator by $(2!x^{-2})^n$, and letting n approach ∞, we have
$$F(x) = f_2(x),\ 2 < x < 3\ ;\ \text{and}\ F(3) = \tfrac{1}{2}[f_2(3) + f_3(3)].$$

And in general
$$F(x) = f_i(x),\quad i < x < i+1,\quad i = 1, 2, 3, \cdots, k-1.$$
$$F(i+1) = \tfrac{1}{2}[f_i(i+1) + f_{i+1}(i+1)].$$
$$F(x) = f_k(x),\quad k < x.$$

AN ELEMENTARY EXAMPLE OF A CONTINUOUS NON-DIFFERENTIABLE FUNCTION*

FRED W. PERKINS, Harvard University

The first example showing that the continuity of a function of the real variable x does not imply the existence of the derivative of this function at a single point is due to Weierstrass.† Other authors have also written upon this subject, but, in view of the importance of the phenomenon, the following discussion, which the writer has endeavored to make as elementary as possible, may not be without interest.

The method by which we propose to construct the function $f(x)$ constituting our example depends upon an interpolation process by means of which we extend the definition of $f(x)$ after we have defined this function for certain pairs of values of x. Let $x = x'$ and $x = x''$ be such a pair of values ($x' \neq x''$) and denote by y' and y'' the numbers $f(x')$ and $f(x'')$ respectively. We set

and
$$f(x' + \tfrac{1}{3}[x'' - x']) = y' + \tfrac{5}{6}[y'' - y']$$
$$f(x' + \tfrac{2}{3}[x'' - x']) = y' + \tfrac{1}{6}[y'' - y'].$$

It will be convenient to represent by P' and P'' the points (x', y') and (x'', y'') respectively, and also to denote by P_1 the point (x_1, y_1) and by P_2 the point (x_2, y_2). Excluding the case that $y' = y''$, we see that y_1 and y_2 lie between y' and y'', and that each of the quantities $|y_1 - y'|$, $|y_2 - y_1|$, $|y'' - y_2|$ is positive, but not greater than $5|y'' - y'|/6$. Furthermore, the lines $P'P_1$, P_1P_2 and P_2P'' have slopes which in each case are at least twice as great, numerically, as the slope of $P'P''$.

We are now ready to define the function $f(x)$ on the interval $0 \leq x \leq 1$. We set $f(0) = 0$ and $f(1) = 1$, and use the process described above to determine $f(1/3)$ and $f(2/3)$. We shall call the two points thus obtained the "interpolated

---

* From AMERICAN MATHEMATICAL MONTHLY, vol. 34 (1927), pp. 476–478.

† Weierstrass gave an example in lectures at Berlin as early as 1861, but it was not published until 1875. Weierstrass' discussion may be found in his *Werke* (Berlin 1895) Vol. 2, page 71ff. Wiener (Journal für Mathematik vol. 90, 1881, p. 221ff.) made a study of the function constituting this example; in connection with this paper, see a note by Weierstrass, (loc. cit. vol. 2, page 228ff.).

Riemann gave a function which enables us to infer that a continuous function may fail to have a derivative on an everywhere dense set of points. (*Gesammelte Mathematische Werke*, p. 225ff.). See also Schwartz, *Gesammelte Mathematische Abhandlungen* (Berlin, 1890), vol. 2, p. 269ff. Among the more recent contributions on this subject may be mentioned two examples of nowhere differentiable functions published in the Jahresbericht der Deutschen Mathematiker Vereinigung, by Faber, vol. 16 (1907), p. 538ff. and by Landsberg, vol. 17 (1908), and a memoir on the Weierstrass' function by Hardy in the Transactions of the American Mathematical Society, vol. 17 (1916), p. 301. See also Hobson, The Theory of Functions of a Real Variable (Cambridge, University Press, 1907), p. 620ff. In the paper by Hardy will be found further references to the literature on this subject.

points of the first order." By the same method we interpolate two more points between each pair of adjacent points now known on the graph of $f(x)$. The six points thus determined we shall call the "interpolated points of the second order," and so on, indefinitely. In this way, the value of $f(x)$ is specified for all values of x of the form $p/3^n$, where n is a positive integer, and p is a non-negative integer, not greater than 3^n.

Let X be any number of the interval $0 \leq x \leq 1$ not of the form $p/3^n$. Since $f(x)$ is defined for values of x differing from X by less than any preassigned positive quantity, and since the function $f(x)$, insofar as it has already been defined, has the lower bound 0 and the upper bound 1, the set of all known points on the graph of $f(x)$ has at least one limit point on the line-segment determined by the relations $x=X$, $0 \leq y \leq 1$. We shall now show that there is in fact just one such limit point. From properties of the interpolation process already stated, it follows that for any positive integer n, and any non-negative integer p such that $p<3^n$ we have

$$\left| f\left(\frac{p+1}{3^n}\right) - f\left(\frac{p}{3^n}\right) \right| \leq \left(\frac{5}{6}\right)^n$$

Any interpolated point with an abscissa lying between $p/3^n$ and $(p+1)/3^n$ has an ordinate lying between $f[p/3^n]$ and $f[(p+1)/3^n]$. Given any $\epsilon>0$, we can choose n so large that $(5/6)^n<\epsilon$, and then choose p so that X is an interior point of the interval bounded by $p/3^n$ and $(p+1)/3^n$. Then, in so far as $f(x)$ is defined on this interval, its least upper and greatest lower bounds here differ by less than ϵ, which shows that the set of all known points on the graph of $f(x)$ cannot have more than one limit point with the abscissa X. Let $f(X)$ be defined as the ordinates of this unique limit point. We extend in this manner the definition of $f(x)$ throughout the entire unit interval. The resulting function is clearly continuous at each of the non-interpolated points.

Further, if $x^* = p/3^n$ is the abscissa of an interpolated point each of the two adjacent interpolated points of the $(n+k)$-th order has an ordinate differing from $f(x^*)$ by not more than $(5/6)^{n+k}$. By choosing the positive integer k sufficiently large, this difference can be made less, in absolute value, than any preassigned positive ϵ. It follows that for such a choice of k, the relation $|f(x^*)-f(x)| \leq \epsilon$ holds for all interpolated points of the interval

$$\frac{3^k p - 1}{3^{n+k}} \leq x \leq \frac{3^k p + 1}{3^{n+k}},$$

and so at their limit points as well. Hence, $f(x)$ is continuous at the interpolated points also. The same sort of reasoning shows that $f(x)$ is continuous at $x=0$ and $x=1$.

We are now ready to show that at no point of the interval $0 \leq x \leq 1$ does the continuous function $f(x)$ possess a derivative. We understand this to mean that for no value of x in the interval in question does the difference quotient $[f(x+h)-f(x)]/h$ approach a limit as h approaches zero.*

Let x_0 be any number of the interval $0 \leq x \leq 1$. We choose any positive integer n, and then determine the positive integer p so that $(p-1)/3^n \leq x_0 \leq p/3^n$. One of the points

$$R: \left[\frac{p-1}{3^n},\ f\left(\frac{p-1}{3^n}\right)\right], \qquad S: \left[\frac{p}{3^n},\ f\left(\frac{p}{3^n}\right)\right]$$

determines with the point $P_0: [x_0, f(x_0)]$ a line with slope at least as great numerically as that of the line RS. But, from a property of the interpolation process, we know that the absolute value of the slope of RS is at least 2^n. Hence, we see that by choosing n sufficiently large we can find a point on the curve as near as we like to P determining with P a line with slope greater numerically than any preassigned quantity. This proves that $f(x)$ has no derivative for $x = x_0$. It does not imply, of course, that the difference quotient $[f(x_0+h)-f(x_0)]/h$ becomes infinite independently of the manner in which h approaches zero. It may be noted, however, that for $x_0 = 0$ or $x_0 = 1$, the difference quotient does become positively infinite independently of the manner in which the point x_0+h of the unit interval approaches x_0.

---

* Some writers, on the contrary, use a definition of differentiability allowing the difference quotient either to approach a limit or to become positively or negatively infinite.

A PECULIAR FUNCTION*

J. P. BALLANTINE, University of Washington

I sometimes show my beginners in calculus the following function:

A pie reposes on a plate of radius R. A piece of central angle θ is cut and put on a separate plate of radius r. How large must r be? Obviously r is a function of θ. It turns out to have the following formula:

$$r = 0 \qquad \theta = 0$$
$$r = \tfrac{1}{2} R \sec \tfrac{1}{2}\theta \qquad 0 < \theta \leq 90°$$
$$r = R \cos \tfrac{1}{2}\theta \qquad 90° \leq \theta \leq 180°$$
$$r = R \qquad 180° \leq \theta \leq 360°.$$

The function has one discontinuity at $\theta = 0$, and its second derivative has various discontinuities.

Note by the Editor

The preceding example by Professor Ballantine seems an uncommonly good one to use when introducing the notion of discontinuous functions to an elementary class as illustrating how naturally such functions arise. It might be of considerable interest and value to make a collection of some more examples of this sort. Can anyone suggest another one equally simple and equally free from an appearance of artificiality?

R. E. Gilman

---

* From AMERICAN MATHEMATICAL MONTHLY, vol. 37 (1930), p. 250.

TWO DISCONTINUOUS FUNCTIONS*

MARIE M. JOHNSON, Oberlin College

1. A rectangular piece of wood is turned horizontally from one corridor into another of the same width and at right angles to the first. Consider pieces with one dimension constant and equal to a. What must be the width w of the corridors as the second dimension x of the wood varies? It is found that the function is defined analytically as follows:

$$w = 0, \qquad x = 0;$$

$$w = \left(x + \frac{a}{2}\right)\frac{\sqrt{2}}{2}, \qquad 0 < x \leq a;$$

$$w = \left(\frac{x}{2} + a\right)\frac{\sqrt{2}}{2}, \qquad x \geq a.$$

This function has a discontinuity at $x=0$, and the first derivative has discontinuities at $x=0$ and $x=a$.

2. Let the first problem be modified in the following manner: The piece of wood is turned around the corner of the corridors only when this permits the width of the corridors to be less than when the piece is not turned. This function is represented by the formula

$$w = 0, \qquad x = 0;$$

$$w = \left(x + \frac{a}{2}\right)\frac{\sqrt{2}}{2}, \qquad 0 < x \leq \frac{2\sqrt{2} - 1}{2}a;$$

$$w = a, \qquad \frac{2\sqrt{2} - 1}{2}a \leq x \leq a;$$

$$w = x, \qquad a \leq x \leq \frac{2}{2\sqrt{2} - 1}a;$$

$$w = \left(\frac{x}{2} + a\right)\frac{\sqrt{2}}{2}, \qquad x \geq \frac{2}{2\sqrt{2} - 1}a;$$

and it is discontinuous at $x=0$. The first derivative has a number of discontinuities.

A Note by the Editor

Another excellent example of a discontinuous function will be found in an article in *The Mathematics Teacher* for December, 1927, entitled, *Functions in General, and the Function [x] in Particular,* by Professor Walter B. Carver of Cornell University.

R. E. Gilman

* From AMERICAN MATHEMATICAL MONTHLY, vol. 37 (1930), p. 497.

A FUNCTION WITH A FINITE DISCONTINUITY*

J. A. WARD, University of Georgia

An illustration of a function with a finite discontinuity that I have found useful in teaching sophomore calculus is the following:

$$y = \text{the degree in } u \text{ of: } (x-1)u^3 + 5u^2 - 3u + 2.$$

Then y is a discontinuous function of x; for at $x=1$, then $y=2$; but for all other values of x we have that $y=3$. This function seems a little less artificial to sophomores than one of the type $y=(x^2-4)/(x-2)$ which has to be re-defined at $x=2$.

Other examples may readily be constructed along these lines, such as:

$$y = \text{the degree in } u \text{ of } (x-1)u^{x^2+4} + 5u^2 - 3u + 2.$$

This gives in the xy-plane a parabola with the single discontinuity at $x=1$. Another variation is a continuous function whose derivative has two discontinuities:

$$y = \text{the degree in } u \text{ of } u^{x^2} + 2u^2 - 5u + 3.$$

In the xy-plane this is a parabola with the lower part of the arc discarded and replaced by a chord.

---

* From AMERICAN MATHEMATICAL MONTHLY, vol. 54 (1947), p. 162.

AN EVERYWHERE CONTINUOUS NOWHERE DIFFERENTIABLE FUNCTION*

JOHN McCARTHY, Princeton University

The following is an especially simple example. It is

$$f(x) = \sum_{n=1}^{\infty} 2^{-n} g(2^{2^n} x)$$

where $g(x) = 1+x$ for $-2 \leq x \leq 0$, $g(x) = 1-x$ for $0 \leq x \leq 2$ and $g(x)$ has period 4.

The function $f(x)$ is continuous because it is the uniform limit of continuous functions. To show that it is not differentiable, take $\Delta x = \pm 2^{-2^k}$, choosing whichever sign makes x and $x + \Delta x$ be on the same linear segment of $g(2^{2^k} x)$. We have

1. $\Delta g(2^{2^n} x) = 0$ for $n > k$, since $g(2^{2^n} x)$ has period $4 \cdot 2^{-2^n}$.
2. $|\Delta g(2^{2^k} x)| = 1$.
3. $\left| \Delta \sum_{n=1}^{k-1} 2^{-n} g(2^{2^n} x) \right| \leq (k-1) \max |\Delta g(2^{2^n} x)| \leq (k-1) 2^{2^{k-1}} 2^{-2^k} < 2^k 2^{-2^{k-1}}$.

Hence $|\Delta f/\Delta x| \geq 2^{-k} 2^{2^k} - 2^k 2^{2^{k-1}}$ which goes to infinity with k.

The proof that the present example has the required property is simpler than that for any other example the author has seen.

Weierstrass gave the example $F(x) = \sum_{n=0}^{\infty} b^n \cos(a^n \pi x)$ for $b < 1$ and $ab > 1 + 3\pi/2$ which is discussed in Goursat-Hedrick, *Mathematical Analysis*.

A complete discussion of the construction of functions with various singular properties is given in Hobson, *Functions of a Real Variable*, volume II, Cambridge, 1926.

---

* From AMERICAN MATHEMATICAL MONTHLY, vol. 60 (1953), p. 709.

A NOTE ON δ AND ϵ*

ATHERTON H. SPRAGUE, Amherst College

Our freshmen, at least some of them, can solve the problem: $\lim_{x \to 2} 3x = 6$, by showing that if $\delta = \frac{1}{3}\epsilon$, then $0 < |x-2| < \delta$ implies $|3x-6| < \epsilon$, for any positive number ϵ. But confront them with $\lim_{x \to 2} x^2 = 4$ and they haven't the remotest idea how to proceed. Of course one can write $|x^2 - 4| = |x-2| \cdot |x+2|$ and start by restricting x to values between 1 and 3 so that $|x+2| < 5$, but this leaves me cold. The purpose of this note is to exhibit a systematic technique for attacking this and similar problems, applicable in fact to all polynomials.

Considering the problem at hand, *i.e.*, $\lim_{x \to 2} x^2 = 4$, we wish to exhibit a positive number δ such that if $0 < |x-2| < \delta$, then $|x^2 - 4| < \epsilon$ for every positive ϵ. The whole trick is to expand $x^2 - 4$ in powers of $x-2$. Thus $x^2 - 4 = (x-2)^2 + 4(x-2)$. Then

$$|x^2 - 4| = |(x-2)^2 + 4(x-2)| \leq |x-2|^2 + 4|x-2|.$$

The latter will be less than ϵ provided that (completing the square)

$$|x-2|^2 + 4|x-2| + 4 < \epsilon + 4$$

or

$$|x-2| + 2 < \sqrt{(\epsilon+4)} \text{ or } |x-2| < \sqrt{(\epsilon+4)} - 2.$$

Thus, choosing $\delta = \sqrt{(\epsilon+4)} - 2$, the problem is solved.

Similarly, to prove that $\lim_{x \to 2} x^3 = 8$, one chooses $\delta = \sqrt[3]{(\epsilon+2^3)} - 2$. In general, for $a > 0$, to prove $\lim_{x \to a} x^n = a^n$, one chooses $\delta = \sqrt[n]{(\epsilon+a^n)} - a$. Obviously, such a δ is always positive.

If $a < 0$ for the case $\lim_{x \to a} x^n = a^n$, we choose $\delta = \sqrt[n]{(\epsilon+|a|^n)} - |a|$; this choice of δ will, of course, work for $a > 0$ as well.

For a problem such as $\lim_{x \to 2} 5x^2 = 20$, since $|5x^2 - 20| < \epsilon$ if $|x^2 - 4| < \frac{1}{5}\epsilon$, we choose $\delta = \sqrt{(\frac{1}{5}\epsilon+4)} - 2$, and so on. In the case of a general polynomial this technique is applied to the individual terms.

As a simple illustration consider $\lim_{x \to 1}(x^3 + 5x^2) = 6$. We wish to exhibit a positive number δ such that if $0 < |x-1| < \delta$, then $|x^3 + 5x^2 - 6| < \epsilon$. Now

$$|x^3 + 5x^2 - 6| = |x^3 - 1 + 5x^2 - 5| \leq |x^3 - 1| + 5|x^2 - 1|.$$

Also, $|x^3 - 1| < \frac{1}{2}\epsilon$ if $0 < |x-1| < \delta_1$, where $\delta_1 = \sqrt[3]{(\frac{1}{2}\epsilon+1)} - 1$; and $5|x^2-1| < \frac{1}{2}\epsilon$ if $0 < |x-1| < \delta_2$, where $\delta_2 = \sqrt{(\frac{1}{10}\epsilon+1)} - 1$. Hence, if we choose δ to be the smaller of δ_1 and δ_2, the problem is solved.

---

* From AMERICAN MATHEMATICAL MONTHLY, vol. 67 (1960), p. 780.

A FURTHER NOTE ON δ AND ε*

ALLAN DAVIS, University of Utah

In Classroom Notes (*A note on δ and ε*, this MONTHLY, vol. 67, p. 780†), Professor A. H. Sprague is concerned with the instructive exercise of finding, for given $\varepsilon>0$, a corresponding $\delta>0$ in order to show from the $\delta-\varepsilon$-definition that for given polynomial function f, $\lim_{x\to a} f(x) = f(a)$. Professor Sprague observes an advantage in expanding $f(x) - f(a)$ in powers of $x-a$, since it is $|x-a|$ which δ restricts. The method of obtaining a suitable δ which he finally presents, however, seems less simple and direct than the one illustrated, through examples, below.

Suppose one wishes to show that $\lim_{x\to 1}(x^3+5x^2) = 6$. Thus, one wishes to demonstrate that for any $\varepsilon>0$, a $\delta>0$ exists such that $|x^3+5x^2-6|<\varepsilon$ whenever $0<|x-1|<\delta$.

Now,

$$
\begin{aligned}
|x^3 + 5x^2 - 6| &= |\{(x-1)^3 + 3x^2 - 3x + 1\} + 5x^2 - 6| \\
&= |(x-1)^3 + 8x^2 - 3x - 5| \\
&= |(x-1)^3 + \{8(x-1)^2 + 16x - 8\} - 3x - 5| \\
&= |(x-1)^3 + 8(x-1)^2 + 13x - 13| \\
&\leq |x-1|^3 + |8||x-1|^2 + |13||x-1|.
\end{aligned}
$$

Choose δ in two stages. First, choose $\delta<1$. Then, if n is any positive integer, when $0<|x-1|<\delta$, so is $|x-1|^n<\delta$. Thus, however, when $0<|x-1|<\delta$,

$$|x-1|^3 + |8||x-1|^2 + |13||x-1| < \delta + 8\delta + 13\delta = 22\delta.$$

Second, choose δ also so that $22\delta<\varepsilon$. Apparently, if $\delta<1$ and $\delta<\varepsilon/22$, then when $0<|x-1|<\delta$, $|x^3+5x^2-6|<\varepsilon$.

As a second example, to show that $\lim_{x\to -2}(-3x^4-8x+2) = -30$, one has,

$$
\begin{aligned}
|(-3x^4-8x+2)-(-30)| &= |-3(x+2)^4 + 24(x+2)^3 - 72(x+2)^2 + 88(x+2)| \\
&\leq |-3||x+2|^4 + |24||x+2|^3 + |-72||x+2|^2 + |88||x+2|.
\end{aligned}
$$

First, choose $\delta<1$, then if $0<|x+2|<\delta$,

$$|-3||x+2|^4 + |24||x+2|^3 + |-72||x+2|^2 + |88||x+2| < 3\delta + 24\delta + 72\delta + 88\delta = 187\delta.$$

Second, choose δ also so that $\delta<\varepsilon/187$. For δ satisfying both choices, $|(-3x^4-8x+2)-(-30)|<\varepsilon$ whenever $0<|x+2|<\delta$.

---

*From AMERICAN MATHEMATICAL MONTHLY, vol. 68 (1961), pp. 567–568.
†*Ed. note*: Reprinted on p. 157, this book.

For any polynomial function f, of degree at least 1, it is apparent that the method for choosing δ which is being used in the above examples, amounts only to taking $\delta < 1$ and also $\delta < \epsilon/K$ where K is the sum of the absolute values of the coefficients in the expansion of $f(x) - f(a)$ in terms of $x - a$. As it turns out, students sometimes make this observation and then interest themselves in how the coefficients may be obtained by a more elegant procedure than the pedagogically rather effective—because simple—device of "insistence and adjustment" which was used in the first example. There have been classroom instances where students have invented or recalled the method for obtaining the coefficients through successive divisions of $f(x) - f(a)$ by $x - a$. Students repeating the course, to be sure, will advocate finding the coefficients via Taylor's formula.

BIBLIOGRAPHIC ENTRIES: CONTINUITY, ϵ AND δ, DISCONTINUITIES

Except for the entries labeled MATHEMATICS MAGAZINE, the references below are to the AMERICAN MATHEMATICAL MONTHLY.

1. Hille, E., and Tamarkin, J. D., Remarks on a known example of a monotone continuous function, vol. 36, p. 255.

   The Cantor singular function.

2. Moritz, R. E., On a totally discontinuous function, vol. 38, p. 394.

   The title is somewhat misleading. The example discussed is a uniformly convergent sequence of functions whose derivatives form a divergent sequence.

3. Hildebrandt, T. H., A simple continuous function with a finite derivative at no point, vol. 40, p. 547.

   Van der Waerden's example of a nowhere differentiable function, with powers of 10 replaced by powers of 2.

4. Begle, E. G., and Ayres, W. L., On Hildebrandt's example of a function without a finite derivative, vol. 43, p. 294.
5. James, Glenn, Remarks on the definition of continuity, vol. 44, p. 235.
6. Longley, W. R., An example of a continuous function with finite discontinuities in its second derivative, vol. 44, p. 467.

   The "wine glass problem" is shown to give rise to a function whose second derivative has two finite discontinuities.

7. Fort, M. K., Jr., The maximum value of a continuous function, vol. 58, p. 32.
8. Bush, K. A., Continuous functions without derivatives, vol. 59, p. 222.
9. Thielman, H. P., Types of functions, vol. 60, p. 156.

   "Neighborly" and "cliquish" functions.

10. Marcus, M. D., Boundedness of a continuous function, vol. 62, p. 580.

11. Pennington, W. B., Existence of a maximum of a continuous function, vol. 67, p. 892.

12. Swift, W. C., Simple constructions of non-differentiable functions and space-filling curves, vol. 68, p. 653.

> The example in this paper seems to be a special case of the example in Bush, vol. 59, p. 222. See also an editorial comment in vol. 69, p. 52.

13. Jungck, G., The extreme value theorem, vol. 70, p. 864.

14. Beardon, A. F., On the continuity of monotonic functions, vol. 74, p. 314.

15. Diaz, J. B., and Metcalf, F. T., A continuous periodic function has every chord twice, vol. 74, p. 833.

16. Bernau, S. J., The bounds of a continuous function, vol. 74, p. 1082.

17. Raisbeck, Gordon, Some examples illustrating continuity and differentiability, MATHEMATICS MAGAZINE, vol. 23, p. 17.

18. Lange, L. H., Successive differentiability, MATHEMATICS MAGAZINE, vol. 34, p. 275.

A function continuous but nowhere differentiable in the interval $0 \leq x \leq \frac{1}{2}$.

5

DIFFERENTIATION

(a)

THEORY

DERIVATIVES OF IMPLICIT FUNCTIONS*

M. R. SPIEGEL, Rensselaer Polytechnic Institute

When presenting a definition of the derivative of a function of a single variable, most textbooks and teachers place emphasis on derivatives of explicit functions using the "Δ-process." Thus the student is informed: Given $y = f(x)$, the derivative of y with respect to x is given by

$$\frac{dy}{dx} = \lim_{\Delta x \to 0} \frac{f(x + \Delta x) - f(x)}{\Delta x},$$

if such a limit exists. After this follows a tremendous number of problems in which the student has to write $dy/dx = \cdots$ a large portion of the time. It seems small wonder then that when presented with an implicit function of x, such as that defined by the relation $4x^2 + y^3 - y = 6x$, the student starts off by writing $dy/dx = \cdots$ and then proceeds to differentiate each term (I trust that many teachers have found the same to be true with their students). In seeking to abolish this sort of thing and to see whether I could instill a better understanding of derivatives of implicit functions, I found that it was instructive to point out to the student, by a few examples, that derivatives of implicit functions need not depend (as many of them believed) on theorems already proved, but that the "Δ-process" could be used directly.

To see how natural such an approach appears to the student, let us consider the following simple example:

If $4x^2 + y^3 - y = 6x$ defines y as a function of x find dy/dx by using the "Δ-process."

Solution: Proceeding according to the "Δ-process" we give x an increment Δx and y "takes on" the increment Δy in such a way that $x + \Delta x$ and $y + \Delta y$ satisfy the given relation. Thus

$$4(x + \Delta x)^2 + (y + \Delta y)^3 - (y + \Delta y) = 6(x + \Delta x)$$

---

* From AMERICAN MATHEMATICAL MONTHLY, vol. 61 (1954), pp. 120–121.

or

(1) $$4x^2 + 8x\Delta x + 4(\Delta x)^2 + y^3 + 3y^2\Delta y + 3y(\Delta y)^2 + (\Delta y)^3 - y - \Delta y = 6x + 6\Delta x.$$

Subtracting

$$4x^2 + y^3 - y = 6x$$

from (1) we obtain

(2) $$8x\Delta x + 4(\Delta x)^2 + 3y^2\Delta y + 3y(\Delta y)^2 + (\Delta y)^3 - \Delta y = 6\Delta x.$$

Upon division of both sides of equation (2) by Δx we have

$$8x + 4\Delta x + 3y^2 \frac{\Delta y}{\Delta x} + 3y\left(\frac{\Delta y}{\Delta x}\right)\Delta y + \left(\frac{\Delta y}{\Delta x}\right)(\Delta y)^2 - \frac{\Delta y}{\Delta x} = 6.$$

Taking the limit as $\Delta x \to 0$, assuming that when $\Delta x \to 0$, $\Delta y \to 0$ and $(\Delta y/\Delta x) \to (dy/dx)$, we find

$$8x + 3y^2 \frac{dy}{dx} - \frac{dy}{dx} = 6$$

from which finally

$$\frac{dy}{dx} = \frac{6 - 8x}{3y^2 - 1}.$$

Similar approaches adopted when dealing with partial derivatives seem to give the student better understanding of explicit and implicit functions of more than one variable. Furthermore the theory of functions of a single variable (explicit and implicit) and the theory of functions of two or more variables seem to become more unified in the minds of the students.

A FUNDAMENTAL THEOREM OF THE CALCULUS*

ISRAEL HALPERIN, Queen's University

Instructors of undergraduate calculus courses frequently use a non-rigorous geometric argument to "prove" that a function $f(x)$ must be constant on an interval if it has a derivative equal to zero at each point of the interval. The usual rigorous proof of this theorem uses the theorem of the mean which is itself based on the real variable theorem that a continuous function on a closed finite interval attains its maximum and minimum values. The difficulty, for the instructor, is that these deeper theorems are not known to the student.

However a completely rigorous proof can be derived at once from the following *weakened* form of the theorem of the mean:

THEOREM W. *If $f(x)$ has a derivative $f'(x)$ for each $a \leq x \leq b$ then for some $a \leq x_0 \leq b$*

$$|f'(x_0)| \geq \left|\frac{f(b) - f(a)}{b - a}\right|.$$

The following proof of Theorem W is simple enough for any undergraduate calculus course!

Without loss of generality, suppose the interval (a, b) is $(0, 1)$. Denote $|f(1) - f(0)|$ by k and let p_1 be the least of the integers $0, 1, \cdots, 9$ for which

$$\left|\frac{f(0.(p_1 + 1)) - f(0.p_1)}{1/10}\right| \geq k.$$

There is such a p_1, for if there were not it would follow that

$$|f(1) - f(0)| \leq |f(1) - f(0.9)| + |f(0.9) - f(0.8)| + \cdots + |f(0.1) - f(0)|$$
$$< \frac{k}{10} + \frac{k}{10} + \cdots + \frac{k}{10} = k,$$

that is, $|f(1) - f(0)| < k$, a contradiction. Designate the interval $(0.p_1, 0.(p_1+1))$ as (a_1, b_1).

Similarly let p_2 be the least of $0, 1, \cdots, 9$ for which

$$\left|\frac{f(0.p_1(p_2 + 1)) - f(0.p_1p_2)}{1/100}\right| \geq k$$

and designate $(0.p_1p_2, 0.p_1(p_2+1))$ as (a_2, b_2).

Repetition of this procedure yields a number x_0, given by its decimal expansion

$$x_0 = 0.p_1p_2\cdots$$

---

* From AMERICAN MATHEMATICAL MONTHLY, vol. 61 (1954), pp. 122–123.

and a sequence of intervals (a_n, b_n) such that for every n

$$\left|\frac{f(b_n) - f(a_n)}{b_n - a_n}\right| \geq k$$

and x_0 is contained in, or is an end point of, the interval (a_n, b_n) and $b_n - a_n \to 0$ as $n \to \infty$. If x_0 is contained in a particular (a_n, b_n) we may preserve the inequality either by replacing a_n by x_0 or by replacing b_n by x_0. Then it follows from the definition of derivative that $|f'(x_0)| \geq k$, as stated.

The sophisticated student will see that this argument is valid for functions $f(x)$ valued in an arbitrary linear, normed space. He will also use the same argument to prove (as variations of Theorem W) that for *real valued* $f(x)$ there are suitable x_1, x_2 on (a, b) with

$$f'(x_1) \leq \frac{f(b) - f(a)}{b - a} \quad \text{and} \quad f'(x_2) \geq \frac{f(b) - f(a)}{b - a}$$

respectively (actually, the real-variable proof of the theorem of the mean shows that there is an ξ in (a, b) for which *both* of these inequalities hold simultaneously).

ON PROVING THE CHAIN RULE

DAVID GANS, New York University

Choose an introductory textbook in calculus at random and examine its proof of the chain rule for differentiating a composite function. The proof will fall into one of three categories according to the amount of care taken, but regardless of the category the proof will be found to be less than satisfactory for the beginning student.

Assuming y to be a differentiable function of u, and u a differentiable function of x, a proof in what we may call category A consists in writing the identity $\Delta y/\Delta x = (\Delta y/\Delta u)(\Delta u/\Delta x)$, letting Δx approach zero, noting that Δu then approaches zero, and inferring that $dy/dx = (dy/du)(du/dx)$. This proof, of course, is defective in that it overlooks the possibility that, regardless of the restriction placed on Δx, Δu in approaching zero may attain the value zero "on the way," in which case $\Delta y/\Delta u$ cannot approach dy/du.

A proof in category B is like the above, but contains the additional remark that Δu must not be zero. Since this remark is always made with the utmost conciseness, often in a brief footnote, unaccompanied by any kind of clarifying discussion, this proof can be regarded as an improvement over the preceding only if it is granted that it is better to leave a student mystified by a difficulty than ignorant of it.

The proofs in category C are rigorous and complete, disposing of the difficulty noted above by an ϵ-procedure of some sort. However, it is hard to see how these comparatively sophisticated proofs can be understood by any but a very small minority of students when it is realized that the latter are usually in contact with the calculus for only about a month when the chain rule is presented to them.

How, then, should we prove the rule to such students? Admittedly, the proof cited above is delightfully simple in outline. That it has not been presented better in our introductory textbooks must be charged, it seems, to the paralyzing force of tradition, *i.e.*, what is good enough for fifty calculus books is good enough for the fifty-first. The fact is, however, that this proof can easily be made rigorous and also understandable, provided that we are willing, as we should be, to make a slight sacrifice in the generality of the theorem. But some simple preliminaries are necessary. First, examples should be given to show that when $u(x)$ is an elementary function of the type familiar to the student Δu can be zero when Δx is not zero, but it cannot be zero if Δx is sufficiently small. Second, it should be shown graphically that $u(x)$ would have to be a most unusual type of function (unless it is a constant) if Δu could be zero regardless how small Δx may be. Third, it should be shown why Δu must not be permitted to equal zero if the limit of $\Delta y/\Delta u$ is to exist and equal dy/du.

---

* From AMERICAN MATHEMATICAL MONTHLY, vol. 62 (1955), pp. 115–116.

With these preliminaries out of the way we can proceed to the proof. If x_1 is the value of x under consideration, and u_1 the corresponding value of u, all that is now necessary beyond what is presented in textbooks is to agree to consider only functions $u(x)$ for which Δu is not zero if x is sufficiently close to x_1 but unequal to it. That is, we state and prove the chain rule only for such functions. Thus, we only exclude functions taking on the value u_1 infinitely many times in the neighborhood of x_1. Except for the trivial case when $u(x)$ is constant in this neighborhood, such excluded functions never occur in first year work, and but rarely in the following few years. The student should be told that the chain rule also applies to these functions but that the proof of this fact is best delayed until such time as he works with them.

ON THE DERIVATIVE OF x^c*

S. LEADER, Rutgers University

For real constants a and c with $a>0$, the differentiation formula

(1) $$Dx^c]_{x=a} = ca^{c-1}$$

can be obtained directly from the special case in which $a=1$ and $c>0$:

(2) $$\lim_{x \to 1} \frac{x^c - 1}{x - 1} = c.$$

To derive (1) from (2) we have only to use the identities

(3) $$\frac{x^c - a^c}{x - a} = \left(\frac{w^c - 1}{w - 1}\right) a^{c-1} = -w^c \left(\frac{w^{-c} - 1}{w - 1}\right) a^{c-1}$$

where $w = x/a$, noting that w approaches 1 as x approaches a. Thus, we need only prove (2).

For the case $c = n$, a positive integer, (2) follows from the identity

(4) $$\frac{x^n - 1}{x - 1} = 1 + x + x^2 + x^3 + \cdots + x^{n-1}.$$

For $c = m/n$, a positive rational, we have, setting $y = x^{1/n}$,

(5) $$\frac{x^{m/n} - 1}{x - 1} = \frac{\dfrac{y^m - 1}{y - 1}}{\dfrac{y^n - 1}{y - 1}}$$

which gives (2) via the preceding case.

For arbitrary $c > 0$ consider any rationals p and q such that $0 < p < c < q$. Then for $0 < x < 1$ we have $x^q < x^c < x^p$ so $x^q - 1 < x^c - 1 < x^p - 1$. Since $x - 1 < 0$,

(6) $$\frac{x^p - 1}{x - 1} < \frac{x^c - 1}{x - 1} < \frac{x^q - 1}{x - 1}.$$

Now (6) also holds for $x > 1$, for then $x^p < x^c < x^q$ and $x - 1 > 0$. Letting x approach 1 in (6) we obtain

(7) $$p \leq \lim_{x \to 1} \frac{x^c - 1}{x - 1} \leq q.$$

Since p and q may be arbitrarily close to c, (7) gives (2).

---

* From AMERICAN MATHEMATICAL MONTHLY, vol. 65 (1958), pp. 364–365.

A SIMPLE PROOF OF A BASIC THEOREM OF THE CALCULUS*

MARY POWDERLY, University of Connecticut

The purpose of this note is to provide an elementary proof of the theorem: *If $f'(x) = 0$ for every x belonging to an interval $[a, b]$, then $f(x) \equiv c$ for all x belonging to $[a, b]$.* The usual proof makes use of the following theorems: A real-valued function continuous over a closed interval takes on its maximum and minimum in that interval, Rolle's Theorem, the Mean Value Theorem, etc. The following proof depends on none of these theorems.

Proof. Assume the contrary, namely, that there exist x_1 and x_2 of $[a, b]$, with $x_1 < x_2$ and $f(x_1) \neq f(x_2)$ and, for definiteness, that $f(x_1) < f(x_2)$ (in case $f(x_1) > f(x_2)$, consider the function $g = -f$). Let $\epsilon = [f(x_2) - f(x_1)]/(x_2 - x_1)$ ($\epsilon > 0$), so that $f(x_2) - f(x_1) = \epsilon(x_2 - x_1)$.

For this ϵ, let $A = \{x \mid x_1 < x \leq b, f(x) - f(x_1) \geq \epsilon(x - x_1)\}$. Then A is nonempty ($x_2 \in A$) and bounded from below. Let x^* be its greatest lower bound: $x^* = \inf A$. Since $f'(x_1) = 0$, there exists a positive $\delta_1 < x_2 - x_1$ such that $x_1 < x < x_1 + \delta_1$ implies $f(x) - f(x_1) < \epsilon(x - x_1)$, and hence $x^* > x_1$. On the other hand, since $f'(x^*) = 0$, there exists a positive δ_2 less than $x^* - x_1$ such that $x^* - \delta_2 < x < x^*$ implies $f(x^*) - f(x) < \epsilon(x^* - x)$.

Let x_3 be any point in $(x^* - \delta_2, x^*)$, so that $x_1 < x_3 < x^*$, and simultaneously:

$$f(x_3) - f(x_1) < \epsilon(x_3 - x_1) \quad \text{(by definition of } x^*\text{)}$$

$$f(x^*) - f(x_3) < \epsilon(x^* - x_3) \quad \text{(by the preceding implication),}$$

and consequently, by addition, $f(x^*) - f(x_1) < \epsilon(x^* - x_1)$.

This inequality implies that $x^* < x_2$. Finally, $f'(x^*) = 0$ implies the existence of a positive $\delta_3 < x_2 - x^*$ such that $x^* < x < x^* + \delta_3$ implies $f(x) - f(x^*) < \epsilon(x - x^*)$ and hence (in conjunction with the last preceding inequality) $f(x) - f(x_1) < \epsilon(x - x_1)$ in violation of the definition of x^* as the *greatest* lower bound of A.

---

* From AMERICAN MATHEMATICAL MONTHLY, vol. 70 (1963), p. 544.

A FALLACY IN DIFFERENTIABILITY *

ALBERT WILANSKY, Lehigh University

The following argument, which at first glance seems to prove that every differentiable function has a continuous derivative, is offered as a challenge for undergraduates.

Let f be differentiable. Fix x. For $k \neq 0$, let

$$\epsilon = \frac{f(x+k) - f(x)}{k} - f'(x).$$

Then

(1) $\qquad f(x+k) = f(x) + kf'(x) + k\epsilon, \qquad \epsilon \to 0 \text{ as } k \to 0.$

In (1) take successively $x=0$, $k=2h$; $x=0$, $k=h$; $x=k=h$. This yields

(2) $\qquad f(2h) = f(0) + 2hf'(0) + 2h\epsilon_1$

(3) $\qquad f(h) = f(0) + hf'(0) + h\epsilon_2$

(4) $\qquad f(2h) = f(h) + hf'(h) + h\epsilon_3.$

Solving (4) yields

$$f'(h) = \frac{f(2h) - f(h)}{h} - \epsilon_3.$$

Substitution from (2), (3) yields $f'(h) = f'(0) + 2\epsilon_1 - \epsilon_2 - \epsilon_3$. Hence $f'(h) \to f'(0)$ as $h \to 0$.

BIBLIOGRAPHIC ENTRIES: THEORY OF DIFFERENTIATION

Except for the entries labeled MATHEMATICS MAGAZINE, the references below are to the AMERICAN MATHEMATICAL MONTHLY.

1. Macnie, J., Introduction to differentiation, vol. 4, p. 172.

    Differentiation of x^r, where r is a positive rational number.

2. Hedrick, E. R., A direct definition of logarithmic derivative, vol. 20, p. 185.
3. Frink, O., Jr., An algebraic method of differentiation, vol. 36, p. 264.
4. Dushnik, B., A generalization of the derivative of a function, vol. 42, p. 414.
5. Joseph, J. A., An extension of a theorem of Leibniz, vol. 45, p. 36.

    Formula for $D^n(uv)$ extended to include negative exponents.

6. Lange, L., Geometrical aspects of the power function, vol. 49, p. 248.
7. Dresden, A., The derivatives of composite functions, vol. 50, p. 9.
8. Ivanoff, V. F., The n-th derivative of a fractional function, vol. 55, p. 491.

---

* From MATHEMATICS MAGAZINE, vol. 38 (1965), p. 108.

9. McKiernan, M., On the n-th derivative of composite functions, vol. 63, p. 331.

10. Fort, M. K., Jr., A fundamental theorem of calculus, vol. 63, p. 334.

If f' is zero on an interval, then f is constant on the interval. See also Halperin, vol. 61, p. 122, reproduced on pp. 149–150, this book.

11. Pandres, D., Jr., On higher ordered differentiation, vol. 64, p. 566.

On the nth derivative of a composite function.

12. Adamson, Iain T., A "static" approach to derivatives, vol. 66, p. 905.

A somewhat philosophical discussion of mathematical terminology.

13. Mead, D. G., The chain rule, vol. 67, p. 787.

14. Porter, G. J., On the differentiability of a certain well-known function, vol. 69, p. 142.

An elementary proof that the "ruler function" is not differentiable.

15. Mazkewitsch, D., The n-th derivative of a product, vol. 70, p. 739.

16. Hight, D. W., and Haddock, A. G., $D_x f[g(x)] = g(x)$, vol. 71, p. 1034.

17. Hunt, Burrowes, Pictures about 0-derivatives, vol. 73, p. 71.

18. Wilansky, Albert, An elementary calculus computation without variables, vol. 73, p. 73.

19. Brookes, C. J., Functions satisfying a certain type of reciprocity of derivatives, vol. 74, p. 578.

Solution of $d^n y / dx^n \cdot d^m x / dy^m = 1$.

20. Aull, C. E., The first symmetric derivative, vol. 74, p. 708.

Study of $\lim_{h \to 0} [f(x+h) - f(x-h)]/2h$.

21. Beesack, P. R., Taylor's formula and the existence of nth derivatives, vol. 74, p. 980.

22. Shieh, P. and Verghese, K., A general formula for the nth derivative of $1/f(x)$, vol. 74, p. 1239.

23. Eaves, J. E., Off the beaten path with some differentiation formulas, MATHEMATICS MAGAZINE, vol. 26, p. 147.

24. Ulam, S. M., and Hyers, D. H., On the stability of differential expressions, MATHEMATICS MAGAZINE, vol. 28, p. 59.

If f and g are uniformly close and if $f'(x)$ changes sign at $x = a$, then $g'(x)$ must be zero for some x close to a.

25. Manheim, J. H., Another look at differentiation, MATHEMATICS MAGAZINE, vol. 39, p. 89.

(b)
APPLICATIONS TO GEOMETRY

ON THE DEFINITION OF INFLECTION POINT*

G. M. EWING, University of Missouri

If we examine the definitions of inflection point given in textbooks on the calculus, certain inconsistencies appear. An inflection point may be defined as a point of a curve at which
  (A) the direction of concavity changes [1];
  (B) the tangent line crosses the curve [2];
  (C) the slope has a maximum or minimum [3], or in other ways [4].

For our present purposes a curve is a set of points of the euclidian plane which can be represented by parametric equations $x = x(t)$, $y = y(t)$, $T_1 \leq t \leq T_2$, with $(x(t_1), y(t_1)) \neq (x(t_2), y(t_2))$ for $t_1 < t_2$, except possibly for $t_1 = T_1$ and $t_2 = T_2$. It is convenient to specify certain curves by equations of the form $y = f(x)$, $x = \phi(y)$, $F(x, y) = 0$, etc.

A curve is said to be concave toward the x-axis (away from the x-axis) if given any three points $(x(t_i), y(t_i))$, $i = 1, 2, 3$, such that $x(t_1) < x(t_2) < x(t_3)$, the determinant

$$D(t_1, t_2, t_3) \equiv \begin{vmatrix} y(t_1) & x(t_1) & 1 \\ y(t_2) & x(t_2) & 1 \\ y(t_3) & x(t_3) & 1 \end{vmatrix}$$

is greater than zero (less than zero).

The equations of certain curves, together with the indicated inflection points are set down for reference:

(1) $\quad y = x^3, \quad (0, 0).$

(2) $\quad y = x^{1/3}, \quad (0, 0).$

(3) $\quad y = x^3, \quad x < 0;$
$\quad\quad = x^2, \quad x \geq 0; \quad (0, 0).$

(4) $\quad y = x^2, \quad x < 0;$
$\quad\quad = x^{1/2}, \quad x \geq 0; \quad (0, 0).$

(5) $\quad y = x^2, \quad x < 0; \quad (0, 1).$
$\quad\quad = -x^2 + 1, \quad x \geq 0.$

(8) $\quad x = y^{2/3}; \quad (0, 0).$

(9) $\quad y = \int_{-1}^{x} \phi(x) dx; \quad (0, 0).$

$\quad \phi(x) = 2x^2 + x^2 \sin \dfrac{1}{x}, \quad x \neq 0;$
$\quad\quad = 0, \quad x = 0.$

---

* From AMERICAN MATHEMATICAL MONTHLY, vol. 45 (1938), pp. 681–683.

(6) $x^2 + y^2 = 1$; (1, 0).

(7) $x = t^2$; (0, 0).
      $y = t^4$, $t < 0$;
      $\quad = 2t^4$, $t \geq 0$.

(10) $y = 2x^3 + x^3 \sin \dfrac{1}{x}$, $x \neq 0$;
      $\quad = 0$, $x = 0$;   (0, 0).

In the following table, check marks indicate which of the properties (A), (B), and (C) are possessed by each of these examples.

Example	A	B	C	Remarks
1	✓	✓	✓	[according to all writers.[5]]
3	✓	✓	✓	$f''(0)$ does not exist. Hence not an inflection point.
2	✓	✓		Slope $\dfrac{dy}{dx}$ is not defined for $x=0$.
8	✓	✓		
4	✓			There is no tangent line at $x=0$.
5	✓			$f(x)$ is discontinuous at $x=0$.
6	✓			
9		✓	✓	
10		✓		
7			✓	

In framing a definition it seems desirable to formulate it so that it will apply to as large a class of curves as possible; and hence properties depending upon certain derivatives are to be avoided. It seems desirable that such points as those of examples (1), (2), and (3) be classed as inflection points whereas those of the remaining examples may well be excluded. The following definition is suggested.

Inflection Point. An inflection point is a point $(x(\bar{t}), y(\bar{t}))$ of a curve $C: x = x(t); y = y(t)$ if (and only if) the following conditions are satisfied:

(1) $x(t)$ and $y(t)$ are continuous for $t = \bar{t}$;

(2) there exist numbers $t_1 < \bar{t}$ and $t_2 > \bar{t}$ such that $D(\tau_1, \tau_2, \tau_3) > 0$ (or < 0) for $t_1 \leq \tau_1 < \tau_2 < \tau_3 \leq \bar{t}$, and $D(\tau_1, \tau_2, \tau_3) < 0$ (or > 0 respectively) for $\bar{t} \leq \tau_1 < \tau_2 < \tau_3 \leq t_2$.

(3) for every linear form $ax + by + c$ which vanishes for $x = x(\bar{t}), y = y(\bar{t})$ there exist numbers $t_3 < \bar{t}$ and $t_4 > \bar{t}$ such that $ax(t) + by(t) + c > 0$ (or < 0) for $t_3 \leq t < \bar{t}$ and $ax(t) + by(t) + c < 0$ (or > 0 respectively) for $\bar{t} < t \leq t_4$.

In descriptive language an inflection point is a point of a curve such that (1)' the curve is continuous at the point, (2)' the direction of concavity changes at the point, and (3)' every straight line through the point has an arc of the curve on each side of it. Condition (2) is not the same as (2)', as shown by example (7); but (2)' is a consequence of (2) and (3). Some of the properties of an inflection point as just defined are as follows:

PROPERTY 1. *Inflection points are unchanged by translations or rotations of axes.*

PROPERTY 2. If the point $(x(\bar{t}), y(\bar{t}))$ is a point of inflection of a curve C, which is of class C'' (i.e., $x''(t)$ and $y''(t)$ are continuous) for all values of t in an interval to which $\bar{t}$ is interior, and if $x'(t)$ and $y'(t)$ are not both zero, then $x'y'' - y'x''$ vanishes for $t = \bar{t}$.

COROLLARY. If C can be represented in the form $y = f(x)$, and if $f''(x)$ is continuous, then $f''(x) = 0$ at the point of inflection.

PROPERTY 3. If C has a tangent line at an inflection point, the tangent line crosses the curve.

The last is immediate from defining Property 3. The proofs of the other stated properties are simple.

Note by the Editor. Doubtless some readers will think that the author has sacrificed simplicity and clarity for generality. We may agree that the discussion is interesting whether the proposed definition of an inflection point should be used in a textbook or not.

R. E. Gilman

References

1. C. E. Love, Differential and Integral Calculus, MacMillan, 1934, p. 48.

N. J. Lennes, Differential and Integral Calculus, Harper, 1931, p. 113.

W. A. Granville, et. al., Elements of the Differential and Integral Calculus, Ginn and Co., 1934, p. 79.

2. W. B. Fite, Advanced Calculus, MacMillan, 1938, p. 76.

R. Courant, Vorlesungen über Differential- und Integralrechnung, Springer, 1930, vol. 1, p. 127.

3. J. V. McKelvey, Calculus, MacMillan, 1937, p. 59.

4. G. H. Hardy, Pure Mathematics, Cambridge Press, 1925, p. 272.

A NOTE ON ENVELOPES*

HOWARD EVES, Syracuse University

At a certain stage in calculus we teach the well known method of finding the envelope of a one-parameter family of curves given by cartesian equations. Occasionally a student wonders how to find the envelope of a one-parameter family of curves given by polar equations (without transforming to cartesian coördinates and then back again). An easily established but general theorem may be stated to the effect that *the same process used for a family of curves given in the cartesian coördinate system may be used for a family of curves given in any point coördinate system.* Inasmuch as many teachers of college mathematics are unaware of this fact and since it does not seem to be stated in any of the ready literature, it might be well to establish it.

Let $f(u, v, c) = 0$ be the one-parameter family of curves given in some (u, v) coördinate system. Imagine a superimposed (x, y) cartesian coördinate system, and let $u = u(x, y)$, $v = v(x, y)$ be the equations of the transformation of coördinates. Then the family of curves is given in the cartesian system by $F(x, y, c) \equiv f[u(x, y), v(x, y), c] = 0$. Since it is obvious that $\partial F/\partial c \equiv \partial f/\partial c$, the theorem is proved.

---

* From AMERICAN MATHEMATICAL MONTHLY, vol. 51 (1944), p. 344.

RULINGS *

C. S. OGILVY, Syracuse University

Unless he has become so mathematically browbeaten by a year or so of the calculus that he receives all new concepts with apathy, a student's curiosity should be stimulated and perhaps a healthy skepticism aroused by his first encounter with ruled surfaces. How can a curved surface like a hyperbolic paraboloid contain straight lines? The purpose of this note is to show, by a method well within the student's grasp at this level, not only that rulings exist on certain surfaces but that in many special cases it is an easy matter actually to find them.

If the instructor is not averse to a bit of showmanship, he can set the stage by the use of a rather striking example. The class has just found approximately by differentials the value of $\sqrt{1+(1.99)(4.02)}$. The instructor remarks that the values of the increments of x and y in the expression $\sqrt{1+xy}$ must be small relative to 2 and 4 in order for the result to be a good approximation. He then takes as a counter example increments of 1 and 1, puts them through the formula for the total differential, and finds to his apparent surprise that the result is not approximately but exactly correct:

$$dz = \left[\frac{y}{2\sqrt{1+xy}} dx + \frac{x}{2\sqrt{1+xy}} dy \right]_{x=2, y=4, dx=1, dy=1} = 1.$$

Thus $z = 3+1 = 4$, the exact value of $\sqrt{1+(2+1)(4+1)}$. In feigned desperation he tries 2, 2 or 3, 3 as increments, only to have the same thing occur: every "approximation" gives the exact square root. If he now asks the class what is happening, someone should volunteer the suggestion that the instructor has been going out not on a limb but along a ruling of the surface $z = \sqrt{1+xy}$.

The process of the example itself suggests a criterion for the existence of rulings: if, starting from an arbitrary point (x, y, z) on a surface $z = f(x, y)$ it is possible for the whole increment Δz to equal the differential element dz, for some fixed ratio $dx:dy$ depending only on the point (x, y, z) and not on the size of the increments, then $f(x, y)$ represents a ruled surface.

To say that $\Delta z = dz$ means

(1) $$f(x+h, y+k) - f(x, y) = \frac{\partial f(x, y)}{\partial x} h + \frac{\partial f(x, y)}{\partial y} k,$$

where $h = \Delta x = dx$ and $k = \Delta y = dy$. (1) is a necessary and sufficient condition for the existence of rulings; it remains, however, to solve for h in terms of k, or in other words to obtain an expression for h/k, to answer the question whether the rulings are real, which they are if and only if h/k is real. Unfortunately (1) is

---

* From AMERICAN MATHEMATICAL MONTHLY, vol. 59 (1952), pp. 547–549.

sometimes an awkward equation to deal with; but we can exhibit a few of the many easy immediate applications.

(a) A proof that $z = Ax + By + C$ is a plane. Equation (1) becomes the identity $0 = 0$, which says that the criterion is fulfilled independent of h/k. Starting at any point (x, y, z) and going in *any* direction on the surface, we are travelling along a ruling. This property characterizes a plane.

(b) $z = x^2 + y^2$. (1) yields $h^2 + k^2 = 0$. This is a typical situation where there are clearly no rulings, since h, k cannot vary, nor is h/k defined, since $h = k = 0$ alone satisfies.

(c) $z = x^2 - y^2$. From (1) we find $h/k = \pm 1$. Since these ratios are real, there are two families of rulings, whose projections on the xy-plane are two orthogonal sets of parallel lines intersecting the axes at 45°.

Testing by means of (1) is often easier said than done. But in seeking a more workable criterion we may have to leave our second year calculus class behind as we turn to the Taylor expansion of $f(x+h, y+k)$ around the point $f(x, y)$:

$$
\begin{aligned}
f(x+h, y+k) = f(x, y) &+ \frac{1}{1!}\left[h\frac{\partial f(x,y)}{\partial x} + k\frac{\partial f(x,y)}{\partial y}\right] \\
&+ \frac{1}{2!}\left[h^2\frac{\partial^2 f(x,y)}{\partial x^2} + 2hk\frac{\partial^2 f(x,y)}{\partial x \partial y} + k^2\frac{\partial^2 f(x,y)}{\partial y^2}\right] \\
&+ \cdots.
\end{aligned}
\tag{2}
$$

If (1) is to hold, the terms of the second and all higher degrees in h, k must vanish identically:

$$h^2 f_{xx} + 2hk f_{xy} + k^2 f_{yy} = 0. \tag{3}$$

For a solution real in h/k, the discriminant of (3) must be non-negative:

$$f_{xy}^2 - f_{xx} f_{yy} \geq 0. \tag{4}$$

Thus (4) is a readily available necessary condition for the existence of rulings: all points must be hyperbolic or parabolic. It is obviously not sufficient; for all higher terms of (2) must also vanish for the values of h/k which satisfy (3) if rulings are to exist. As an example, take $z = xy + x^4 - y^4$. Equation (4) is satisfied:

$$f_{xy}^2 - f_{xx} f_{yy} = 1 + 144 x^2 y^2 > 0 \qquad \text{for all } x, y.$$

Hence h/k has two real solutions in (3). However, neither of them annuls identically the third order term of (2), and hence the surface is not ruled.

Equation (3) may be obtained in another way by starting with the assumption that rulings exist. One can go further and observe that eliminating the quantity h/k between (3) and the equation obtained by setting the third order term of (2) equal to zero results in a single partial differential equation of third order, namely the differential equation of the general ruled surface. This implies that if for a given function a solution of (3) annuls also the third order terms of (2), then the surface is ruled. This is indeed the necessary and sufficient condition. Its application is tedious, however, and certain shortcuts can be obtained; but to develop them here is beyond the scope of this note.

All treatment of this topic is omitted from most modern textbooks. De Morgan deals briefly with equation (2) *et seq.* in his 1842 calculus book (pp. 424–425), where he credits the idea to Monge. The differential equation of the general ruled surface is of course well known; (see for instance Salmon, *Analytic Geometry of Three Dimensions*, V Ed., Vol. II, p. 19). Equation (1) appears not to have been exploited, although it has the advantage of enlightening the student who is not yet ready for (2).

WHAT IS A POINT OF INFLECTION?*

A. W. WALKER, University of Toronto

Has the curve $y=x^4$ a point of inflection at the origin? At present, the answer depends on whether we consider the matter from the point of view of calculus or algebraic geometry.

As Ewing [1†] has remarked, considerable care is required in expressing the necessary and sufficient conditions for an inflection point in the language of the calculus, and inconsistencies are found when various texts are compared. However, the generally accepted view is surely the one expressed by von Mangoldt [2] in the section of the *Encyklopädie* on the application of calculus to curves and surfaces, namely, that a point of inflection is any point where a plane curve crosses both its tangent and its normal. (The curve may consist of two parts with different equations, joining, without discontinuity in the direction of the tangent line, at the point in question. An inflection point is not necessarily a point of zero curvature, and conversely.) According to this definition, the curve $y=x^4$ has, of course, no inflection points.

In algebraic geometry we are concerned to a considerable extent with complex or abstract fields, in which the above definition has no meaning.‡ The general viewpoint here (expressed with lack of clarity and consistency by many texts in English) seems to be the one clearly stated by R. J. Walker [3]. After remarking that the terms *flex* and *point of inflection* are identical, he defines a flex of a plane curve as a non-singular point at which the tangent line has n coincident intersections with the curve, where $n \geq 3$. Berzolari [4], in the section of the *Encyklopädie* on higher plane algebraic curves, refers to an *ordinary* inflection point when $n=3$, and to a *higher order* inflection point when $n>3$; in particular, the term *undulation point* is used when $n=4$. This is so for the curve $y=x^4$ at the origin. (It is instructive to consider $y=x^4-k^2x^2$ and let k approach zero.)

The only known reference to the fact that there are (in the real field) these two conflicting definitions is a short footnote in de la Vallée Poussin [5], which does not appear in the first edition but was added later. If the dual interpretation is to continue, it would avoid confusion if writers of future texts on both subjects followed this lead, preferably including a reference to [2] and [4] or their equivalents. An alternative would be the universal adoption in algebraic geometry of the concise term *flex*, defined as above. A flex would be either an inflection point (n odd) or an undulation point (n even), extending the scope of this latter term; the distinction would be significant only in the real field. (Salmon [6] remarks: "Cramer calls those points at which the tangent meets the curve in an odd number of consecutive points, points of visible inflection, to

---

*From AMERICAN MATHEMATICAL MONTHLY, vol. 63 (1956), pp. 182–183.
Ed. note: Reprinted on pp. 171–173, this book.
‡**The writer is indebted to the referee for this significant remark.**

distinguish them from points of undulation, which do not, to the eye, differ from ordinary points on the curve.")

References

1. This MONTHLY, vol. 45, 1938, p. 681.
2. Encyk. der Math. Wiss., III, 3, p. 9.
3. R. J. Walker, Algebraic Curves, Princeton, 1950, p. 71.
4. Encyk. der Math. Wiss., III, 2(1), p. 320.
5. De la Vallée Poussin, Cours d'Analyse Infinitésimale, 7th ed., 1930, Vol. 1, p. 242.
6. Salmon, Higher Plane Curves, 3rd ed., 1879, p. 37.

BIBLIOGRAPHIC ENTRIES: APPLICATIONS TO GEOMETRY

The references below are to the AMERICAN MATHEMATICAL MONTHLY.

1. Wilson, E. B., Relating to infinitesimal methods in geometry, vol. 24. p. 241.
2. Tamarkin, J. D., Some geometric illustrations for the elementary course in differential equations, vol. 35, p. 27.

> Properties of integral curves of first-order linear differential equations and some related equations.

3. Olds, E. G., A note on the linear differential equation of the first order, vol. 35, p. 306.

> A follow-up to Tamarkin's paper.

4. Kline, M., Note on elementary vector analysis and on an application to differential geometry, vol. 43, p. 555.

> Proof of equivalence of various definitions of limit of a vector-valued function. Also, a discussion of order of contact of curves.

5. Taylor, A. E., Derivatives in the calculus, vol. 49, p. 631.
6. Lars, L. S., A simplification of the second derivative test, vol. 54, p. 543.
7. Green, J. W., On the envelope of curves given in parametric form, vol. 59, p. 626.
8. Valentine, F. A., The motion of a particle constrained to move on a rough convex curve, vol. 63, p. 16.
9. Nash, Stanley W., The higher derivative test for extrema and points of inflection, vol. 66, p. 709.
10. Yates, Robert C., Curvatures of $r^n = \cos n\theta$, vol. 67, p. 275.
11. Thurston, H. A., On the definition of a tangent line, vol. 71, p. 1099.

(c)
APPLICATIONS TO MECHANICS

NORMAL AND TANGENTIAL ACCELERATION*

R. C. YATES AND C. P. NICHOLAS, USMA

As a first approach to acceleration in the study of plane motion both student and teacher would be happy to avoid the customary routine of converting from components in an arbitrary rectangular reference system to the more useful normal and tangential ones. The following discussion, essentially intrinsic in character, has this appeal.

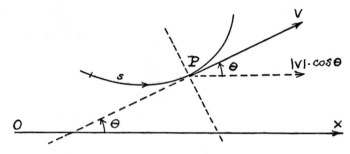

At a point P of the path let θ be the angle made by the velocity vector V with any direction OX. This angle, measured *from* the direction line OX *to* the velocity vector, is counterclockwise-positive. The component of V in this direction is

$$V_\theta = |V| \cdot \cos \theta.$$

The acceleration component in this same direction is the time derivative:

$$a_\theta = \frac{dV_\theta}{dt} = \frac{d|V|}{dt} \cdot \cos \theta - |V| \cdot (\sin \theta) \cdot \frac{d\theta}{dt}$$

$$= \frac{d|V|}{dt} \cdot \cos \theta - |V| \cdot (\sin \theta) \cdot \frac{d\theta}{ds} \cdot \frac{ds}{dt},$$

or

$$a_\theta = \frac{d|V|}{dt} \cos \theta - |V|^2 \sin \theta \frac{d\theta}{ds}.$$

---

* From AMERICAN MATHEMATICAL MONTHLY, vol. 58 (1951), pp. 255–256.

If we select the direction OX parallel to the tangent by taking $\theta = 0$, we obtain the *tangential acceleration*:

$$a_T = \frac{d|V|}{dt}.$$

Supposing that $d\theta/ds \neq 0$ at P, we now select OX along the normal at P pointing toward the concave side of the curve. If $d\theta/ds > 0$ at P, we have: $\theta = -\pi/2$ and $d\theta/ds = 1/R$ where R is the radius of curvature. If $d\theta/ds < 0$ at P, we have $\theta = \pi/2$ and $d\theta/ds = -1/R$. Then in either case:

$$a_N = \frac{|V|^2}{R},$$

the *normal* component of the acceleration measured in the direction of the center of curvature.

INVERSE SQUARE ORBITS: A SIMPLE TREATMENT*

D. E. RICHMOND, Williams College

The determination of the orbit of a body attracted to a center by an inverse square force is of great historical importance. Its rôle in astronomy and atomic physics makes the derivation of continuing interest. However, it is usually accessible to the student only after a fairly extensive knowledge of differential equations. There are available a number of vector treatments but they too require a considerable degree of mathematical maturity for their understanding.

This paper presents an extremely simple vector derivation of the equation of the orbit and of the relation between velocity and radial distance. It also includes a simple construction for the velocity. This treatment presupposes a knowledge of vector differentiation but uses neither scalar nor vector products. It is believed that the approach is well within the understanding of a college sophomore.

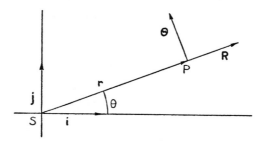

FIG. 1

Let a planet P be located at any time t by the vector $\mathbf{r}(t)$ from the sun S to P. Let $r(t)$ be the length of $\mathbf{r}(t)$ and let $\theta(t)$ specify its direction. Hereafter, the argument t will be omitted. In terms of the unit vectors $\mathbf{i}$ and $\mathbf{j}$,

$$\mathbf{r} = r(\cos\theta\,\mathbf{i} + \sin\theta\,\mathbf{j}) = r\mathbf{R},$$

where $\mathbf{R}$ is a unit vector in the direction of $\mathbf{r}$.

The velocity $\mathbf{v} = d\mathbf{r}/dt$, hence

(1) $$\mathbf{v} = r'(\cos\theta\,\mathbf{i} + \sin\theta\,\mathbf{j}) + r(-\sin\theta\,\mathbf{i} + \cos\theta\,\mathbf{j})\theta' = r'\mathbf{R} + r\theta'\mathbf{\Theta}$$

where $\mathbf{\Theta}$ is a unit vector perpendicular to $\mathbf{R}$ in the direction of increasing θ, and where primes denote differentiation with respect to t.

Implicit in (1) is the relation $d\mathbf{R}/dt = \theta'\mathbf{\Theta}$. A companion relation is $d\mathbf{\Theta}/dt = -\theta'\mathbf{R}$, which is immediately verified by differentiating $\mathbf{\Theta} = -\sin\theta\,\mathbf{i} + \cos\theta\,\mathbf{j}$.

One other preliminary is necessary. For central forces, $r^2\theta' = h$, a constant. The derivation is familiar and will be be omitted here.

---

* From AMERICAN MATHEMATICAL MONTHLY, vol. 59 (1952), pp. 694–696.

The inverse square law is expressed by

$$\mathbf{a} = \frac{d\mathbf{v}}{dt} = \frac{-\mu}{r^2}\mathbf{R} \tag{2}$$

where μ is a constant. To integrate (2), it is desirable to express the right member as a time derivative. This suggests substituting $\mathbf{R} = -(1/\theta')d\mathbf{\Theta}/dt$. Using $r^2\theta' = h$,

$$\frac{d\mathbf{v}}{dt} = \frac{-\mu}{r^2}\left(-\frac{1}{\theta'}\frac{d\mathbf{\Theta}}{dt}\right) = \frac{\mu}{r^2\theta'}\frac{d\mathbf{\Theta}}{dt} = \frac{\mu}{h}\frac{d\mathbf{\Theta}}{dt}. \tag{3}$$

Then

$$\frac{h}{\mu}\mathbf{v} = \mathbf{\Theta} + \mathbf{C} \tag{4}$$

where $\mathbf{C}$ is a constant vector.

The value of $\mathbf{C}$ depends upon the initial conditions. Let us assume that at $t=0$, P is closest to S, that is, at perihelion, and that $\mathbf{i}$ is drawn in this direction. Since $r'=0$ at $t=0$, $\mathbf{v}_0$ is in the direction of $\mathbf{\Theta}_0 = \mathbf{j}$. Hence $\mathbf{C} = \epsilon \mathbf{j}$ where ϵ is constant. Thus

$$\frac{h}{\mu}\mathbf{v} = \mathbf{\Theta} + \epsilon \mathbf{j}. \tag{5}$$

Figure 2 gives the corresponding geometrical construction.

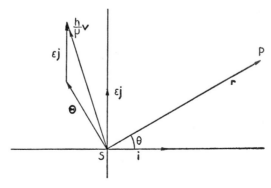

FIG. 2

Let us now take components of the vector $(h/\mu)\mathbf{v}$ in the direction of $\boldsymbol{\Theta}$. The result is $(h/\mu)r\theta'$ or, since $r^2\theta' = h$, $h^2/(\mu r)$. By (5), this must equal the projection of $\boldsymbol{\Theta} + \epsilon \mathbf{j}$ on $\boldsymbol{\Theta}$, namely, $1 + \epsilon \cos \theta$. Hence

(6) $$\frac{h^2}{\mu r} = 1 + \epsilon \cos \theta$$

or

(7) $$r = \frac{p}{1 + \epsilon \cos \theta}, \quad p = \frac{h^2}{\mu},$$

which is the polar equation of a conic. We have assumed that r is a minimum at $\theta = 0$. This implies that $\epsilon > 0$, excluding the trivial case of the circular orbit ($\epsilon = 0$).

From the law of cosines applied to the triangle of sides $\boldsymbol{\Theta}$, $\epsilon \mathbf{j}$, and $(h/\mu)\mathbf{v}$ (figure 2),

$$\frac{h^2 v^2}{\mu^2} = 1 + \epsilon^2 + 2\epsilon \cos \theta.$$

Substituting for $\epsilon \cos \theta$ from (6)

$$v^2 = \frac{2\mu}{r} - \frac{\mu^2(1 - \epsilon^2)}{h^2}.$$

In the important case of an elliptical orbit ($\epsilon < 1$), the semi-major axis

$$a = \frac{p}{1 - \epsilon^2} = \frac{h^2}{\mu(1 - \epsilon^2)}.$$

Then

(8) $$v^2 = \mu \left[\frac{2}{r} - \frac{1}{a} \right],$$

a widely used relation.

BIBLIOGRAPHIC ENTRIES: APPLICATIONS TO MECHANICS

The references below are to the AMERICAN MATHEMATICAL MONTHLY.

1. Hamilton, H. J., Uniform circular motion is singular, vol. 63, p. 109.
2. Dyer-Bennet, J., A problem on escape velocity, vol. 67, p. 1014.
3. Stein, S. K., Kepler's second law and the speed of a planet, vol. 74, p. 1246.

(d)
DIFFERENTIALS

DIFFERENTIALS*

MARK KAC and J. F. RANDOLPH, Cornell University

As easy and desirable as it is to get along without differentials, they seem to be with us to stay. The fact that differentials enable one to arrive at correct results when he thinks of dx and dy as "little pieces of x and y" makes him believe he is thinking correctly and he therefore will not give up his concept of differentials because some pedantic mathematician tells him he is just lucky.

Granted, then, that we are not going to dispense with differentials, can we not do something to help students understand what they are? The unsophisticated undergraduate who tries to believe everything he reads and his instructor tells him is hopelessly confused. One day he tries to believe (but does not succeed) that dy/dx is *not* dy divided by dx. The next day he may have momentary comfort when he learns that it is true after all that dy/dx is dy divided by dx and that $\lim_{\Delta x \to 0} \Delta y/\Delta x = dy/dx$. However, he is then told or, heaven forbid, sees "proved" that $dx = \Delta x$. He may admire the cleverness of dx for being able to "remain constant on one side of an equation and (since $dx = \Delta x$) approach zero or the other," but it bothers him just the same.

In the usual way of representing a function (please do not fight with us about this functional notation) let x be the independent variable and $y = f(x)$ the dependent variable. Then let Δx be a number $\neq 0$ and such that $f(x + \Delta x)$ is defined, but otherwise let Δx be arbitrary. Consider the number $f(x + \Delta x) - f(x)$ (if x insists on living up to its reputation as a variable, replace x by x_0), let

$$\Delta y, \Delta f, \Delta f(x)$$

be alternative notations for this number, and define each of the symbols

$$D_x y, D_x f, D_x f(x), y', f', f'(x)$$

to be the following limit, assumed to exist,

$$\lim_{\Delta x \to 0} \frac{f(x + \Delta x) - f(x)}{\Delta x}.$$

Above all things do not define dy/dx to be this limit.

Now let dx be an absolutely arbitrary number and define

$$dy = f'(x)dx.$$

---

* From AMERICAN MATHEMATICAL MONTHLY, vol. 49 (1942), pp. 110–112.

Clearly if $dx=0$ then $dy=0$, and if $dx\neq 0$ then $dy/dx = f'(x)$. Then from the very

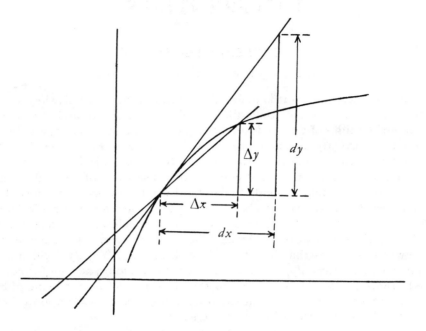

beginning dy/dx, $dx\neq 0$, is dy divided by dx.

Since Δx and dx are quite arbitrary, there is no inherent reason why they should be the same, and we therefore first take them different. Consequently, instead of the usual picture illustrating the relation of dy and Δy we have the geometric interpretation given in the figure. Since the slope of the chord approaches the slope of the tangent as $\Delta x \to 0$ we clearly have

$$\lim_{\Delta x \to 0} \frac{\Delta y}{\Delta x} = \frac{dy}{dx},$$

and there is no worry about a strain on dx as $\Delta x \to 0$.

If now we wish to use differentials as approximations to increments, it should be clear to the student that dy is an approximation to Δy if we choose, as we may, $dx = \Delta x$ and "small." Also if it is thought that infinitesimals are desirable in this connection, then write

$$\lim_{dx=\Delta x\to 0} \frac{dy - \Delta y}{\Delta x} = \lim_{dx=\Delta x\to 0} \frac{dy}{dx} - \lim_{dx=\Delta x\to 0} \frac{\Delta y}{\Delta x}$$
$$= \lim_{dx=\Delta x\to 0} f'(x) - f'(x) = f'(x) - f'(x) = 0,$$

and thus have that $dy - \Delta y$ is an infinitesimal of higher order than $\Delta x (= dx)$.

This definition of dx and dy amounts to what is essentially the Leibnitz definition of differentials; namely, dx is arbitrary and
$$dy = \lim_{t \to 0} \frac{f(x + t\,dx) - f(x)}{t}.$$

Differentials are, to be sure, not a great help when dealing with a function of one variable, but they become more important in the theory of functions of two or more variables. However, if differentials are introduced early in the course, their use will seem more natural to the student in later more complicated situations. It should, from the very beginning, be emphasized that a differential of an independent variable is itself independent, but a differential of the dependent variable is not independent but is determined.

In a course in mechanics when "virtual displacement", i.e., displacement along the tangent to a path, is introduced, the symbols δx and δy are generally used for the x- and y-components of this displacement. The reason why two symbols, δ's and d's, should be introduced when the algebric manipulations are the same is never clear to the student. The student is right in this case; the δ's are only differentials and this symbol is therefore superfluous.

DIFFERENTIALS*

ALONZO CHURCH, Princeton University

1. I am interested in the note of Kac and Randolph in the February number of the MONTHLY (pp. 110–112†), because I agree with them that the usual definition of the differential which they criticize is unsound, and that this unsoundness is within the understanding of the more intelligent beginning student in the calculus—or at least that it is sufficiently near the threshold of his understanding so that the definition causes him difficulty (even if he cannot make explicit the reason for his difficulty).

I would urge, however, that the objection which they make to the usual definition is not sufficient to reveal an unsoundness in it. In effect their objection is that, in the equation

$$\frac{dy}{dx} = \lim_{\Delta x \to 0} \frac{\Delta y}{\Delta x},$$

if dx is identified with Δx, the same variable Δx appears in the equation in two different rôles, on the left as a free variable and on the right as a bound variable.‡ But this must not be considered an error. For example, in a context where π has been defined as

$$\int_{-1}^{1} \frac{dx}{\sqrt{1-x^2}},$$

no one would think of objecting to the equation,

$$\tan(x + \pi) = \tan x,$$

on the ground that π stands for an expression containing x as a bound variable. Likewise, in a context where $D_x y$ has been defined as

$$\lim_{\Delta x \to 0} \frac{\Delta y}{\Delta x},$$

there should be no objection to writing an equation which contains $D_x y$ and at the same time contains a separate occurrence of Δx as a free variable.

The modification of Kac and Randolph in the definition of the differential must therefore be considered as designed to remove a difficulty for the student, rather than as correcting an actual error. Its advantage is that it avoids the possible necessity for an added explanation, which would, at least in effect, have to reproduce the distinction between free and bound variables.

---

*From AMERICAN MATHEMATICAL MONTHLY, vol. 49 (1942), pp. 389–392.

†*Ed. note:* Reprinted on pp. 185–187, this book.

‡ A variable is *free* in a given expression (in which it occurs) if the meaning or value of the expression depends upon determination of a value of the variable; in other words, if the expression can be considered as representing a function with that variable as argument. In the contrary case the variable is called a *bound* (or *apparent*, or *dummy*) variable.

2. On the other hand, there is a more serious objection which applies alike to the definition of Kac and Randolph and to the more usual definition which they wish to replace.

This objection may be formulated in the following terms. Both definitions agree in defining the differential of the independent variable, say x, by taking dx to be a new independent variable—hence it should with equal correctness be possible to use a single letter, say z, to represent this new variable. Then the differential of a dependent variable, say y, is defined by taking dy to be $(D_x y)dx$ —i.e., $(D_x y)z$. But a survey of the more usual purposes for which differentials are employed in the calculus will show that not all of these are adequately served by taking dx and dy to be z and $(D_x y)z$ respectively.

In particular, the use of differentials in connection with integration fails to be provided for. Thus $\int x dx$ becomes simply $\int xz$, and the whole significance of the notion is lost. (What plausibly is the operation $\int$ which, applied to the product of two independent variables x and z, yields $\frac{1}{2}x^2 + C$?)—It should be emphasized that the facility which comes with the use of differentials is more marked in the integral calculus than in the differential calculus, and that any definition of differentials which fails to account for their use with the sign $\int$ has therefore lost more than half their value. Compare, e.g., any one of the following processes in terms of differentials with the clumsier parallel process which regards the notation $\int \cdots dx$ as indivisible and employs only derivatives without differentials: (1) the integration $\int x\sqrt{4x+3}\, dx$ by the substitution $t^2 = 4x+3$, using the equation between differentials, $dx = \frac{1}{2}t dt$; (2) the solution of the differential equation $dy/dx = xy$ by multiplying both sides by dx/y and then applying the operation $\int$ to both sides; (3) the discovery by inspection of an integrating factor for a differential equation of the first order and first degree in two (or more) variables.

Another aspect of the foregoing objection lies in the point that the same notation, dx or dy, is given different meanings according as x is independent or dependent variable. This is especially unfortunate because it is often precisely one of the advantages in the use of differentials that various variables are symmetrically treated, without arbitrarily singling out one or more of them as the independent variable or variables (compare, e.g., the equation $ds^2 = dx^2 + dy^2$ with the corresponding equation for $(ds/dx)^2$, or for $(ds/dy))^2$.

Sometimes a student will ask why the result,

$$\frac{dy}{dx} = \frac{dy}{du}\frac{du}{dx},$$

cannot be obtained by simple cancellation of du against du, and the more difficult argument which employs properties of limits thus avoided. On the basis of the usual definition of a differential, the reply is to point out that du has different meanings in its two occurrences; but this immediately reveals the weakness of this usual definition.

If desired, the objection that the usual definition of a differential does not treat the variables symmetrically can be regarded as the fundamental objection. The operation $\int$ must, of course, be the inverse of the operation d, however the latter operation is defined; and if the operation d fails to treat the variables symmetrically, the same lack of symmetry must affect the inverse operation. The difficulty in connection with integration reduces in part to the point that the embarrassment occasioned by the lack of symmetry is more acute in the case of the inverse operation. But the matter is further complicated by the way in which the usual definition of dy introduces a new independent variable z.

Unless some solution of these difficulties can be found, it seems that it would be preferable to introduce differentials in a frankly inaccurate and heuristic manner as small values of the increment or "little bits" of the "variable quantity" involved, rather than to clothe the idea with the deceptive appearance of logical accuracy.

3. There is a method of introducing differentials which suggests itself as a possible remedy, but unfortunately it may not be suitable for use in an elementary course except by devoting a disproportionate amount of time to the study of parametric equations The statement of it which follows is at all events not intended to be in form for presentation to the student.

This method is simply to define dx and dy to be $D_\tau x$ and $D_\tau y$ respectively, where τ is an arbitrary parameter.

This does not contradict the usual statement that differentials are direction numbers of the tangent line. On the contrary it implies that statement. But it also adds a supplement to it which is needed to provide for certain ordinary uses of differentials. In particular the sign $\int$ taken by itself is then naturally understood to mean integration with respect to τ.

The extension of this idea to the differential du of a function u of two independent variables x and y is possible but somewhat cumbrous. It would perhaps be preferable to interpret equations involving du as relative to an arbitrary functional relationship between x and y (i.e., as holding for every such functional relationship which satisfies appropriate conditions).

No very convenient method is provided of introducing second differentials, or of associating differentials with small values of the increment. In fact it seems that, if this definition of differentials is adopted, the notion of a differential should be kept entirely separate from considerations connected with small values of the increment. A special notation to represent $f'(x)\Delta x$ may be unnecessary; if such a notation is introduced, it should be something like $\delta_x f(x)$ or $\delta_x y$ (the subscript being dropped only in cases where it is irrelevant which variable is independent).

BRINGING IN DIFFERENTIALS EARLIER*

W. R. RANSOM, Tufts College

Suppose we skip the usual chapter on functions and continuity, and start with the concept of variables connected by equations:

$$x^2 + y^2 = r^2, \quad x = r \cos \theta, \cdots . \tag{1}$$

Not variables as "symbols to which numerical values can be assigned," but variables as Newton (and beginners in calculus) might conceive them—the measures of quantities in nature and geometry. And not continuity as unnaturally defined for a point only, but the sort of continuity implied in the ancient doctrine "natura non agit per saltum."

A set of variables connected by equations will have sets of simultaneous numerical values, and such a set constitutes a "state." In an interval from one state to another, the variables receive increments (denoted by $\Delta x, \Delta y, \cdots$ etc.) and the new values, $x+\Delta x, \cdots$ are connected by the same relations as before. The intervals between states are taken small enough so that no question of multiple values arises.

By subtractions a new set of equations is obtained which describe relations between the variables and their increments. If these equations are all divided by the same increment, we get relations between the variables for some state and some mean rates of increase for an interval from that state to another. We define exact rate (now simply called rate) as the limit which the mean rate approaches when the interval shrinks so that the second state approaches the first. These rates are then adjoined to the set of variables, thus enlarging the state.

Now we come to differentials. In the equations which connect the variables with the rates, express the rates as fractions with a common denominator. This denominator is called the differential of that variable whose increment was used as a divisor, and the numerators are called the differentials of the variables in whose rates they appear. That is, we have symbolized the rates so that $D_x y = dy/dx$, $D_x r = dr/dx$, $\cdots$. This defines the differentials, one as a denominator of rates, and the others as numerators of rates. Now adjoin these differentials to the state containing variables and rates.

By the argument which appears in current texts under the title "function of a function," it is then to be shown that not only are

$$D_x y = dy/dx, \quad D_x r = dr/dx, \cdots$$

but also:

$$D_y x = dx/dy, \quad D_r y = dy/dr, \cdots .$$

Accordingly the distinction relating to one differential as denominator of rates may, and should forthwith be laid aside: any two differentials in a state may be taken as numerator and denominator for the corresponding rate.

---

* From AMERICAN MATHEMATICAL MONTHLY, vol. 58 (1951), pp. 336–337.

It is desirable to frame rules for writing down the equations that connect variables with their differentials, as, from (1):

$$2xdx + 2ydy = 2rdr, \quad dx = dr \cos\theta + r \sin\theta d\theta, \cdots.$$

From these we get the information about dy/dx, dr/dy, $d\theta/dy$, $\cdots$.

The definition commonly given in current texts for dy is $dy = (D_x y)\Delta x$, whence $dx = \Delta x$. This does not make sense, since Δx is not defined for the state to which the *rates* belong, but only for the interval to which mean rates belong, Graphically dy and dx are the rise and run of a piece of the tangent, and have nothing to do with two points on the curve.

Since differentials are determined numerically only to within a common factor, they cannot themselves be differentiated. There is no d^2y. The symbol d^2y/dx^2 is shorthand for $d(dy/dx)/dx$ just as $D_x^2 y$ is shorthand for $D_x(D_x y)$.

By using differentials from the start, we get rid of the clumsy phrase "with respect to" in both differentiation and integration. The dx in the integral tables becomes a differential, a factor of the differential to be integrated, and not a part of a symbol "$\int(\ )dx$." It seems scandalous to the writer that the dx in dy/dx should be equal to Δx, while the dx in $\int dx$ should not be equal to dx.

Editorial Note: This paper by Professor Ransom is the outgrowth of extended discussion between him and the editor arising from a discussion group at Tufts College in August, 1950 on the "Teaching of Calculus." Although the editor does not agree with Professor Ransom at many points, he is publishing this paper as a stimulus to discussion of an area in which there seems to be great confusion among textbook writers and teachers. Classroom Notes will welcome a limited number of replies to Professor Ransom's paper, particularly those which discuss the questions:

(1) What is a differential?

(2) Does the "dx" of differential calculus have the same meaning as the "dx" in the integral: $\int_a^b f(x)\,dx$?

Prospective commentators will be interested in the various articles on this subject in earlier volumes of this MONTHLY. These are listed under "Differentiation" in the Index which was recently issued to readers of the MONTHLY.*

C. B. Allendoerfer

---

* *Ed. note*: See AMERICAN MATHEMATICAL MONTHLY, vol. 57, part II.

DIFFERENTIALS*

M. K. FORT, JR., University of Illinois

1. Introduction. The beginning student usually has more trouble understanding differentials than any other topic in Calculus. One reason for this, as has often been noted, is the use of the "dy/dx notation" instead of the "$f'(x)$ notation" for derivatives. However, even the use of the superior "$f'(x)$ notation" does not help the student to understand what the author of his text is trying to get across when he comes to the notorious proof that $dx = \Delta x$. In order to eliminate the confusion concerning differentials, it is sufficient (and probably necessary) to first give an adequate treatment of the concept of function. With regard to functions, the average Calculus text is open to the following criticisms:

(1) "Function" is never actually defined, but instead a definition is given for the logically different concept "the variable y is a function of the variable x."

(2) The symbol "$f(x)$" is used to denote both a function and the value of that function at x.

(3) No distinction is made between a function and an equation which is used to define the function. (*e.g.* Text books frequently speak of "the function $y = x^2 + 3$" rather than "the function f for which $f(x) = x^2 + 3$.")

We now outline a treatment of functions and differentials which the author feels is suitable for presentation at the Calculus level.

2. Functions. F is a *function* if F is a rule which associates with each member x of a certain set A of objects a unique member $F(x)$ of a set B of objects. The set A is the *domain* of the function F and the set B is the *range* of F. If x is in the domain of F, then $F(x)$ is the *value of F at x*. The sets A and B are often sets of numbers, but this is not always the case.

Example 1. Let S be the rule which associates with each real number x its square x^2. Then $S(x) = x^2$ for each real number x. The domain of S is the set of all real numbers and the range of S is the set of all non-negative real numbers.

Example 2. Let d be the rule which associates with each point (x, y) of the plane the distance from (x, y) to $(0, 0)$. The domain of d is the set of all ordered number pairs and the range of d is the set of all non-negative real numbers. A function whose domain is a set of ordered number pairs is often called a *function of two variables* and the value of such a function f at (x, y) is denoted by $f(x, y)$ instead of $f((x, y))$. Thus for d we have $d(x, y) = (x^2 + y^2)^{1/2}$

Example 3. Let C be the function which associates with each triangle T its inscribed circle $C(T)$. The domain of C is the set of all triangles and the range of C is the set of all circles.

It is convenient to use the expression "ordinary function" to describe a function whose domain and range are both sets of numbers. As soon as derivatives

---
* From AMERICAN MATHEMATICAL MONTHLY, vol. 59 (1952), pp. 392–395.

are defined (carefully avoiding the dy/dx notation) the following example is instructive.

Example 4. Let D associate with each differentiable function f the function f'. Then D is a function whose domain and range are both sets of ordinary functions.

3. Differentials. Let f be a function which has a derivative at a number x. We define the *differential of f at x* to be the function g for which $g(h) = f'(x) \cdot h$ for each real number h. Thus the differential of f at x is an ordinary function which is linear and whose domain is the set of all real numbers. We shall denote the differential of f at x by df_x. The *differential of f* is a function of two variables whose domain is the set of all pairs (x, h) for which x is a number at which f has a derivative and h is a real number; the differential of f is denoted by df and is defined by the equation $df(x, h) = df_x(h)$. (If the reader chooses, he may replace h by Δx throughout this paper.)

If one defines in the usual way the addition of functions, the multiplication of functions and multiplication of functions by numbers, then it is easy to prove that

$$d(f + g) = df + dg$$
$$d(f \cdot g) = df \cdot g + f \cdot dg$$
$$d(c \cdot f) = c \cdot df$$

whenever f and g are functions and c is a number.

We define the identity function I by letting $I(x) = x$ for each real number x. It is easy to prove that $dI_x(h) = h$ for all numbers x and h. Hence, for any function f which has a derivative at x,

$$df_x(h) = f'(x) \cdot dI_x(h)$$

for all h. It follows that the function df_x is equal to the product of the function dI_x by the number $f'(x)$ and thus

$$df_x = f'(x) \cdot dI_x.$$

The above equation is true for each x at which f has a derivative, and hence

(i) $$df = f' \cdot dI.$$

Formula (i) is equivalent to the theorem "$dy = f'(x)dx$" which states that the differential of a function f is equal to the product of the derivative of f and the differential of the identity function.

The differential df is a function of two variables. In order to check his understanding of our notation the reader should prove the following statement. If f'' exists, then the partial derivative of df with respect to the first variable exists and is equal to the differential of f'.

4. Integration. The Editor has requested discussion of the following question:*

(ii) Does the "dx" of differential calculus have the same meaning as the "dx" in the integral $\int_a^b f(x)dx$?

The answer to this question is both yes and no.

If $\int_a^b f(x)dx$ is the Riemann integral (*i.e.* defined as the limit of a finite sum), then there is no connection between the "dx" in the integral and "dx" as used in the differential calculus. In this case the "dx" in the integral is used to make the "x" a bound variable. It serves the same purpose that the first "j" does in the expression "$\sum_{j=3}^{5} k^j$." (Notice that if we omitted the j and instead wrote $\sum_{3}^{5} k^j$ then we would be in doubt as to whether $3^j+4^j+5^j$ or $k^3+k^4+k^5$ was intended.) While $\int_a^b f(x)$ would not be a good notation for the integral (since x appears in this expression as a free variable), there is no logical objection to a notation such as $\int_a^b f$. However, it would be difficult to express $\int_2^3(x^2-5)dx$ in this latter notation since there is no commonly used notation which denotes the function whose value at x is x^2-5. This function can be denoted by an expression such as $\lambda x(x^2-5)$ [see The Calculi of Lambda-Conversion†], and if this notation were used for the function then we could denote our integral by $\int_a^b \lambda x(x^2-5)$ instead of $\int_a^b(x^2-5)dx$. This concludes the discussion of our answer of "no" to (ii).

We now examine the circumstances under which the answer to (ii) is "yes." Let us suppose that f is a continuous function. Then it is possible to prove that there exist functions g such that $g'=f$. It follows from (i) that $dg=f\cdot dI$ for such functions g. We may now define $\int f\cdot dI$ to be the class of all functions g for which $dg=f\cdot dI$. (The "dI" in $f\cdot dI$ is actually superfluous since it is just as reasonable to denote this class of functions by $\int f$.) Finally, we can define $\int_a^b f\cdot dI = g(b)-g(a)$ where g is a member of the class $\int d\cdot dI$. Thus, if integration is introduced via the indefinite integral or primitive method, and "dx" represents the differential of the identity function, then it is possible to interpret the "dx" in $\int_a^b f(x)dx$ as being the same as the "dx" of the differential calculus.

5. Conclusion. The author feels that most of the confusion concerning differentials disappears if the instructor:

(1) uses the word "variable" only to mean a symbol (*i.e.* an ink spot or chalk mark, etc.,) and avoids the usual implication that a variable is a mysterious Dr. Jekyll and Mr. Hyde type of gadget;

(2) carefully distinguishes between a function f and the number $f(x)$ which is the value of f at the number x;

(3) defines the differentials of functions and not the differentials of variables.

---

*This MONTHLY, vol. 58, 1951, p. 337. [*Ed. note:* See pp. 205–207, this book.]

† Alonzo Church, *The Calculi of Lambda-Conversion*, Princeton University Press, 1941.

THE RELATION OF DIFFERENTIAL AND DELTA INCREMENTS*

C. G. PHIPPS, University of Florida

Do mathematicians in general have as much trouble understanding the use and meaning of the differential notation as is indicated by the sporadic references to it which have appeared in this MONTHLY? It is to be hoped not, for there should be no confusion *when the symbolism is properly defined.*

This note is written in response to the Editor's two queries: (1) what is a differential and (2) how is the differential notation related to that for integration?† One consistent set of definitions is given here which answers the questions in the order they are proposed.

Let $y = F(x)$ be a continuous function with continuous derivatives of all orders. The derivative of this function is found by the usual limiting process and the result may be indicated by any one of the usual symbols, say $D_x y$.

Let (x, y) and (x', y') be two different solutions of the above equation. We define the delta-increments as

(1a) $$\Delta x = x' - x \quad \text{and} \quad \Delta y = y' - y.$$

Then we define the differentials of these variables as

(1b) $$dx = \Delta x \quad \text{and} \quad dy = (D_x y)\Delta x.$$

The first definition in (1b) applies specifically to the independent variable and the second to the dependent one. Furthermore, these definitions are consistent for, if we write $y = x$, the second definition yields the same result which is stated in the first.

From the latter two definitions we obtain the usual forms

(2a) $\quad dy = (D_x y)dx \quad$ and $\quad$ (2b) $\quad \dfrac{dy}{dx} = D_x y.$

The second of these forms supplies the reason for writing the derivative in its usual fractional form. It also shows the change from Δx to dx gives a symmetry to the notation.

The first of these forms has two interpretations:

(i) it is a *rule* for carrying out the operation of "taking the differential"; and

(ii) it is a *formula* for computing the value of the differential of the dependent variable.

The right member of (2a) is a function of two independent quantities, x and dx. If dx is held constant, dy becomes a function of x alone to which we can apply the rule in (i) and define the second differential of y as

---

* From AMERICAN MATHEMATICAL MONTHLY, vol. 59 (1952), pp. 395–398.
† Ed. note: See pp. 205–207, this book.]

(3a) $$d(dy) = d^2y = [(D_x^2 y)dx]dx = (D_x^2 y)dx^2.$$

Further, we define the higher differentials of y as the result of successive applications of this rule under the same conditions and obtain

(3b) $$d^n y = (D_x^n y)dx^n.$$

These differential equations may then be transformed into the usual fractional notation for derivatives of higher orders.

If the function $F(x)$ is expanded into a Taylor's series, we may write

(4a) $$\Delta y = (D_x y)dx + \frac{1}{2!}(D_x^2 y)dx^2 + \cdots + \frac{1}{n!}(D_x^n y)dx^n + \cdots.$$

By the aid of our definitions this series may be transformed into

(4b) $$\Delta y = dy + \frac{1}{2!}d^2 y + \frac{1}{3!}d^3 y + \cdots + \frac{1}{n!}d^n y + \cdots.$$

Thus it is seen that our definitions are convenient as well as consistent.

It is well-known that within the interval of convergence the partial sums of the terms of this series may be used as successive approximations to Δy. For instance, the first term alone gives the value Δy would have if its rate of change were constant while x changed from x to x'. The first two terms yield the value Δy would have if this rate of change itself had changed at a constant rate. The first three terms have the constant rate removed one step further. And partial sums with more terms have corresponding interpretations.

If the function $F(x)$ does not possess continuous derivatives beyond the n-th order, we can of course obtain no more than the n-th differential. At the same time, the series approximation would likewise not extend beyond the n-th order.

The extension of these definitions to functions of more than two variables is illustrated by two examples. Let $z = G(x, y)$. As independent variables, we define the differentials of x and y as equal to their delta-increments. Then

(5) $$dz = G_x dx + G_y dy,$$

where the subscript denotes partial differentiation. This form gives both a *rule* for "taking the differential" and a *formula* for computing its value. The definition may be extended to any order. Furthermore, the relation between Δz and its differentials of all orders is given by a form identical with that in (4b).

If $z = G(x, y)$ where $y = F(x)$, the value and form of dz is still given by (5) wherein the value of dy is given by (2a). The second differential, however, is more complicated; it is

(6) $$d^2 z = G_{xx}dx^2 + 2G_{xy}dxdy + G_{yy}dy^2 + G_y d^2 y.$$

The differentials of higher orders are correspondingly more complicated but can with care be written down. Also, the form in (4b) still holds true.

Turning now to the notation for integration, we find its connection with differential notation dependent upon the meaning attached to the integral.

(a) We may begin by defining the integral sign as the indefinite operator defined in the equation

$$\int dy = y + C.$$

If $y = F(x)$ and $D_x y = f(x)$, we have by substitution that

$$\int f(x)dx = F(x) + C.$$

By this definition, integration is defined (except for the constant C) as the inverse operation of "taking the differential." Hence, the integral sign as an operator in this sense has no meaning unless it is applied to a perfect differential.

(b) Alternately, we may define Δy exactly by the expression

$$\Delta y = \int_{x_0}^{x_0+\Delta x} f(x)dx,$$

where $y = F(x)$ and $D_x y = f(x)$ as above.

The value of this integral may be approximated by taking $f(x) = f(x_0)$ throughout the interval. This approximate value is $(dy =) f(x_0)\Delta x$. A better approximation may be obtained by dividing the interval into more and smaller parts and using the above approximation for each part. The limit as the number of parts becomes infinite is the exact value of the integral.

This method of approximating Δy does not correspond, except for the first step, with the method which uses either form of (4). Hence the dx in this integral has no relation either to Δx or to the previous form of dx.

The only obvious reason for using the same symbolism in (b) as in (a) is that the definition in (a) may be used to evaluate the limit represented by (b), namely as

$$F(x_0 + \Delta x) - F(x_0).$$

Thus by writing the definite integral in the same form as the indefinite, the known *properties of the latter form* may be used to find the *value of the former*. As with differentials, one interpretation gives the form, the other gives the substance.

TOWARD UNDERSTANDING DIFFERENTIALS*

H. J. HAMILTON, Pomona College

1. Introduction. The mathematics teacher hardly needs to excite himself today about identification of differentials with "nascent" or "evanescent" increments or with "vanishingly small" quantities, or even with "infinitesimals." If physics has not entirely abandoned these mystical ideas, at least mathematics has done so. It is now standard mathematical pedagogy to introduce differentials as follows. Let $y=f(x)$, where x is the independent variable. Then we define $dx \equiv \Delta x$ and $dy \equiv f'(x)dx$, where Δx is an arbitrary, non-zero increment in x. A picture goes with these definitions, and we explain to our students that a perfectly good value for dx is the distance in millimeters to the remotest known star; dy/dx is still equal to $f'(x)$.

(Of course, if we are as careful as we should be, we say that dx is a *variable* increment rather than an *arbitrary* increment, for we need to "see" the "dx" in most cases—certainly in integrands, where to give dx an arbitrary value would be to introduce ambiguity: $\int 6x^2 = 2x^3+c$ because I took $dx=1$; $\int 6x^2 = x^3+c$ because Professor Smith took $dx=2$.)

But it is time that we stopped patting ourselves on the backs for having invented a good definition. We need now to attack the problems of *understanding*, which our logical definition has by no means solved—even, I think, for most of us teachers. These problems are the problems of concreteness; they are, frankly, psychological; and they arise when we interchange dependent and independent variables, introduce a new variable, or become involved with relations between several variables—especially when more than one are independent.

To be sure, our elementary text-books prove that dy/dx is invariant under each of the first two of these changes. (The proofs, however, could be improved: If x is the independent variable, we should logically write $(dx)_x$ and $(dy)_x$ for the symbols defined above, and we should use corresponding notations when other variables are chosen as independent. We should then show that the ratio, $(dy)_t/(dx)_t$ does not depend on the choice of the independent variable t—including of course the special cases $t=x$ and $t=y$. Thereafter we could feel logically secure in writing dy/dx without subscripts.) But there remains the haunting if purely psychological question of just which "d" is the "Δ" in a given differential expression. (And are we not indeed expecting our freshmen to comprehend something a shade worse than a variable *functional*?) Since concreteness seems to be a *sine qua non* for mental tranquility, we might perhaps agree to identify the differential with the increment for, say, the *last* variable whose differential appears in any specific problem—and keep it that, no matter what other variables we may later introduce. This is quite justifiable for problems involving a single independent variable, and something like it may be optimum pedagogy. *Note, however, that it actually dispels the mystery of the unspecified differential increment only at the expense of introducing the inelegance of the specified differential increment.* Also, it necessitates further artificialities in approximation

* From AMERICAN MATHEMATICAL MONTHLY, vol. 59 (1952), pp. 398–403.

problems in which one would normally identify some *other* increment with a value of the corresponding differential.

The problem of concreteness is more difficult in the case of several independent variables. For example, if $z = f(x, y)$, where x and y are independent variables, we customarily define (or *should* define) $(dx)_{x,y} \equiv \Delta x$, $(dy)_{x,y} \equiv \Delta y$, and $(dz)_{x,y} \equiv f_x(x, y)(dx)_{x,y} + f_y(x, y)(dy)_{x,y}$, where Δx and Δy are independently variable increments in x and y, respectively. If we now choose to regard two other parameters, u and v (whether they are among x, y, and z or not), as the independent variables, we arrive, by means of certain simple and elegant theorems to be found in any respectable calculus text-book, at a formal identity which may be expressed as $dz = f_x(x, y)dx + f_y(x, y)dy$. (How many of us teachers have stopped to think that this means, in particular, that $\Delta z = f_x(x, y)(dx)_{v,z} + f_y(x, y)\Delta y$?) Again we may answer the yearning for concreteness by a device: we may agree that the last *two* differentials which appear in any specific problem shall be interpreted throughout that problem as independent increments—*if there are precisely two independent variables in the problem*. But if x and y are functions of two, or one, or a million independent variables, just so long as $z = f(x, y)$, those same admirable text-book theorems show us that $dz = f_x(x, y)dx + f_y(x, y)dy$. And when not only *what* the independent variables are, but *how many of them* there are, is unspecified, I am afraid that our wish for concreteness will have to remain not entirely satisfied. (As a test of our *own* need for concreteness, we might read critically the paragraphs in current elementary differential equations text-books devoted to the "total" equation $P(x, y, z)dx + Q(x, y, z)dy + R(x, y, z)dz = 0$ and to its geometric applications.)

It seems to me that, in seeking a way out of our dilemma, we should frame our definitions with reference to neither *increments* of variables nor any *particular set* of independent variables. I offer the following general, yet skeletal and condensed development with the hope that it may achieve for differentials (A) continued dissociation from notions of smallness, (B) symmetry of definition, (C) stability of definition under transformations, (D) a degree of compatability with the psychological desire for concreteness of interpretation, and (E) creatively critical consideration at levels of one and two independent variables by teachers who wish to make the concept more meaningful.

2. An algebraic development of differentials. (The justification for this title will appear in due course.) In an attempt to minimize prejudice in this theoretical discussion, we avoid the "d" notation; this we accomplish by reserving Latin letters for the primary variables and using the corresponding Greek letters for their differentials. We assume that all of our functions are "sufficiently well-behaved."

Let all of the relations* between a set of variables† $x_i (i = 1, 2, \cdots, l)$ be implied by a set of r independent equations

---

* By "relations" we shall mean "functional relations."

† When each variable literal subscript is introduced, we state its range; thereafter this range for that subscript is assumed.

(1) $$f_m(x_1, x_2, \cdots, x_l) = 0 \qquad m = 1, 2, \cdots, r.$$

Since the rank r of the matrix

(2) $$(f_{m\,x_i})$$

equals both* the number of independent equations in (1) and the number of the x_i which are defined by (1) as functions of the others, we have $r < l$. Hence the r equations in the new variables ξ_i,

(3) $$f_{m\,x_1}\xi_1 + f_{m\,x_2}\xi_2 + \cdots + f_{m\,x_l}\xi_l = 0,$$

have solutions other than the trivial one. Indeed, since the subsets x'_j ($j = 1, 2, \cdots, r$) of the x_i for which determinant $(f_{m\,x'_j}) \neq 0$ coincide with the subsets* of the x_i which are defined by (1) as functions of the others, their positions coincide with those of the subsets ξ'_j of the ξ_i which are defined by (3) as functions of the remaining ones of the ξ_i.

The differentials of the x_i relative to the system (1) we define to be the (corresponding) ξ_i as linearly interrelated by (3).

Let

(4) $$f_0(x_1, x_2, \cdots, x_l) = 0$$

represent *any* relation between the x_i. Since (4) is implied by (1), the rank of the matrix

$$(f_{n\,x_i}) \qquad n = 0, 1, 2, \cdots, r$$

is the same as that of (2), namely r. Hence all sets of values of the ξ_i which satisfy (3) also satisfy

(5) $$f_{0\,x_1}\xi_1 + f_{0\,x_2}\xi_2 + \cdots + f_{0\,x_l}\xi_l = 0.$$

In particular, if all of the relations between the x_i are implied by the set of $\bar{r}$ independent equations

(6) $$f_p(x_1, x_2, \cdots, x_l) = 0 \qquad p = r+1, r+2, \cdots, r+\bar{r},$$

then

(7) $$f_{p\,x_1}\xi_1 + f_{p\,x_2}\xi_2 + \cdots + f_{p\,x_l}\xi_l = 0.$$

Now if we require of the ξ_i only that they satisfy (7), the arguments by which we deduced (7) from (1) and (3) permit us to deduce (3) from (6) and (7). Thus, the differentials of a set of variables are dependent only on the relations between those variables and not on any particular system which expresses those relations; and so *we may omit the phrase* "relative to the system (1)" *in our earlier definition.*

Certain other invariant properties of systems which define the relations between the x_i may be deduced from the reciprocal implications of (1) and (6).

---

*P. Franklin, A Treatise on Advanced Calculus, New York, 1940, pp. 340–347.

Thus: *we have $\bar{r}=r$, since the rank of (f_{qx_i}) $(q=1, 2, \cdots, r+\bar{r})$ is equal at once to r and $\bar{r}$; if determinant $(f_{mx'_j}) \neq 0$, then determinant $(f_{px'_j}) \neq 0$, and conversely,* since these conditions express the solvability of (1) and (6), respectively, for the x'_j; hence also *if (3) is solvable for a set ξ'_j of the ξ_i, then (7) is solvable for the ξ'_j,* and conversely. We repeat here that, *if (1) is solvable for a set x'_j of the x_i, then (3) is solvable for the corresponding ξ'_j and conversely.*

Thus for any set of related variables x_i $(i=1, 2, \cdots, l)$ there are a number $r<l$ and a set of subsets $x'_j (j=1, 2, \cdots, r)$ which are defined as functions of the remaining ones of the x_i, say x''_k $(k=1, 2, \cdots, l-r)$; the differentials ξ'_j of the x'_j are then necessarily defined as linear functions of the differentials ξ''_k of the x''_k. These subsets and these definitions are provided by *any* system of the form (1) which expresses all of the relations between the x_i and by the associated system of the form (3).

3. Transformation and constraint of variables. Transformation of variables amounts to (i) the addition to (1) of certain equations which involve new variables as well as the old ones and (ii) the addition to (3) of corresponding equations which involve new differentials and primary variables as well as the old ones, with (iii) the rank of the new matrix corresponding to (2) maintained equal to r. Neither the previous definitions of the original differentials, nor the previously established relations between them, nor any previous concept or interpretation (see Section 5) of them need be changed in any way. (To be sure, we must envision spaces of more dimensions *when* we wish to regard the augmented system of primary variables and differentials as a whole.)

Constraint of variables, whether with or without addition of new ones, involves the addition to (1) and to (3) of new equations with an increase in the rank of the matrix corresponding to (2). The differentials are thus also constrained. But (1) and (3) continue to hold, and we remain, although restricted, within the configurations in terms of which we may have interpreted those relations (see Section 5).

4. Two explicit formulas. For simplicity in the following we shall use the x'_j and ξ'_j notations of Section 2 to indicate the typical sets of r "dependent" variables and of their differentials (that is, those for which (1) and (3) can be theoretically solved), and the x''_k and ξ''_k notations to indicate the corresponding $l-r$ "independent" variables and their differentials (those as functions of which (1) and (3) determine the x'_j and ξ'_j respectively). For further brevity we put $s=l-r$. From (3) we have

(8) If $x_i = \phi_i(x_1, x_2, \cdots, x_l)$, then $\xi_i = \phi_{ix_1}\xi_1 + \phi_{ix_2}\xi_2 + \cdots + \phi_{ix_l}\xi_l$.

In particular,

(9) If $x_i = \psi_i(x_1'', x_2'', \cdots, x_s'')$, then $\xi_i = \psi_{ix_1''}\xi_1'' + \psi_{ix_2''}\xi_2'' + \cdots + \psi_{ix_s''}\xi_s''$.

5. A vectorial interpretation. The equations (1) represent r surfaces in l-space, and the equations (3) identify the differentials ξ_i with the components of a vector in that space which is tangent to each of these surfaces and is therefore tangent to the s-dimensional surface which is their common intersection. (This vector one would of course call *the differential vector of the x_i*.) Equation (4) represents a surface which passes through this intersection, and (5) is merely a statement of the fact that the differential vector of the x_i is tangent to this surface.

6. Differentials and "vanishing smallness." In the setting up of differential equations the mathematical scientist often derives an approximate equation of the form

(10) $\quad A_1\Delta x_1 + A_2\Delta x_2 + \cdots + A_l\Delta x_l \approx 0$ for small increments,

where $A_i = A_i(x_1, x_2, \cdots, x_l)$, and then states that

(11) $\quad A_1 dx_1 + A_2 dx_2 + \cdots + A_l dx_l = 0$

—the procedure whose legitimacy is doubtless responsible for the sometime feeling that differentials are quantities, and in fact quantities which are pretty awfully small. Let us show that the procedure is indeed legitimate. First we note that (10) means that

(12) $\quad A_1\Delta x_1 + A_2\Delta x_2 + \cdots + A_l\Delta x_l = o(\sqrt{\Delta x_1^2 + \Delta x_2^2 + \cdots + \Delta x_l^2}).$

Now, holding all of an arbitrary set of independent variables $x_1'', x_2'', \cdots, x_s''$ fixed except one, say x_k'', dividing by $\Delta x_k''$, and letting $\Delta x_k''$ tend to zero, we have

(13) $\quad A_1\psi_{1\,x_k''} + A_2\psi_{2\,x_k''} + \cdots + A_l\psi_{l\,x_k''} = 0,$

where $x_i = \psi_i(x_1'', x_2'', \cdots, x_s'')$. Multiplying (13) by ξ_k'', summing the results with $k = 1, 2, \cdots, s$, and using (9), we get (11).

Conversely, (11) implies (10)—a fact of great value in the theory of approximation. This is an immediate consequence of the substitution into (11) of the approximate values $\xi_i \approx \Delta x_i$ which are obtained from (9) when we put $\xi_k'' = \Delta x_k''$ (which, note, we are privileged but not compelled by definition to do) and take account of the formula for approximate increments.

7. Differentials and integration. The integration problem relevant to our discussion is this: Given ξ_i, to find x_i. The solution is denoted by $x_i = \int \xi_i$. Let us illustrate.

First we examine the most familiar case, $y = \int f(x)\xi$. This means that $\eta = f(x)\xi$, and the solution is $y = F(x)$ where $F'(x) = f(x)$, by (9). The method of change of variable amounts to introducing a new relation $x = \phi(t)$, which gives rise to $\xi = \phi'(t)\tau$. In terms of t, our problem becomes that of finding y for which $\eta = f[\phi(t)]\phi'(t)\tau$, and the solution is $y = \Phi(t)$ where $\Phi'(t) = f[\phi(t)]\phi'(t)$.

Consider next the line integral $l = \int (y\xi + x\eta)$ along a path $x = \phi(t)$, $y = \psi(t)$. We are to find l for which $\lambda = y\xi + x\eta$. The normal procedure in line integration would be to find ξ and η from the equations of the path and express our problem in the form $\lambda = f(t)\tau$, but because of the special nature of the integrand (it is an "exact differential") we may expedite matters by introducing a new variable $z = xy$. For then $\zeta = y\xi + x\eta$ and our problem becomes merely to solve $\lambda = \zeta$, with the answer $l = z + c$.

The preceding illustrations involve only one independent variable, but perhaps they are adequate in as tentative a sketch as this one.

EDITORIAL [DIFFERENTIALS]

C. B. ALLENDOERFER

After reading the numerous papers submitted to Classroom Notes on differentials, and after discussions with other mathematicians, your editor is convinced that there is no commonly accepted definition of differential which fits all uses to which this notation is applied.

In the theory of functions of a single variable, $y = f(x)$, a reasonable case can be made for the customary definition: $dx = \Delta x$; $dy = f'(x)dx$. This breaks down, however, when one extends it to functions of several variables and considers double integrals of the form $\iint f(x, y)\, dxdy$. Students are rightly baffled when they attempt to convert such an integral to polar coordinates and are told that no longer is it permissible to put $dx = -r\sin\theta\, d\theta + \cos\theta\, dr$, etc. The Jacobian must be used instead, and at this point the logical structure which was built so carefully collapses entirely. If we wish to make calculus an intellectually honest subject and not a collection of convenient tricks, it is time we made a fresh start.

Consideration of the uses of differentials in various parts of calculus indicates that mathematicians really do use these symbols to mean quite different things in different circumstances. It seems to the writer that we should admit this from the outset and not try to give a single definition. But surely we must give our students some explanation of differentials or they will be more confused than ever. The following remarks are an attempt to do this. The best description of a differential that I have been able to formulate is:

DIFFERENTIALS: *The presence of a differential in a mathematical expression is a sign to the reader that this expression is obtained by a limiting process from a second expression in which the differentials dx, dy, etc. are replaced by finite increments Δx, Δy, etc. The nature of this limiting process depends upon the particular expression involved and must be inferred from the context.*

This description applies nicely to derivatives where

$$\frac{dy}{dx} = \lim_{\Delta x \to 0} \frac{\Delta y}{\Delta x}.$$

There seems to be no good reason to give dy and dx independent meanings, and no need to write expressions such as $dy = f'(x)dx$. An important use of this formula occurs in error problems, but these are more clearly presented by noting that for small values of Δx,

$$\frac{\Delta y}{\Delta x} \approx f'(x). \quad [\approx \text{means approximately equal}]$$

Hence
$$\Delta y \approx f'(x)\Delta x.$$

For integrals the description again applies to tell us that $\int f(x)dx$ is by definition the $\lim \sum_{i=1}^{n} f(x_i)\Delta x_i$. Similarly the notation $\int f(x)\, dg(x)$ for a Stieltjes integral tells us what limit has been taken, but does not ascribe an independent meaning to $dg(x)$. The meaning of differentials as used here has no logical connection with their meaning in derivatives. Such a connection is often obtained incorrectly in teaching by failure to state clearly the theorem regarding change of variable in an integral:

$$\int_{x_0}^{x_1} f(x)dx = \int_{t_0}^{t_1} f[x(t)]\frac{dx}{dt}dt; \quad x_0 = x(t_0);\ x_1 = x(t_1).$$

The proof of this theorem *does not* consist of the observation that: $dx = (dx/dt)dt$; even though this collection of nonsense may be an aid to the memory, a crutch for poor students, and a convenience for the experienced mathematician.

In multiple integrals differentials play a similar role and serve to indicate the type of limiting process which is under consideration. Further applications of this sort include line integrals such as $\int_c P(x, y)dx + Q(x, y)dy$ in which the differentials have a similar meaning but one which is still logically separate from those mentioned above.

Another appearance of these strange creatures is in differential equations such as:

(1) $$xdx + ydy + zdz = 0$$

whose "solution" is said to be: $x^2 + y^2 + z^2 = c$. Here their interpretation is even more obscure. Any (good or bad) physicist will tell you that (1) means that for small $\Delta x, \Delta y, \Delta z$;

(2) $$x\Delta x + y\Delta y + z\Delta z \approx 0,$$

where $\Delta x, \Delta y, \Delta z$ are arbitrary changes of these variables consistent with the problem at hand. But clearly (1) is not the straight limit of (2). The problem becomes clearer if we connect the point $P(x, y, z)$ to the point $Q(x+\Delta x, y+\Delta y, z+\Delta z)$ by a curve of parameter t, whose value at P is t_0 and at Q is $t_0+\Delta t$. Then by division of (2) by Δt we obtain:

(3) $$x\frac{\Delta x}{\Delta t} + y\frac{\Delta y}{\Delta t} + z\frac{\Delta z}{\Delta t} \approx 0.$$

Now we take limits as $\Delta t \to 0$ and obtain:

(4) $$x\frac{dx}{dt} + y\frac{dy}{dt} + z\frac{dz}{dt} = 0.$$

The problem required that (2) hold for all permissible displacements (Δx, Δy, Δz) and so for all curves of this type. The omission of dt in the passage from (4) to (1) is then the statement that (4) is true for all such curves. To show that $x^2+y^2+z^2=c$ is a solution of (1) we assume that x, y, and z are arbitrary differentiable functions of a parameter t which satisfy this relation. We then find that:

$$\frac{d}{dt}(x^2 + y^2 + z^2) = 0$$

is the equation (4); and hence conclude that we have a solution of (1).

In differential geometry expressions such as $ds^2 = E\ du^2 + 2F\ du\ dv + G\ dv^2$ are considered frequently as if ds, du, and dv were well defined quantities. However, a careful inspection of the use of these forms reveals that in every case such a form is really an integrand with the integral sign omitted for convenience. The meanings of the differentials are to be interpreted as if an integral were written explicitly. A similar understanding exists in probability theory when one speaks of the "probability $f(x)dx$" where $f(x)$ is a probability density.

There is a discredited view that differentials are some sort of "infinitely small quantities." Of course, this is nonsense, but there is a germ of truth in it. For differentials do mark the places in mathematical expressions which are held by small, but finite, quantities while some type of limiting process is carried out in which these small quantities tend to zero. To the writer they have no other consistent meaning.

BIBLIOGRAPHIC ENTRIES: DIFFERENTIALS

The references below are to the AMERICAN MATHEMATICAL MONTHLY.

1. Campbell, J. W., On exact differentials, vol. 28, p. 66.

    A necessary and sufficient condition for exactness in terms of line integrals.

2. Jackson, D., A comment on "differentials," vol. 49, p. 389.

    A comment on the paper by Kac and Randolph, vol. 49, p. 110, reproduced on pp. 167–169 in this volume.

3. Thurston, H. A., The definition of $dy:dx$, vol. 70, p. 539.

    A different definition of differential which has utility in handling certain classes of problems.

4. Thurston, H. A., Tangents and differentials, vol. 71, p. 660.

    A sequel to the previous paper.

(e)

PARTIAL DERIVATIVES

TRANSFORMATIONS OF THE LAPLACIAN*

R. P. AGNEW, Cornell University

It is the purpose of this note to make some simple remarks about the familiar tedious procedure involved in using a transformation of coordinates to convert the Laplacian operator from rectangular to spherical coordinates. Let the rectangular, cylindrical, and spherical coordinates of a point be, respectively, (x, y, z), (ρ, ϕ, z), and (r, ϕ, θ). The problem is to start with the Laplacian Δu defined by

$$\text{(1)} \qquad \Delta u = \frac{\partial^2 u}{\partial x^2} + \frac{\partial^2 u}{\partial y^2} + \frac{\partial^2 u}{\partial z^2}$$

and to show that

$$\text{(2)} \qquad \Delta u = \frac{\partial^2 u}{\partial r^2} + \frac{2}{r}\frac{\partial u}{\partial r} + \frac{1}{r^2}\frac{\partial^2 u}{\partial \theta^2} + \frac{\cos\theta}{r^2 \sin\theta}\frac{\partial u}{\partial \theta} + \frac{1}{r^2 \sin^2\theta}\frac{\partial^2 u}{\partial \phi^2}.$$

Direct use of the formulas

$$\text{(3)} \qquad x = r\sin\theta\cos\phi, \qquad y = r\sin\theta\sin\phi, \qquad z = r\cos\theta$$

to convert (1) into (2) is the familiar and tedious procedure noted above. It is very much easier to use the simpler formulas

$$\text{(4)} \qquad x = \rho\cos\phi, \qquad y = \rho\sin\phi, \qquad z = z$$

relating rectangular and cylindrical coordinates to convert (1) into the expression

$$\text{(5)} \qquad \Delta u = \frac{\partial^2 u}{\partial \rho^2} + \frac{1}{\rho}\frac{\partial u}{\partial \rho} + \frac{1}{\rho^2}\frac{\partial^2 u}{\partial \phi^2} + \frac{\partial^2 u}{\partial z^2}$$

in cylindrical coordinates.

When (5) has been obtained, the formulas

$$\text{(6)} \qquad z = r\cos\theta, \qquad \rho = r\sin\theta, \qquad \phi = \phi$$

relating cylindrical and spherical coordinates may be used to convert (5) into (2); but a simple observation eliminates nearly all of the work involved in this step. We note that, except for the names of the variables involved, the trans-

---

* From AMERICAN MATHEMATICAL MONTHLY, vol. 60 (1953), pp. 323–325.

formation (6) is identical with the transformation (4). Since (4) converts the sum of the first two terms of the right member of (1) into the sum of the first three terms in the right member of (5), it follows that (6) converts the sum of the first and last terms of the right member of (5) into

$$\frac{\partial^2 u}{\partial r^2} + \frac{1}{r}\frac{\partial u}{\partial r} + \frac{1}{r^2}\frac{\partial^2 u}{\partial \theta^2}. \tag{7}$$

After obtaining the relatively much simpler fact that (6) converts the sum of the two central terms of the right member of (5) into

$$\frac{1}{r}\frac{\partial u}{\partial r} + \frac{\cos\theta}{r^2 \sin\theta}\frac{\partial u}{\partial \theta} + \frac{1}{r^2 \sin^2\theta}\frac{\partial^2 u}{\partial \phi^2}, \tag{8}$$

we see without laborious calculation that (6) converts (5) into (2). Thus we are able to obtain both (5) and the more complicated (2) with a total amount of labor very little greater than that required to obtain (5) alone.

Since derivations of (2) appear in so many books and lectures in pure and applied mathematics, it would seem to be impossible to be certain that the above remarks are new. In any case, two facts indicate that the remarks have never been adequately publicized. In the first place, (2) is very often derived by tedious use of (3). In the second place, there have been times when (5) was derived by use of (4), but the derivation of (2) was omitted on the ground that the derivation was too tedious.

Finally, we may ask whether there are other occasions in which one can make significant use of the fact that, except for the names of the variables involved, the transformations (4) and (6) are identical.

A PROOF OF A THEOREM ON JACOBIANS*†

HIDEHIKO YAMABE, Institute of Technology, University of Minnesota

1. Let x, y denote points in a d-dimensional Euclidean space E^d, and let F be a C^1-mapping from E^d into another d-dimensional space.

We denote by $x^1, \cdots, x^d$, the coordinates of x and by $f^1(x), \cdots, f^d(x)$ those of $F(x)$. It is well known that if the Jacobian det. $J(z) = \det. (\partial f^i(x)/\partial x^j)_{x=z} \neq 0$, then $f(x)$ is a one-to-one and open mapping around z. *The purpose of this note is to give a simple proof of this theorem.*

2. **Proof of the theorem.** We may assume without loss of generality that $z = 0$. We shall use the vector and matrix notations, and shall consider matrices as linear operators on this vector space. Since $J(x)$ is continuous in x, there exists a small positive ϵ such that if $|x| = \text{Max}_i |x^i| < \epsilon$ then for any vector u

(1) $$|J(x)u - J(0)u| \leq (1 - \epsilon)|J(0)u|.$$

This is possible because $J(0)$ is nonsingular. Now consider a curve $F(x+ty)$ where t ranges over the unit interval $[0, 1]$. We assume that $|x+ty| < \epsilon$ for all t in $[0, 1]$.

Define

(2) $$(\partial/\partial t)F(x + ty) = \lim_{h \to 0} (1/h)(F(x + (t + h)y) - F(x + ty)).$$

Then by easy computations,

(3) $$(\partial/\partial t)F(x + ty) = J(x + ty)y.$$

Integrate the left hand side from 0 to 1, and we have

(4) $$F(x + y) - F(x) = \int_0^1 J(x + ty)y dt.$$

Now suppose that $F(x+y) = F(x)$. Then

$$0 = \left| \int_0^1 J(x + ty) y dt \right|$$

$$= \left| \int_0^1 J(0) y dt + \int_0^1 (J(x + ty) - J(0)) y dt \right|$$

$$\geq |J(0)y| - (1 - \epsilon)|J(0)y| = \epsilon |J(0)y|.$$

Hence $J(0)y = 0$, which implies $y = 0$ because $J(0)$ is nonsingular.

---

* From AMERICAN MATHEMATICAL MONTHLY, vol. 64 (1957), pp. 725–726.
† This work was done with the support of the National Science Foundation and the Sloan Foundation.

Next, we assume $F(0) = 0$ and prove that the image of $|x| < \epsilon/2$ contains the sphere $|z| < \delta = \epsilon/2 \inf_{|x|=\epsilon/2} |J(0)x|$. Let z_0 be given, $|z_0| < \delta$. Find x_0 such that

$$|F(x_0) - z_0| = \inf_{|x| \leq \epsilon/2} |F(x) - z_0|.$$

For any x, $|x| = \epsilon/2$, by (4) and (1)

$$|F(x)| = \left|J(0)x + \int_0^1 (J(tx) - J(0))x\,dt\right| \geq \epsilon |J(0)x| \geq 2\delta,$$

$$|F(x) - z_0| \geq \delta > |F(0) - z_0| \geq |F(x_0) - z_0|.$$

Hence $|x_0| < \epsilon/2$, and y can be chosen so that $|x_0+ty| \leq \epsilon/2$ for $0 \leq t \leq 1$, and $J(0)y = -\eta(F(x_0) - z_0)$ for some small positive η. Then, by (4) and (1) $|F(x_0+y) - F(x_0) - J(0)y| \leq (1-\epsilon)|J(0)y|$, $|F(x_0+y) - z_0| \leq |F(x_0) - z_0 + J(0)y| + (1-\epsilon)|J(0)y| \leq (1-\eta\epsilon)|F(x_0) - z_0|$. From the definition of x_0 it follows that $F(x_0) - z_0 = 0$. The theorem is hence proved.

Remark. We can extend this result to a certain more general infinite dimensional space if we can give a proper formulation to the continuity and the non-singularity of $J(x)$.

BIBLIOGRAPHIC ENTRIES: PARTIAL DERIVATIVES

The references below are to the AMERICAN MATHEMATICAL MONTHLY.

1. Karapetoff, V., Formal unification of gradient, divergence and curl, by means of an infinitesimal operational volume, vol. 35, p. 14.

2. Brand, Louis, Inversion of Jacobian matrices, vol. 68, p. 281.

   Special emphasis on the case of an orthogonal transformation of coordinates.

3. Weston, J. D., Some remarks about the curl of a vector field, vol. 68, p. 359.

6

MEAN VALUE THEOREM FOR DERIVATIVES, INDETERMINATE FORMS

(a)

MEAN VALUE THEOREM

PROOF OF THE FIRST MEAN VALUE THEOREM OF THE INTEGRAL CALCULUS*

T. PUTNEY, State College of Washington

Professor A. A. Bennett in Volume 31 (1924), page 40, of this MONTHLY called attention to the didactic interest which attaches to a particular form of proof of the first mean value theorem of the differential calculus. A proof of the same character is to be found in the well-known book by Theodore Chaundy *The Differential Calculus* (Oxford University Press, 1935), p. 84. A precisely analogous argument can be employed to obtain a proof of the first mean value theorem of the integral calculus, and may be of interest in classroom instruction.

Define two functions $G(x)$ and $H(x)$ in $a \leq x \leq b$ by

$$G(x) \equiv \int_a^x f(t)g(t)dt, \qquad H(x) \equiv \int_a^x g(t)dt$$

where $f(x)$ and $g(x)$ are continuous in $a \leq x \leq b$ and $g(x)$ does not vanish in $a < x < b$. In the determinant

$$D(x) \equiv \begin{vmatrix} G(x) & H(x) & 1 \\ G(a) & H(a) & 1 \\ G(b) & H(b) & 1 \end{vmatrix} = \begin{vmatrix} G(x) & H(x) & 1 \\ 0 & 0 & 1 \\ G(b) & H(b) & 1 \end{vmatrix},$$

---

* From AMERICAN MATHEMATICAL MONTHLY, vol. 60 (1953), pp. 113–114.

since $G(a)=H(a)=0$. Furthermore, $D(a)=D(b)=0$. Hence, by Rolle's theorem, $D'(\xi)=0$ for some ξ, $a<\xi<b$ or

$$g(\xi)\begin{vmatrix} f(\xi) & 1 & 0 \\ 0 & 0 & 1 \\ G(b) & H(b) & 1 \end{vmatrix} = 0.$$

Finally, since $g(\xi)\neq 0$, this yields the desired result, namely $G(b)=f(\xi)H(b)$.

AN APPLICATION OF THE MEAN VALUE THEOREM*

DAVID ZEITLIN, Remington Rand UNIVAC

Given $y = u^v$, where u and v are differentiable functions of x, the derivative $D_x y$ is found by logarithmic differentiation. Thus,

(1) $$D_x(u^v) = v u^{v-1} \cdot D_x u + u^v \log u \cdot D_x v$$

is derived in this manner. In most textbooks, no attempt is made to derive (1) using the definition of the derivative, i.e., $\lim_{\Delta x \to 0} \Delta y/\Delta x$. For in doing so, one runs into difficulties involving a binomial expansion.

The mean value theorem is usually introduced in the study of indeterminate forms. At this point, the derivative of u^v may be easily found. Thus

(2) $$\Delta y = \{(u + \Delta u)^{v+\Delta v} - u^{v+\Delta v}\} + \{u^{v+\Delta v} - u^v\}.$$

If we apply the mean value theorem to the first term in (2), treating $v + \Delta v$ as fixed, we find that

$$(u+\Delta u)^{v+\Delta v} - u^{v+\Delta v} = (\Delta u)(v+\Delta v) \cdot r^{v+\Delta v - 1},$$

where $u + \Delta u < r < u$ or $u < r < u + \Delta u$. Thus

(3) $$\frac{\Delta y}{\Delta x} = (v + \Delta v) \cdot r^{v+\Delta v - 1} \cdot \left(\frac{\Delta u}{\Delta x}\right) + \left(\frac{u^{v+\Delta v} - u^v}{\Delta v}\right) \cdot \left(\frac{\Delta v}{\Delta x}\right)$$

As $\Delta x \to 0$, $\Delta u \to 0$, $\Delta v \to 0$, $r \to u$, and (1) is now established.

---

* From AMERICAN MATHEMATICAL MONTHLY, vol. 64 (1957), p. 427.

THE LAW OF THE MEAN*

R. C. YATES, The College of William and Mary

In discussing the Law of the Mean, we consider the function

$$\phi(x) = f(x) - f(a) - \frac{f(b) - f(a)}{b - a}(x - a).$$

This is a formidable expression whose origin is puzzling until it is pointed out as the difference PQ of the ordinate of a point on the graph of $f(x)$ and the ordinate to the secant line for the same x.

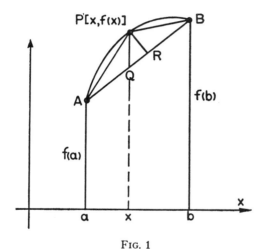

Fig. 1

A different approach results from selecting PR, the perpendicular from P to the secant (Fig. 1). Now $\sin \angle PQR = PR/PQ$ and, since $\angle PQR$ is a constant angle for all values x in the interval, then PR is proportional to PQ throughout and thus has all the desired properties of $\phi(x)$. But this variable quantity PR is the altitude of the triangle APB whose base is the chord AB. Thus $\phi(x)$ is also proportional to the area K of triangle APB. Accordingly, instead of $\phi(x)$, we might as well consider:

$$2K = \begin{vmatrix} x & f(x) & 1 \\ a & f(a) & 1 \\ b & f(b) & 1 \end{vmatrix}.$$

---

* From AMERICAN MATHEMATICAL MONTHLY, vol. 66 (1959), pp. 579–580.

The derivative of this variable-area function must vanish by Rolle's theorem at some $x = X$ between a and b. That is,

$$\begin{vmatrix} 1 & f'(X) & 0 \\ a & f(a) & 1 \\ b & f(b) & 1 \end{vmatrix} = 0,$$

or

$$f(a) - f(b) - f'(X)(a - b) = 0.$$

Thus

$$(b - a)f'(X) = f(b) - f(a).$$

THE EXTENDED LAW OF THE MEAN BY A TRANSLATION-ROTATION OF AXES*

JACQUELINE P. EVANS, Wellesley College

In the Classroom Note *Proof of the mean value theorem* (this MONTHLY, vol. 65, 1958, pp. 362–364), C. L. Wang uses a translation-rotation of the Cartesian coordinate system to obtain the Mean Value Theorem from Rolle's Theorem. The chief merit of this proof is that it provides a natural geometric motivation for the Mean Value Theorem. However, as the author points out, there are cases where his method of proof breaks down. This occurs whenever the graph of the given curve doubles back upon itself and is not the graph of a *function* in terms of the XY-coordinate system. (See Fig. 1 for example.)

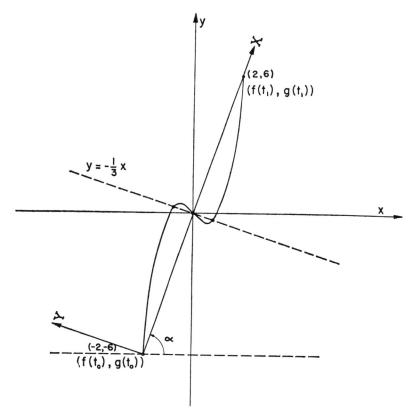

FIG. 1. $y = x^3 - x$, $-2 \leq x \leq 2$; $x = f(t)$, $y = g(t)$, $t_0 \leq t \leq t_1$.

---

* From AMERICAN MATHEMATICAL MONTHLY, vol. 67 (1960), pp. 580–581.

The method of proof used by Wang can be adapted to cover all cases. When this is done we obtain the Extended Law of the Mean. The proof retains the virtue of yielding a natural motivation and interpretation of this theorem and avoids the use of artificially concocted functions to which Rolle's Theorem is usually applied.

Let $x=f(t)$, $y=g(t)$, $t_0 \leq t \leq t_1$, where f and g are continuous for $t_0 \leq t \leq t_1$, and $f'(t)$, $g'(t)$ exist for $t_0 < t < t_1$. If $g(t_0) = g(t_1) = 0$, there exists, by Rolle's Theorem applied to g, at least one T, $t_0 < T < t_1$, such that $g'(T) = 0$. Let the given equations be regarded as the parametric representation of a given curve. If $f'(T) \neq 0$, this curve must have a horizontal tangent. If $f'(T) = 0$, we can draw no conclusion without further investigation.

Now assume that $g(t_0) \neq g(t_1)$ but retain the other conditions on f and g. The geometric situation described above can be shifted to a new coordinate system by a translation and rotation of the xy-system into the XY-system. Using the results of Wang, the parametric equations of the given curve, in terms of the new coordinate system, are:

$$F(t) = X = [g(t) - g(t_0)] \sin \alpha + [f(t) - f(t_0)] \cos \alpha,$$
$$G(t) = Y = [g(t) - g(t_0)] \cos \alpha - [f(t) - f(t_0)] \sin \alpha, \qquad t_0 \leq t \leq t_1.$$

By Rolle's Theorem applied to G there exists at least one T, $t_0 < T < t_1$, such that $G'(T) = g'(T) \cos \alpha - f'(T) \sin \alpha = 0$, from which it follows that $g'(T)[f(t_1) - f(t_0)] = f'(T)[g(t_1) - g(t_0)]$. This result when interpreted geometrically in terms of the xy-coordinate system simply says that either there is one point on the given curve such that the tangent there is parallel to the secant through the two endpoints of the curve or else there is an indeterminate point. If we make the added assumption that $f'(t) \neq 0$, $t_0 < t < t_1$, the first case occurs and we can write:

$$\frac{g'(T)}{f'(T)} = \frac{g(t_1) - g(t_0)}{f(t_1) - f(t_0)},$$

thus obtaining the Extended Law of the Mean.

EXTENDED LAWS OF THE MEAN*

LOUIS C. BARRETT AND RICHARD A. JACOBSON, South Dakota School of Mines and Technology

The usual proofs of the first and extended mean value theorems involve the process of applying Rolle's theorem to functions happily designed to yield the desired conclusions. Frequently, no mention is made of how these functions are discovered. Indeed, except in isolated instances [1], it is customary to introduce the determinants thus employed without comment. In this note we illustrate how these determinants may be motivated quite naturally through geometrical reasoning applied to areas of parallelograms and to volumes of parallelopipeds.

We begin with the extended mean value theorem:

THEOREM 1. *If $x(t)$ and $y(t)$ are continuous on $a \leq t \leq b$, and differentiable on $a < t < b$, then there exists at least one value T such that $a < T < b$ and*

(1) $$x'(T)[y(b) - y(a)] = y'(T)[x(b) - x(a)].$$

A determinant, which upon the application of Rolle's theorem leads to equation (1), comes to our attention in connection with the observation that if $x(t)$ and $y(t)$ satisfy the hypotheses of Theorem 1, then the area $A(t)$ of parallelogram $PQRS$ (Fig. 1), is a function that satisfies Rolle's theorem. For, when the variable vertex $P: [x(t), y(t)]$ coincides with either of the fixed vertices $Q: [x(a), y(a)]$ or $S: [x(b), y(b)]$, $A(t)$ is zero. Moreover, $A(t)$ is continuous on $a \leq t \leq b$ and differentiable on $a < t < b$. But, from analytic geometry,

(2) $$\pm A(t) = \begin{vmatrix} x(a) & y(a) & 1 \\ x(b) & y(b) & 1 \\ x(t) & y(t) & 1 \end{vmatrix}$$

Upon applying Rolle's theorem to this determinant, (1) follows at once. When $x(t)$ is replaced by t, Theorem 1 reduces to the familiar law of the mean.

A more general theorem of the mean, including Theorem 1 as a special case, is

THEOREM 2. *If $x(t)$, $y(t)$, and $z(t)$ are continuous on $a \leq t \leq b$ and differentiable on $a < t < b$, then there exists at least one value T such that $a < T < b$ and*

(3) $$x(a)[y(b)z'(T) - z(b)y'(T)] + z(a)[x(b)y'(T) - y(b)x'(T)]$$
$$= y(a)[x(b)z'(T) - z(b)x'(T)].$$

The functions $x(t)$, $y(t)$, and $z(t)$ in this theorem may be thought of as parametric representations of a curve C in space having the points $M: [x(a), y(a), z(a)]$ and $N: [x(b), y(b), z(b)]$ as fixed points and $P: [x(t), y(t), z(t)]$ as a variable

---

* From AMERICAN MATHEMATICAL MONTHLY, vol. 67 (1960), pp. 1005–1007.

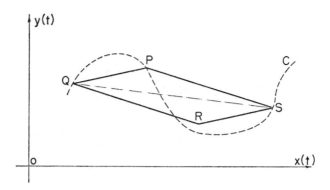

FIG. 1

point (Fig. 2). These three points determine, in turn, the constant vectors **A** and **B** and the variable vector **U**, all emanating from a common point 0. Let $V(t)$ denote the volume of the parallelopiped having these three vectors as concurrent edges. When **U** coincides with either **A** or **B**, i.e. when $t=a$ or $t=b$, $V(t)$ vanishes. Furthermore, if $x(t)$, $y(t)$, and $z(t)$ fulfill the hypotheses of Theorem 2, then $V(t)$ is continuous on $a \leq t \leq b$ and differentiable on $a < t < b$ and, therefore, satisfies Rolle's theorem.

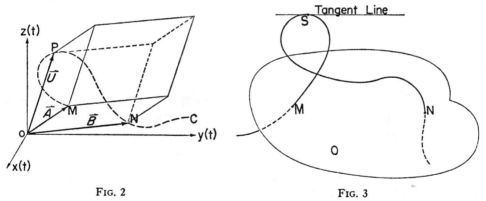

FIG. 2   FIG. 3

But, $V(t)$ is given by the scalar triple product (also called the triple scalar product)

(4) $$\pm V(t) = \mathbf{A} \times \mathbf{B} \cdot \mathbf{U} = \begin{vmatrix} x(a) & y(a) & z(a) \\ x(b) & y(b) & z(b) \\ x(t) & y(t) & z(t) \end{vmatrix}.$$

This determinant may now be utilized to establish Theorem 2 in the usual way.

A well-known geometrical interpretation of Theorem 1 is that there exists at least one point on the curve C (Fig. 1) between Q and S at which the tangent to the curve is parallel to the secant line QS.

In order to interpret Theorem 2 geometrically, we differentiate (4) and apply Rolle's theorem to obtain

$$\mathbf{A} \times \mathbf{B} \cdot \mathbf{U}'(T) = 0, \qquad a < T < b \tag{5}$$

which, as is readily verified by expanding, is simply (3) written in vector form. The vector $\mathbf{U}'(T)$ lies along the tangent to C at a point $S: [x(T), y(T), z(T)]$ distinct from M and N (Fig. 3), and is perpendicular to the vector $\mathbf{A} \times \mathbf{B}$. But, $\mathbf{A} \times \mathbf{B}$ is perpendicular, in turn, to both $\mathbf{A}$ and $\mathbf{B}$. Consequently, $\mathbf{U}'(T)$ must, in general, be parallel to the plane of $\mathbf{A}$ and $\mathbf{B}$. In fact, by applying a similar analysis to the vectors $\mathbf{A} - \mathbf{K}$, $\mathbf{B} - \mathbf{K}$, and $\mathbf{U} - \mathbf{K}$, where $\mathbf{K}$ is an arbitrary constant vector, it can be shown that corresponding to any plane through the points M and N there is at least one point on C, distinct from M and N, where the tangent line to C is parallel to the given plane.

These observations reveal that even more general laws of the mean are generated when Rolle's theorem is applied to determinants of higher order than three, and that these determinants may be readily motivated by an analysis of the content of parallelotopes in n-dimensions.

Reference

1. R. C. Yates, The law of the mean, this MONTHLY, vol. 66, 1959, pp. 583–584. [*Ed. note:* Reprinted on pp. 215–216.

A NATURAL AUXILIARY FUNCTION FOR THE MEAN VALUE THEOREM*

M. J. POLIFERNO, Trinity College, Hartford, Connecticut

The usual classroom proof of the mean value theorem is based on the auxiliary function $g: g(x) = f(x) - f(a) - m(x-a)$, where $m = \{f(b) - f(a)\}/(b-a)$, to which Rolle's theorem is applied. Consideration of g is motivated by the observation that x_0 is likely to occur as a value of x at which the vertical directed distance over x from the chord containing $(a, f(a))$ and $(b, f(b))$ to the graph of f is an extreme (Fig. 1).

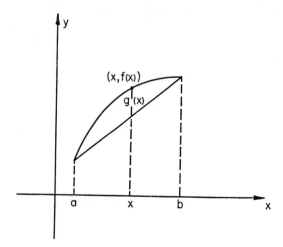

FIG. 1

There are a number of pedagogical disadvantages to this approach. In the first place, it seems more natural to rotate the xy-coordinate system than to look at the space between the graph and the chord. Consider the result of rotating the xy-system so that the x'-axis (the new x-axis) is parallel to the chord. Then Rolle's theorem applies to the function $h: h(x) =$ the directed distance from the x'-axis to the point $(x, f(x))$ (Fig. 2).

It is easy to express h in terms of f: the directed distance from a line $y = mx + b$ to a point (x_1, y_1) is $(y_1 - mx_1 - b)/\sqrt{(m^2+1)}$, and an equation of the x'-axis is $y = mx$, so that $h(x) = f(x) - mx/\sqrt{(m^2+1)}$.

---

* From AMERICAN MATHEMATICAL MONTHLY, vol. 69 (1962), pp. 45–47.

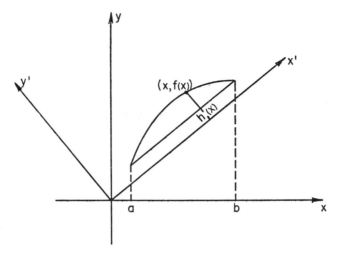

Fig. 2

The student will find it obvious geometrically (and easy enough algebraically) that $h(a) = h(b)$. Note also that he will encounter no difficulty in differentiating h, since the denominator is a constant (and he will not have to contend with $f(a)$, which occurs in g and gives him pause). The main pedagogical advantage in h, I think, is that it gives a very easy rigorous proof along lines that parallel perfectly the natural intuitive approach.

Although $\sqrt{(m^2+1)}$ is not really needed, retention of this denominator presents no complication whatever, and gives the natural geometric interpretation to h. Some students will realize this, and might be encouraged to investigate alternative auxiliary functions; others are apt to be confused if the dispensability of $\sqrt{(m^2+1)}$ is discussed before the proof is completed.

A suitable modification H of h can be used to similar advantage as an auxiliary function for Cauchy's theorem: *If f and g are continuous on $[a, b]$ and differentiable on (a, b), and $g'(t) \neq 0$ on (a, b) then there is a number t_0 in (a, b) such that*

$$\frac{f(b) - f(a)}{g(b) - g(a)} = \frac{f'(t_0)}{g'(t_0)}.$$

Here $H(t) = \{f(t) - mg(t)\}/\sqrt{(m^2+1)}$ and, with the parametric equations $y = f(t)$, $x = g(t)$, the geometric interpretation is the same.

ON AVOIDING THE MEAN VALUE THEOREM*

LIPMAN BERS, Columbia University

The arrangement (hardly a new one†), which I had in mind when talking to Leon Cohen [1], is as follows.

THEOREM. *If $f'(x) > 0$ for $a < x < b$, then f is increasing.*

This is intuitively obvious and easy to prove. Indeed, assume there is a p, $a < p < b$, such that the set S of all x, $a < x < p$, with $f(x) \geq f(p)$ is not empty. Set $q = \sup S$; since $f'(p) > 0$, we have $a < q < p$. If $f(q) \geq f(p)$, then since $f'(q) > 0$, there are points of S to the right of q. If $f(q) < f(p)$, then q is not in S and, by continuity, there are no points of S near and to the left of q. Contradiction.

COROLLARY 1. *If $f'(x) \geq 0$ for $a < x < b$, then f is nondecreasing.* (Apply the theorem to $f(x) + \epsilon x$, $\epsilon > 0$.)

This implies that a function with an identically vanishing derivative is constant.

COROLLARY 2. *If $f'(x) \leq K$ for $a \leq x \leq b$, then $f(b) \leq f(a) + K(b-a)$.* (Apply Corollary 1 to $K(x-a) - f(x)$.)

Using the intermediate value theorem and either Corollary 2 or the fundamental theorem of calculus, one obtains at once the mean value theorem for continuously differentiable functions. That's all one needs in calculus.

The "full" mean value theorem, for differentiable but not continuously differentiable functions, is a curiosity. It may be discussed together with another curiosity, Darboux' theorem that every derivative obeys the intermediate value theorem.

Reference

1. Leon Cohen, On being mean to the mean value theorem, the preceding article in this MONTHLY [vol. 74, pp. 581–582].

---

* From AMERICAN MATHEMATICAL MONTHLY, vol. 74 (1967), p. 583.
† I would be grateful to any reader who may know of a reference.

MEAN VALUE THEOREMS AND TAYLOR SERIES*

M. R. SPIEGEL, Rensselaer Polytechnic Institute

In all of the textbooks on elementary and advanced calculus with which the author is acquainted, the various mean-value theorems and Taylor series with a remainder are arrived at by setting up a function judiciously and then applying Rolle's Theorem. In many cases the student justifiably may get the feeling that this "suitable" function is pulled out of the proverbial hat. The purpose of this article is to present a basis for arriving at these "suitable" functions. The method employed does not seem to be well known, but appears to be valuable from a pedagogical as well as theoretical viewpoint.

We begin by stating the theorem due to Rolle which will be referred to throughout the paper.

Rolle's Theorem: Let $F(x)$ be defined, single-valued and continuous in the interval $a \leq x \leq b$ and be such that $F(a) = F(b) = 0$. Furthermore, let $F(x)$ have a derivative (finite or infinite) in the open interval $a < x < b$. Then there is at least one number ξ such that $a < \xi < b$ and for which $F'(\xi) = 0$.

This theorem is adequately proved in many texts although they seem to avoid unnecessarily the case where $F(x)$ has an infinite derivative.

Let us now consider a function $f(x)$ defined, single-valued and continuous in $a \leq x \leq b$ and having a derivative in $a < x < b$. We seek to approximate $f(x)$ in the interval $a \leq x \leq b$ by a linear function $A + Bx$ where A and B are constants which we shall determine. Let us consider the difference between $f(x)$ and the linear function $A + Bx$ and write

(1) $$F(x) \equiv f(x) - (A + Bx), \qquad a \leq x \leq b.$$

We now choose the constants A and B so that $F(a) = F(b) = 0$. That is,

(2) $$F(a) \equiv f(a) - (A + Ba) = 0,$$
(3) $$F(b) \equiv f(b) - (A + Bb) = 0.$$

Since $F(x)$ now satisfies the conditions of Rolle's Theorem, it follows that

(4) $$F'(\xi) \equiv f'(\xi) - B = 0, \qquad a < \xi < b,$$

so that

(5) $$f'(\xi) = B, \qquad a < \xi < b.$$

By solving equations (2) and (3) simultaneously for B, equation (5) becomes

(6) $$f'(\xi) = \frac{f(b) - f(a)}{b - a}, \qquad a < \xi < b,$$

i.e.,

(7) $$f(b) = f(a) + (b - a)f'(\xi).$$

---

* From MATHEMATICS MAGAZINE, vol. 29 (1956), pp. 263–266.

It is now natural to ask whether we can generalize the procedure adopted above by considering the difference between $f(x)$ and a quadratic approximation to $f(x)$. For this purpose we define

(8) $$F(x) \equiv f(x) - (A_1 + B_1 x + C_1 x^2),$$

where A_1, B_1, and C_1 are constants which we wish to determine. Again we ask that the conditions $F(a) = F(b) = 0$ be imposed so that once more we may make use of Rolle's Theorem. There is certainly no loss in generality if we write (8) in the form

(9) $$F(x) \equiv f(x) - [A + B(x - a) + C(x - a)^2],$$

which will be a little more convenient in simplifying the algebra involved. Upon imposing the condition that $F(a) = 0$ we find from (9) that $A = f(a)$, so that (9) becomes

(10) $$F(x) \equiv f(x) - f(a) - B(x - a) - C(x - a)^2.$$

If $F(b)$ is to vanish we see from (10) that we must require the condition

(11) $$f(b) = f(a) + B(b - a) + C(b - a)^2.$$

By applying Rolle's theorem we have

(12) $$F'(\xi) \equiv f'(\xi) - B - 2C(\xi - a) = 0 \qquad a < \xi < b.$$

However, this is not enough to determine the constants B and C. We do this by imposing a further condition. Since we have $F'(\xi) = 0$ it is natural to ask that $F'(a) = 0$, for if both $F'(a) = 0$ and $F'(\xi) = 0$ we will be able to apply Rolle's Theorem to the function $F'(x)$ and say that there exists a number m between a and ξ such that $F''(m) = 0$. Of course, to do this we must be sure that $F'(x)$ is continuous in $a \leq x \leq b$ and has a derivative in $a < x < b$. This is accomplished by requiring that $f'(x)$ have similar properties. By asking that $F'(a) = 0$ we have from (10)

(13) $$F'(a) \equiv f(a) - B = 0 \quad \text{or} \quad B = f'(a).$$

Thus (11) becomes

(14) $$f(b) = f(a) + f'(a)(b - a) + C(b - a)^2.$$

Since $F'(a) = F'(\xi) = 0$ we apply Rolle's theorem to obtain

(15) $$F''(m) = f''(m) - 2C = 0, \qquad a < m < \xi,$$

or

(16) $$C = \tfrac{1}{2} f''(m), \qquad a < m < \xi.$$

Thus (14) becomes

(17) $$f(b) = f(a) + f'(a)(b - a) + \tfrac{1}{2} f''(m)(b - a), \qquad a < m < \xi < b.$$

This is merely a statement of Taylor's theorem with the remainder given in the form due to Lagrange. It is interesting to note that the number m in (17) is closer to a than the number ξ in (7).

Upon extending the above procedure to a consideration of the function

(18) $\qquad F(x) \equiv f(x) - [A + B(x - a) + C(x - a)^2 + D(x - a)^3],$

where $f(x), f'(x), f''(x)$ are continuous in $a \leq x \leq b$ and $f'''(x)$ has a finite or infinite derivative in $a < x < b$, the reader may easily obtain, by successive applications of Rolle's theorem in a manner exactly analogous to that given above

(19) $\qquad f(b) = f(a) + f'(a)(b - a) + \dfrac{f''(a)}{2}(b - a)^2 + \dfrac{f'''(\lambda)}{6}(b - a)^3,$

where $a < \lambda < m < \xi < b$.

Further extensions to the Taylor series with the remainder after any number of terms may easily be supplied by the reader.

One further remark may not be out of place here. Let us consider the function

(20) $\qquad F(x) \equiv f(x) - [A + B\theta(x)], \qquad a \leq x \leq b,$

where the given function $\theta(x)$ is used in place of x. By imposing the conditions $F(a) = F(b)$ we find

(21) $\qquad A + B\theta(a) = 0,$

(22) $\qquad A + B\theta(b) = 0.$

Hence by Rolle's Theorem [with the suitable restrictions on $f(x)$ and $\theta(x)$] we have

(23) $\qquad F'(\xi) = f'(\xi) - B\theta'(\xi) = 0.$

Upon requiring that $\theta'(x) \neq 0$ in $a \leq x \leq b$ we have

(24) $\qquad \dfrac{f'(\xi)}{\theta'(\xi)} = B.$

Hence from (21) and (22) if $\theta(a) \neq \theta(b)$ we have

(25) $\qquad \dfrac{f(b) - f(a)}{\theta(b) - \theta(a)} = \dfrac{f'(\xi)}{\theta'(\xi)}, \qquad a < \xi < b.$

Further generalizations may occur to the reader.

BIBLIOGRAPHIC ENTRIES: MEAN VALUE THEOREM

Except for the entries labeled MATHEMATICS MAGAZINE, the references below refer to the AMERICAN MATHEMATICAL MONTHLY.

1. Downing, H. H., Note on the law of the mean, vol. 27, p. 467.

    Studies the set of θ and h satisfying $f(x+h)-f(x)=hf'(x+\theta h)$ for fixed x when f is a polynomial of degree ≤ 3.

2. Bennett, A. A., The consequences of Rolle's theorem, vol. 31, p. 40.
3. Krall, H. L., On the mean value theorem, vol. 42, p. 604.
4. Shain, J., The method of cascades, vol. 44, p. 24.

    Rolle's method of approximating the roots of a polynomial equation by locating zeros of the derivative.

5. Bush, K. A., On an application of the mean value theorem, vol. 62, p. 577.
6. Goodner, D. B., Extensions of the law of the mean, vol. 64, p. 185.
7. Wang, Chung Lie, Proof of the mean value theorem, vol. 65, p. 362.

    The proof comes from Rolle's theorem by translation and rotation of axes.

8. Barrett, L. C., Methods of proving mean value theorems, vol. 69, p. 50.
9. Cohen, L. W., On being mean to the mean value theorem, vol. 74, p. 581.
10. Levi, H., Integration, anti-differentiation and a converse to the mean value theorem, vol. 74, p. 585.
11. Osborn, Roger, Some geometric considerations related to the mean value theorem, MATHEMATICS MAGAZINE, vol. 33, p. 271.

    Discusses and interprets various forms of the "manufactured" functions used to derive the mean value theorem from Rolle's theorem.

12. Lan, Chih-Chin, An extension of the mean-value theorem in E_n, MATHEMATICS MAGAZINE, vol. 39, p. 91.

(b)
INDETERMINATE FORMS

A GEOMETRIC EXAMPLE OF AN INDETERMINATE FORM*

ARTHUR C. LUNN, University of Chicago

If the prolate ellipsoid of revolution generated by turning the ellipse $x^2/a^2 + y^2/b^2 = 1$ around the x-axis be cut by the plane $x = c$ through one focus, the area of the smaller part, between this focal plane and the nearer vertex, is given by the integral

$$S = \frac{2\pi b}{a^2} \int_c^a \sqrt{a^4 - (a^2 - b^2)x^2} \, dx.$$

When the constants are expressed in terms of the eccentricity e and the distance q between the focus and vertex, by means of the relations

$$c = ae, \quad b = a\sqrt{1 - e^2}, \quad q = a(1 - e),$$

the integral reduces to

(1) $$S = \pi q^2 \sqrt{\frac{1+e}{(1-e)^3}} \left\{ \sqrt{1-e^2} - e\sqrt{1-e^4} + \frac{\sin^{-1} e - \sin^{-1} e^2}{e} \right\}.$$

If now the distance q be kept constant, the limiting surface (for $e = 1$) is the paraboloid of revolution generated by $y^2 = 4qx$ (with a convenient change of axes); and the corresponding area is

$$S_1 = \frac{\pi}{q} \int_0^{2q} y\sqrt{y^2 + 4q^2} \, dy$$

which reduces to

(2) $$S_1 = \frac{8\pi q^2}{3} (\sqrt{8} - 1).$$

It is thus apparent that as e approaches unity the limit S should be S_1. But the expression for S is then formally indeterminate, and verification is not directly practicable by the customary method of successive differentiation with respect to e, the reason being the presence of the factor $(1-e)$ raised to a fractional power, which cannot be reduced to a constant by repeated differentiation. It may be noticed also that the point $e = 1$ is a branch-point for $\sin^{-1} e$ and for $\sin^{-1} e^2$, where two values of a many-valued function come together.

---
* From AMERICAN MATHEMATICAL MONTHLY, vol. 19 (1912), pp. 116–117.

This suggests the change of variable $1-e=t^2$; then S as a function of t may be treated by successive differentiation; or, what is at bottom the same thing, let the various terms in the brace be expanded in integral powers of t, with the values at $t=0$ suitably chosen thus:

$$\sqrt{1-e^2} = \sqrt{2}\, t - \tfrac{1}{4}\sqrt{2}\, t^3 \cdots$$
$$e\sqrt{1-e^4} = 2t - \tfrac{7}{2}t^3 \cdots$$
$$\sin^{-1} e = \frac{\pi}{2} - \sqrt{2}\, t - \frac{\sqrt{2}}{12} t^3 \cdots$$
$$\sin^{-1} e^2 = \frac{\pi}{2} - 2t + \tfrac{1}{6}t^3 \cdots$$

then in the expression for S, the factors t^3, to which the indeterminateness is due, may be cancelled, and the leading coefficients are then such as to show that the limit of S is S_1 as t approaches zero.

A contrasting case occurs with respect to the entire area of the ellipsoid, which is

$$2\pi a^2 \sqrt{1-e^2} \left\{ \frac{\sin^{-1} e}{e} + \sqrt{1-e^2} \right\},$$

the limit of which, as e approaches zero, is the area of a sphere of radius a. Here the indeterminateness occurring in $\sin^{-1} e/e$ is of integral order.

Many similar examples could be cited, for instance, that of the center of gravity or center of pressure of part of an ellipse and the corresponding part of the limiting parabola, where the limit of an indeterminate form, discovered by geometric or physical considerations, is to be analytically verified.

A NOTE ON INDETERMINATE FORMS*

ROGER OSBORN, University of Texas

Recently, in the discussion of indeterminate forms in an elementary calculus class, one of the students posed the question, "Why isn't ∞^0 equal to one?" He based this question not on a consideration of the inapplicability of the algebraic rule $a^0 = 1$ to the concept ∞, but on the fact that he had yet to see a limit problem of the form ∞^0 for which the answer was different from 1.

The first two examples that I chose for illustration had the same fault as those in the textbook being used by the class—the answer in each case was 1. We found that it was not difficult to construct an example of a limit problem of the form ∞^0 which had an arbitrarily chosen answer, but that very few were given in textbooks. Even though the subject did not warrant the time spent on it, we found and worked examples of this type given in fifteen standard texts on elementary calculus which were readily available in my office. These ranged from old, well accepted treatises on the subject to some of the newer books which combine calculus with analytic geometry. The books examined had from five to no problems of the type ∞^0 apiece, and in only one of the books† was there a problem given of the form ∞^0 for which the answer was different from 1. It was the conclusion of the class (and of the instructor) that if it is worthwhile to consider such problems, the cases in which they differ from ordinary algebra should be included by the instructor since they are generally not included by the textbook.

---

* From AMERICAN MATHEMATICAL MONTHLY, vol. 59 (1952), pp. 549–550.
† Peterson, Thurman S., Elements of Calculus, New York, Harper and Brothers, 1950, p. 155, problem 11 is: $\lim_{x\to\infty} (e^x + x)^{1/x} = e$.

A NOTE ON INDETERMINATE FORMS*

L. J. PAIGE, University of California at Los Angeles

It is not surprising that many students suspect the indeterminate form 0^0 to be equal to 1, believing that the elementary rules of algebra will apply. The example $x^{\alpha/\log x}$ immediately dispels this myth.

In attempting to construct examples of the indeterminate form 0^0, we immediately try $\lim_{x \to 0} x^{f(x)}$, where $f(0) = 0$ and the derivative $f'(x)$ is assumed to exist in a neighborhood of the origin. But here one is led to an interesting limit and the surprising result that if $\lim_{x \to 0} x^{f(x)}$ exists it must be 1. Perhaps this word of caution will prevent lost time in looking for examples to remedy the students' suspicions.

We attempt to evaluate $\lim_{x \to 0} x^{f(x)}$ by following the text and take the logarithm, obtaining $\lim_{x \to 0} \dfrac{\log x}{1/f(x)}$. Here an application of l'Hospital's rule yields for consideration

$$\lim_{x \to 0} \frac{f(x)}{x} \cdot \frac{f(x)}{f'(x)}.$$

If, at this point, we assume $f'(0) \neq 0$ and that $f'(x)$ is continuous, the $\lim_{x \to 0} f(x)/f'(x)$ exists and has the value 0; proving, of course, that $\lim_{x \to 0} x^{f(x)} = 1$. The interesting observation to be made is the following:

Let $f(x)$ satisfy $f(0) = 0$ and possess a derivative in a neighborhood of the origin. Then if $\lim_{x \to 0} f(x)/f'(x)$ exists it must have the value 0.

The italicized statement of course justifies the students' suspicions that $\lim_{x \to 0} x^{f(x)} = 1$, under the hypothesis on $f(x)$. The proof proceeds as follows:

Assume $\lim_{x \to 0} f(x)/f'(x)$ exists. Then 0 cannot be a limit point of zeros of $f'(x)$ or we are confronted with 0 being the limit point of points where $f(x)/f'(x)$ is either infinite or undefined. Hence we conclude that $f'(x) \neq 0$ for $0 < |x| < \delta$. We immediately conclude that $f(x) \neq 0$ in the interval $0 < x < \delta$, since otherwise $f'(\xi) = 0$, for some ξ, $0 < \xi < \delta$. Thus, without loss of generality, let $f(x) > 0$ in the interval $0 < x < \delta$.

If we now assume $\lim_{x \to 0} f(x)/f'(x) = L > 0$, there exists a neighborhood δ' such that

$$\left. \begin{array}{c} \dfrac{L}{2} < \dfrac{f(x)}{f'(x)} < \dfrac{3L}{2} \\[6pt] \dfrac{2}{L} > \dfrac{f'(x)}{f(x)} > \dfrac{2}{3L} \end{array} \right\} \quad \text{for } 0 < x < \delta'.$$

---

* From AMERICAN MATHEMATICAL MONTHLY, vol. 61 (1954), pp. 189–190.

Applying the Mean Value Theorem to $f(x)$ we find

$$\frac{f(x)}{f'(x)} = x \frac{f'(\xi)}{f'(x)} = x \cdot \frac{f(\xi)}{f(x)} \cdot \frac{f(x)}{f'(x)} \cdot \frac{f'(\xi)}{f(\xi)},$$

for x on the interval $0 < x < \delta''$ ($\delta'' \equiv$ smaller of δ, δ') and for some ξ, $0 < \xi < x$. We now observe that $f(x)/f'(x)$ is bounded by $3L/2$, $f'(\xi)/f(\xi)$ by $2/L$, and $f(\xi)/f(x)$ by 1 since $f'(x)$ must be >0. Therefore

$$\frac{f(x)}{f'(x)} < x \cdot 3$$

on $0 < x < \delta''$; and consequently $\lim_{x \to 0} f(x)/f'(x) = 0$, a contradiction to the assumption that $L \neq 0$. Hence $L = \lim_{x \to 0} f(x)/f'(x)$ must equal 0, and the student should be pleased since $\lim_{x \to 0} x^{f(x)} = 1$ as desired.

L'HÔPITAL'S RULE AND EXPANSION OF FUNCTIONS IN POWER SERIES*

M. R. SPIEGEL, Rensselaer Polytechnic Institute

When a student is first introduced to the concept of expansion of functions in power series, it is instructive to provide methods for evaluating the coefficients preliminary to using the conventional one of computing successive derivatives according to the rules of Maclaurin and Taylor. One such method, which will be described here, is useful when the student has had L'Hôpital's rule for the evaluation of indeterminate forms. For illustrative purposes the case of the expansion of sin x in powers of x will be discussed.

One begins with the customary assumption that

$$(1) \qquad \sin x = a_0 + a_1 x + a_2 x^2 + a_3 x^3 + \cdots$$

is to be an identity in x. It is desired to determine what possible values of the a's are consistent with this assumption.

To evaluate a_0 let $x=0$. If (1) is to be an identity it follows that a_0 ought to be equal to zero. Hence

$$(2) \qquad \sin x = a_1 x + a_2 x^2 + a_3 x^3 + \cdots .$$

Dividing both sides of (2) by x yields

$$(3) \qquad \frac{\sin x}{x} = a_1 + a_2 x + a_3 x^2 + \cdots .$$

Taking the limit of both sides of (3) as $x \to 0$ it follows that $a_1 = 1$. Hence

$$(4) \qquad \frac{\sin x}{x} = 1 + a_2 x + a_3 x^2 + \cdots ,$$

or

$$(5) \qquad \frac{\sin x - x}{x} = a_2 x + a_3 x^2 + \cdots .$$

Dividing by x,

$$(6) \qquad \frac{\sin x - x}{x^2} = a_2 + a_3 x + \cdots .$$

Taking the limit of both sides of (6) as $x \to 0$, making use of L'Hôpital's rule in the evaluation of the indeterminate form on the left, we find $a_2 = 0$. Hence

$$(7) \qquad \frac{\sin x - x}{x^2} = a_3 x + a_4 x^2 + \cdots .$$

---

* From AMERICAN MATHEMATICAL MONTHLY, vol. 62 (1955), pp. 358–360.

Dividing by x and letting $x \to 0$ using L'Hôpital's rule again we have $a_3 = -\frac{1}{6}$ and so

(8) $$\frac{\sin x - x}{x^3} = -\frac{1}{6} + a_4 x + a_5 x^2 + \cdots,$$

i.e.,

(9) $$\frac{x^3 + 6 \sin x - 6x}{6x^3} = a_4 x + a_5 x^2 + \cdots$$

or

(10) $$\frac{x^3 + 6 \sin x - 6x}{6x^4} = a_4 + a_5 x + \cdots,$$

from which the limit as $x \to 0$ gives $a_4 = 0$. A continuation of the procedure yields $a_5 = 1/120$. It is then found that

(11) $$\frac{x^3 + 6 \sin x - 6x}{6x^5} = \frac{1}{120} + a_6 x + a_7 x^2 + \cdots.$$

In a similar way further coefficients may be found. By stopping at any desired place as, for example, at (11) and solving for $\sin x$, or by using the coefficients already found in the equation (1), the first few terms of the series are obtained. In this case the result is

(12) $$\begin{aligned}\sin x &= x - \frac{x^3}{6} + \frac{x^5}{120} + a_6 x^6 + a_7 x^7 + \cdots \\ &= x - \frac{x^3}{3!} + \frac{x^5}{5!} + a_6 x^6 + a_7 x^7 + \cdots\end{aligned}$$

from which the general pattern is indicated but, of course, not proved.

The author has found by approaching the problem in this way that students greatly welcome (and consequently remember better) the conventional method which is so much less tedious. The above procedure is useful also in that it provides a probably needed review of indeterminate forms and L'Hôpital's rule which by this time may have disappeared from the student's mathematical vocabulary. It also prepares him to see later a possible application of series in the evaluation of indeterminate forms, thus shortening the labor of applying L'Hôpital's rule too frequently in certain cases.

In addition this approach enables the student to realize that there may be many different ways of evaluating the coefficients in the power series expansion of a given function. In fact one possible method which could be suggested is the somewhat natural approach of letting x equal certain values on both sides of (1), such as for example $x = \pi/2^n$, $n = 1, 2, 3, \cdots$ for which sin x may always be found exactly. In this manner, the concept of an infinite number of equations involving an infinite number of unknowns and consequently determinants of infinite order could be mentioned as possibilities for further investigation.

At this stage it is even a good idea to inquire as to the possibility of expansion of functions into series other than power series, perhaps presenting just a brief mention of Fourier series.

It turns out that such remarks as given above often provide for some lively discussion in class and seem to add more to the meaning of mathematics and consequently to the better understanding of mathematics in general. It is certainly worth the little time spent.

BIBLIOGRAPHIC ENTRIES: INDETERMINATE FORMS

Except for the entry labeled MATHEMATICS MAGAZINE, the references below are to the AMERICAN MATHEMATICAL MONTHLY.

1. Taylor, A. E., L'Hospital's rule, vol. 59, p. 20.

    Rigorous proofs of various cases of L'Hospital's Rule.

2. Furstenberg, H., Note on one type of indeterminate form, vol. 60, p. 700.

    The type is $[\phi(n)]^{n^{-r}}$, where $\phi(n) \to \infty$.

3. Watson, G. C., A note on indeterminate forms, vol. 68, p. 490.

    Examples in which 0^0 does not lead to a limit of 1.

4. Rickert, N. W., A calculus counterexample, vol. 75 p. 166.
5. Schaefer, Paul, A note on the limit of $f(x)/f'(x)$, MATHEMATICS MAGAZINE, vol. 34, p. 268.

7

POLYNOMIALS AND POLYNOMIAL APPROXIMATIONS

(a)

TAYLOR POLYNOMIALS

ELEMENTARY DEVELOPMENT OF CERTAIN INFINITE SERIES*

J. P. BALLANTINE, University of Washington

We find it is possible to obtain the usual series for certain fundamental functions rigorously without presupposing any knowledge of infinite series or of convergence. The method will be illustrated by developing $\log(1+x)$; other functions which could be so treated include $(1+x)^n$ when n is fractional or negative, arc tan x, sin x, cos x, and e^x. We assume a knowledge of integration of x^n, of differentiation of x^n and the particular function being developed, and of one further theorem, namely:

THEOREM 1. *If $f(x) > 0$ for values of x in the interval $a > x > b$, then*

$$\int_a^b f(x)dx > 0.$$

From Theorem 1, we may prove immediately three other theorems cited below. The present development calls only for Theorem 2, but the proofs here given of Theorems 3 and 4 may be of interest.

THEOREM 2. *If $f(x) < g(x)$ for the interval $a < x < b$, then*

$$\int_a^b f(x)dx < \int_a^b g(x)dx.$$

---

* From AMERICAN MATHEMATICAL MONTHLY, vol. 44 (1937), pp. 470–472.

THEOREM 3. If $f(x) < M$ for the interval $a < x < b$, then

$$\int_a^b f(x)dx < M(b-a).$$

THEOREM 4. If $f(x) < M$ and $g(x) > 0$ for the interval $a < x < b$, then

$$\int_a^b f(x)g(x)dx < M \int_a^b g(x)dx.$$

Each theorem follows directly from Theorem 1 by replacing the function $f(x)$ of Theorem 1 by a particular positive function, namely:

$g(x) - f(x)$ in the case of Theorem 2,
$M - f(x)$ in the case of Theorem 3,
$\{M - f(x)\}\{g(x)\}$ in the case of Theorem 4.

Development of $\log(1+x)$. The derivative of $\log(1+x)$ is $1/(1+x)$. It is easy to find simple polynomials larger or smaller than $1/(1+x)$. Thus, for $0 < x < b$,

$$\frac{1}{1+x} < 1, \quad \frac{1}{1+x} > 1-x, \quad \frac{1}{1+x} < 1-x+x^2,$$

and so on. These inequalities are readily proved by clearing of fractions. Therefore, by Theorem 2, integrating from 0 to x, where $0 < x$, we have

$$\log(1+x) < x, \quad \log(1+x) > x - \frac{x^2}{2},$$

$$\log(1+x) < x - \frac{x^2}{2} + \frac{x^3}{3}, \quad \text{etc.}$$

These inequalities hold, even if x is given a large value, but for practical purposes, x should be small. Then the expressions that are smaller than $\log(1+x)$ are very close to those that are larger, and a close determination can be made.

Since these relations have not been established for negative values of x, we shall develop $\log(1-x)$. Its derivative is $-1/(1-x)$. As before, we have, for $0 < x < b < 1$,

$$\frac{1}{1-x} > 1, \quad \frac{1}{1-x} > 1+x, \quad \frac{1}{1-x} > 1+x+x^2.$$

and so on; from which we readily obtain (by integration and change of sign)

$$\log(1-x) < -x, \quad \log(1-x) < -x - \frac{x^2}{2},$$

$$\log(1-x) < -x - \frac{x^2}{2} - \frac{x^3}{3}, \quad \text{etc.}$$

Thus we have any number of expressions which are greater than $\log(1-x)$. We now seek expressions which are smaller. By long division with remainder, we find that

$$\frac{1}{1-x} = 1 + \frac{x}{1-x}, \quad \frac{1}{1-x} = 1 + x + \frac{x^2}{1-x}, \quad \text{etc.}$$

Consider x limited to the interval $0 < x < b < 1$. Then using the relation

$$\frac{1}{1-x} < \frac{1}{1-b} = c \qquad 0 < x < b < 1$$

in the equations just considered, we obtain the inequalities

$$\frac{1}{1-x} < 1 + \frac{x}{1-b} = 1 + cx, \quad \frac{1}{1-x} < 1 + x + cx^2,$$

$$\frac{1}{1-x} < 1 + x + x^2 + cx^3, \quad \text{etc.}$$

If we integrate from 0 to b, change signs, replace c by $1/(1-b)$, and b by x, we have

$$\log(1-x) > -\frac{x}{1-x}, \quad \log(1-x) > -x - \frac{x^2}{2(1-x)},$$

$$\log(1-x) > -x - \frac{x^2}{2} - \frac{x^3}{3(1-x)}, \quad \text{etc.}$$

Consider one of the expressions which is larger than $\log(1-x)$ and the corresponding one which is smaller; we have, for example,

$$-x - \frac{x^2}{2} - \frac{x^3}{3(1-x)} < \log(1-x) < -x - \frac{x^2}{2} - \frac{x^3}{3}.$$

They are precisely the same, except for the extra factor $1-x$ in the denominator of the last term of one of them. For these expressions to be of practical use, x should be small, in which case $1-x$ is near 1, and the two final terms are nearly equal. If enough terms have been taken so that the final terms are negligible, then the factor $1-x$ makes very little difference, and either member of the above inequality may be taken as the desired value of $\log(1-x)$.

While the above series for $\log(1-x)$ is not new, the remainder term is better than the usual one. For example, if $\log(1-.9)$ is computed by taking ten terms of the series, the error is found to be less than

$$\frac{.9^{10}}{10(1-.9)} = .9^{10} = .35.$$

The classical remainder formula, $R_n(x) = f^n(\xi)x^n/n!$, $0 < \xi < x$, in this case shows only that

$$R_{10}(.9) = \frac{9!}{(1-\xi)^{10}} \frac{.9^{10}}{10!} < \frac{.9^{10}}{10(.1)^{10}} = 350{,}000{,}000.$$

A NOTE ON TAYLOR'S THEOREM*

C. L. SEEBECK, JR., University of Alabama

Although there are many proofs of Taylor's Theorem in mathematical literature, most of these are difficult for a not-too-advanced Calculus student to master. The following proof requires no knowledge of infinite series, or of convergence. It is direct, readily illustrated geometrically, and provides the remainder using only elementary or easily explained theorems.

Let $f(x)$ be a function continuous with its derivatives through order $n+1$ in a suitable neighborhood of $x=a$. Moreover, assume $f^{[k]}(a)$ may be evaluated for $k=0, 1, 2, \cdots, n$. We desire the value of $f(x)$ at a point in the given neighborhood. Since the evaluation process may call for operations other than $(+, -, \times, \div)$, we first approximate its value by means of

$$P(x) = \sum_{k=0}^{n} c_k(x-a)^k,$$

a polynomial. We choose c_k so that $P(x)$ will be identified with $f(x)$ as closely as possible when $x=a$, or such that

(1) $\qquad P^{[k]}(a) = f^{[k]}(a), \qquad k = 0, 1, 2, \cdots, n,$

a condition that implies

$$c_k = \frac{f^{[k]}(a)}{k!}.$$

Now let $R(x)$ be the amount by which the approximation fails.

(2) $\qquad R(x) = f(x) - P(x).$

Equation (1) implies

(3) $\qquad R^{[k]}(a) = 0, \qquad k = 0, 1, 2, \cdots, n.$

Differentiating equation (2) n times gives

$$R^{[n]}(x) = f^{[n]}(x) - P^{[n]}(x).$$

But since $P(x)$ is a polynomial of degree n, its nth derivative is a constant,

$$P^{[n]}(x) = P^{[n]}(a) = f^{[n]}(a), \quad \text{and}$$
$$R^{[n]}(x) = f^{[n]}(x) - f^{[n]}(a).$$

By the law of the mean

(4) $\qquad R^{[n]}(x) = f^{[n+1]}(\theta)(x-a), \qquad \theta \text{ between } a \text{ and } x.$

---

* From AMERICAN MATHEMATICAL MONTHLY, vol. 57 (1950), pp. 32–34.

Let M and m be respectively the maximum and minimum values of $f^{[n+1]}(\theta)$ in the interval (a, x). Then if $a \leq t \leq x$,

$$m(t - a) \leq R^{[n]}(t) \leq M(t - a).$$

It is clear geometrically that if an ordinate of one curve is never greater than the corresponding ordinate of a second curve in a given interval, then the area under the first curve cannot exceed the area under the second for this interval. Accordingly, we integrate this inequality from a to x and obtain

$$\tfrac{1}{2}m(x - a)^2 \leq R^{[n-1]}(x) \leq \tfrac{1}{2}M(x - a)^2.$$

Continuity of the $(n+1)$st derivative of $f(t)$ now implies

(5) $$R^{[n-1]}(x) = f^{[n+1]}(\theta_1) \frac{(x - a)^2}{2}, \qquad \theta_1 \text{ between } a \text{ and } x.*$$

If $x \leq t \leq a$, the inequalities are reversed and (5) is again true.

Repeated integration of the inequalities from a to x will produce

$$R(x) = \frac{f^{[n+1]}(\phi)}{(n + 1)!} (x - a)^{n+1}, \qquad \phi \text{ between } a \text{ and } x.$$

Equation (2) now gives

(6) $$f(x) = P(x) + R(x)$$

which is Taylor's Theorem with remainder.

It may be noted that the question of convergence has been completely sidestepped. $P(x)$ is a good approximation of $f(x)$ whenever the least upper bound for $|R(x)|$ is as small as desired. When n is finite, equation (6) is identically true in x, and the question of the series representing the function does not arise.

---

* This follows more directly and more rigorously from equation (4) by the first law of the mean for integrals if this is available.

A CONNECTION BETWEEN TAYLOR'S THEOREM AND LINEAR DIFFERENTIAL EQUATIONS*

D. C. LEWIS, Johns Hopkins University

The following amusing elementary derivation and generalization of Taylor's theorem (with the integral form for the remainder) was accidentally encountered while teaching a calculus course in which, for the convenience of the engineering students, the introduction of Taylor's theorem was postponed until after they had had some elementary theory of differential equations.

We consider the general homogeneous linear differential equation of the nth order with coefficients $a_1(x), \cdots, a_n(x)$ continuous on a certain interval I of the x-axis,

(1) $$y^{(n)} + a_1(x) y^{(n-1)} + \cdots + a_{n-1}(x) y' + a_n(x) y = 0,$$

and suppose we consider the system of independent solutions $y = y_k(x, t)$, $k = 0, 1, 2, \cdots, n-1$, determined by the conditions

(2) $$(d^k/dx^k) y_k(x, t) \big|_{x=t} = 1, \quad (d^h/dx^h) y_k(x, t) \big|_{x=t} = 0, \quad h \neq k.$$

Here the parameter t may take on any value in I. Then the solution of the corresponding non-homogeneous equation,

(3) $$Y^{(n)} + a_1(x) Y^{(n-1)} + \cdots + a_n(x) Y = F(x)$$

(where $F(x)$ is continuous on I), satisfying the initial conditions

(4) $$Y^{(k)}(c) = b_k, \quad k = 0, 1, 2, \cdots, n-1, c \in I,$$

can be written in the form

(5) $$Y(x) = \sum_{k=0}^{n-1} b_k y_k(x, c) + \int_c^x F(t) y_{n-1}(x, t) dt.$$

This essentially well known result can be verified *a posteriori*, by differentiation of (5) and substitution in (3), using, among the other elements in the definition of the $y_k(x, t)$, the fact that $y_{n-1}(x, t)$ satisfies (1) for every value of t. It is also possible to deduce (5) as the end result of the Lagrange method of variation of parameters.

But *every* function $f(x)$ of class C^n on I can be regarded as being a solution of (3) no matter how the coefficients $a_1(x), \cdots, a_n(x)$ may have been chosen, so long as we take

(6) $$F(x) = f^{(n)}(x) + \sum_{k=1}^{n} a_k(x) f^{(n-k)}(x).$$

---

* From AMERICAN MATHEMATICAL MONTHLY, vol. 59 (1952), pp. 692–693.

Hence, with this definition of $F(x)$, we find that every function $f(x)$ of class C^n on I satisfies the identity,

$$\text{(7)} \qquad f(x) = \sum_{k=0}^{n-1} f^{(k)}(c) y_k(x, c) + \int_c^x F(t) y_{n-1}(x, t) dt,$$

which we of course deduce from (5) by taking appropriate values for the b's.

Formula (7) may be regarded as a generalization of Taylor's theorem. In fact, if $a_1 \equiv a_2 \equiv \cdots \equiv a_n \equiv 0$, we must have, from (6), $F(x) = f^{(n)}(x)$, and, from (1) and (2), we have

$$y_k(x, t) = (x - t)^k / k! \qquad k = 0, 1, \cdots, n - 1.$$

Hence (7) becomes in this special case

$$f(x) = \sum_{k=0}^{n-1} f^{(k)}(c) \frac{(x - c)^k}{k!} + \frac{1}{(n - 1)!} \int_c^x f^{(n)}(t)(x - t)^{n-1} dt,$$

which is Taylor's theorem with integral form of the remainder, usually proved directly by an inductive process employing integration by parts.

Obviously there are many other special cases of (7) which can be obtained by special choices of the a's.

MORE ON TAYLOR'S THEOREM IN A FIRST COURSE*

C. P. NICHOLAS, U.S. Military Academy

This MONTHLY (vol. 58, 1951, pp. 559–562) carried a derivation of Taylor's Theorem which I offered as suitable for simplifying the subject in a first course. The derivation can be further simplified, as shown below. Like the earlier derivation, the new one presupposes that the student can evaluate a successive integral; and also that he has been led into the problem sufficiently to know that we wish to find an approximate value of $f(x)$, based on known values of $f(x)$ and its first n derivatives at a neighboring point $x=a$.

The simplification occurs at the start where, instead of developing two special formulas for the value of an increment, we use the already familiar relationship:

$$\int_a^x f'(x)dx = f(x) - f(a).$$

This may be transposed to read

(I) $$f(x) = f(a) + \int_a^x f'(x)dx.$$

Now, (I) may be used as an iteration formula to expand any function which has successive derivatives. Specifically, it may be used to expand the function under the integral sign of (I) itself. Thus, applying (I) to $f'(x)$ we have:

$$f(x) = f(a) + \int_a^x \left[f'(a) + \int_a^x f''(x)dx \right] dx$$

$$= f(a) + \int_a^x f'(a)dx + \int_a^x \int_a^x f''(x)(dx)^2.$$

Continuing, we apply (I) successively to $f''(x)$ under the double integral sign, then to $f'''(x)$ under the triple integral sign, and so on. Thus:

$$f(x) = f(a) + f'(a)\int_a^x dx + f''(a)\int_a^x \int_a^x (dx)^2 + f'''(a)\int_a^x \int_a^x \int_a^x (dx)^3$$

$$+ \cdots + f^{n-1}(a) \int_a^x \int_a^x \cdots \int_a^x (dx)^{n-1}$$

$$+ \int_a^x \int_a^x \int_a^x \cdots \int_a^x f^{(n)}(x)(dx)^n.$$

---

* From AMERICAN MATHEMATICAL MONTHLY, vol. 60 (1953), pp. 329–331.

Evaluating each integral except the last, we obtain Taylor's formula:

$$f(x) = f(a) + f'(a) \cdot (x - a) + f''(a) \frac{(x - a)^2}{2!} + f'''(a) \frac{(x - a)^3}{3!} + \cdots$$
$$+ f^{(n-1)}(a) \frac{(x - a)^{n-1}}{(n - 1)!} + R_n$$

where R_n, the remainder after n terms, is given by

$$R_n = \int_a^x \int_a^x \cdots \int_a^x f^{(n)}(x)(dx)^n.$$

The Lagrange form of the remainder may now be derived as explained in my previous article. Professor H. J. Hamilton followed with an excellent derivation of the Cauchy form in this MONTHLY (vol. 59, 1952, p. 320).

It should be noted that the foundation of this proof includes the Fundamental Theorem of Integral Calculus, which in turn presupposes the Mean-Value Theorem for Derivatives, usually derived with the aid of Rolle's Theorem. Thus, the proof does not fundamentally avoid Rolle's Theorem. Its advantage lies in the fact that it avoids the elaborate auxiliary function usually introduced when Rolle's Theorem is employed to prove Taylor's Theorem or the Extended Theorem of the Mean.

A PROOF OF TAYLOR'S FORMULA

JAMES WOLFE, University of Utah

The following proof of Taylor's formula with remainder may seem more natural than the proofs ordinarily offered in calculus texts.

Suppose $f(x)$ has a continuous $(n-1)$st derivative in the closed interval between a and b and an nth derivative in the open interval between a and b. Then $f(x)$ can be approximated by a polynomial $p(x)$ of degree n which agrees in value with $f(x)$ at a and b and such that the first $n-1$ derivatives of p and f agree at a:

$$p(x) = f(a) + f'(a)(x-a) + \frac{f''(a)}{2!}(x-a)^2 + \cdots$$
$$+ \frac{f^{(n-1)}(a)}{(n-1)!}(x-a)^{n-1} + k(x-a)^n,$$

where k is determined so that $p(b)=f(b)$. Let $g(x)=f(x)-p(x)$. Then $g(x)$ and its first $n-1$ derivatives vanish at $x=a$ and also $g(b)=0$. Using successive applications of Rolle's theorem, $g'(x_1)=0$ for some x_1 between a and b, $g''(x_2)=0$ for some x_2 between a and x_1, etc. and finally $g^{(n)}(x_n)=0$ for some x_n between a and x_{n-1}. But $g^{(n)}(x) = f^{(n)}(x) - n!k$, consequently $k = f^{(n)}(x_n)/n!$ and

$$f(b) = p(b) = f(a) + f'(a)(b-a) + \cdots$$
$$+ \frac{f^{(n-1)}(a)}{(n-1)!}(b-a)^{n-1} + \frac{f^{(n)}(x_n)}{n!}(b-a)^n.$$

For $n=1$, this proof reduces to the usual proof of the mean value theorem.

TAYLOR'S THEOREM AND NEWTON'S METHOD[†]

F. D. PARKER, University of Alaska

Most texts in calculus study Newton's method long before the study of Taylor's series and Taylor's theorem. While studying the latter it may be worth while to stop to refine Newton's method and to show the relation between the two.

If we use only the first two terms of a Taylor series of a function $f(x)$ expanded about a point x_1 near a zero of $f(x)$, then $f(x) = f(x_1) + f'(x_1)(x - x_1)$. Setting $f(x)$ equal to zero we get an approximation to the root $x = x_1 - f(x_1)/f'(x_1)$, which is familiar to the student as Newton's method. If now we take three terms of the Taylor series, we get an approximation

$$x = x_1 - \frac{f'(x_1) \pm \{[f'(x_1)]^2 - 2f(x_1)f''(x_1)\}^{1/2}}{f''(x_1)}$$

provided, of course, $f''(x_1) \neq 0$. While Newton's method fits a line to the graph of the function, this refinement fits a parabola with the same slope and curvature to the graph of the function. The choice of the sign is dictated by the problem. Further refinements are possible of course, but not practical.

In solving $x^4 + 2x^3 + x^2 - 6x - 12 = 0$ with a first approximation $x = 2$, Newton's method yields 1.7778, the suggested refinement yields 1.7265, whereas the correct root is $\sqrt{3} = 1.7321$.

[†] From AMERICAN MATHEMATICAL MONTHLY, vol. 66 (1959), p. 51.

REMAINDER FORMULAE IN TAYLOR'S THEOREM*

WILLIAM J. FIREY, Washington State University

It is not uncommon in an elementary calculus course to develop Taylor's formula with a remainder described on the one hand in integral form and on the other by use of an intermediate value of a derivative of the function being represented. Thus if we write

$$f(x) = \sum_{k=0}^{n} \frac{f^{(k)}(a)}{k!} (x-a)^k + R_n,$$

we have

(1) $$R_n = \int_a^x \frac{(s-t)^n}{n!} f^{(n+1)}(t) dt$$

or

(2) $$R_n = \frac{(x-a)^{n+1}}{(n+1)!} f^{n+1}(X),$$

where $(X-a)(X-x) < 0$. The first form is often arrived at by a succession of integrations by parts; the second by an application of Rolle's theorem to a suitable function.

The comparison of these two representations of the remainder provides an opportunity for introducing by example the generalized mean-value theorem for integrals (to use the terminology of Courant [1]), but we wish to show that it can also be used in an elementary course to motivate a consideration of distinctions between types of discontinuities.

For convenience (1) will be called the integral form of R_n, (2) Lagrange's form. Further we shall assume $a < x$; the necessary changes in the discussion will be obvious in the alternative case.

In an elementary treatment, it is usual to assume in arriving at the integral form that $f^{(n+1)}$ is continuous in the closed interval from a to x. At the same time, the generalized mean-value theorem for integrals, viz.,

(3) $$\int_a^x p(t)\phi(t) dt = \phi(X) \int_a^x p(t) dt, \qquad (a < X < x),$$

is also usually proven under the assumptions: (a) $p(t) \geq 0$, (b) p and ϕ continuous in the closed interval from a to x. Then one establishes directly Lagrange's remainder from the integral form. But in applying Rolle's theorem, Lagrange's

---

* From AMERICAN MATHEMATICAL MONTHLY, vol. 67 (1960), pp. 903–905.

form of the remainder is obtained under a weaker assumption on $f^{(n+1)}$—namely its existence in the open interval from a to x. This suggests that Taylor's formula with Lagrange's remainder applies to a wider class of functions.

However, Taylor's formula with the integral remainder could be established by integration by parts assuming, say, only Riemann integrability for $f^{(n+1)}$. This puts that form of Taylor's theorem in a little more favorable light; but the question now arises as to whether, if we wish, we can establish Lagrange's remainder by an integral mean-value theorem. To do this we extend the generalized mean-value theorem for integrals.

We shall say ϕ assumes intermediate values over an interval from a to b if for each pair x_0, x_1 of numbers selected from this interval and for any number y between $\phi(x_0)$ and $\phi(x_1)$, there is an x between x_0 and x_1 such that $\phi(x) = y$. This was, implicitly, the definition of continuity in less sophisticated times. It is easy to prove in the usual way that if in the closed interval from a to x: (i) $p(t) \geq 0$, (ii) ϕ assumes intermediate values, and (iii) the indicated integrals in (3) exist, then (3) is true.

Using this form of the mean-value theorem, Lagrange's remainder can be established as a consequence of the integral remainder under the assumptions that $f^{(n+1)}$ satisfies the conditions (ii) and (iii) just stated. But the interesting point which can be made is that (ii) is *automatically* satisfied when ϕ is a derivative, e.g., $f^{(n+1)}$. That is, a derivative, even though discontinuous, must assume intermediate values, or put another way, a derivative cannot have jump discontinuities, (*cf.* [2]). To summarize, Taylor's theorem with integral remainder is valid if $f^{(n+1)}$ is, say, Riemann integrable; with Lagrange's remainder it is valid under the weaker condition that $f^{(n+1)}$ exist in the open interval from a to x.

The pedagogical point of the discussion is this. It is often observed that the integral remainder is better suited to some purposes and this furnishes an obvious motivation to the above extension of its range of validity. At the same time, it is an opportunity to show reasons for introducing finer distinctions in discussions of discontinuities. And finally there is perhaps some attraction in the device of displaying an apparent discrepancy of results followed by at least a partial resolution of the discrepancy.

References

1. R. Courant, Differential and Integral Calculus, New York.
2. W. Rudin, Principles of Mathematical Analysis, New York, 1953.

GEOMETRIC INTERPRETATIONS OF POLYNOMIAL APPROXIMATIONS OF THE COSINE FUNCTION*

E. R. HEINEMAN, Texas Technological College

The polynomial approximations obtained from the power series expansion

(1) $$\cos x = 1 - \frac{x^2}{2!} + \frac{x^4}{4!} - \frac{x^6}{6!} + \cdots$$

converge to the cosine function as shown.

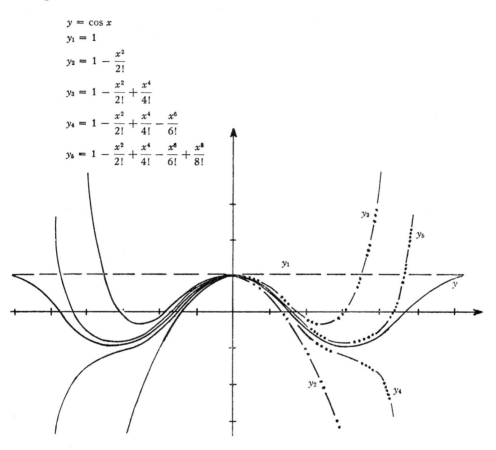

$y = \cos x$
$y_1 = 1$
$y_2 = 1 - \frac{x^2}{2!}$
$y_3 = 1 - \frac{x^2}{2!} + \frac{x^4}{4!}$
$y_4 = 1 - \frac{x^2}{2!} + \frac{x^4}{4!} - \frac{x^6}{6!}$
$y_5 = 1 - \frac{x^2}{2!} + \frac{x^4}{4!} - \frac{x^6}{6!} + \frac{x^8}{8!}$

It is interesting to notice that if $y_i(x)$ is the ith polynomial approximation obtained from (1), then

$$\frac{d^2 y_i}{dx^2} = - y_{i-1},$$

---

* From AMERICAN MATHEMATICAL MONTHLY, vol. 73 (1966), pp. 648–649.

which shows that the inflection points of the graph of y_i have the same abscissas as the x-intercepts of y_{i-1}.

In one sense, it is clear that the successive approximations become "better." The student, however, must be warned against the pseudo-truth: "Every time you add another term, the approximation gets better." The horizontal line $y_1 = 1$ is a shabby approximation for the cosine curve, but it is never in error by more than 2, which is more than can be said about the later approximations.

The addition of another term to y_2 gives y_3, which is a better approximation for cos x, but *only if* $|x| < \sqrt{12 + \sqrt{96}} \approx 4.67$. For $x > 2\sqrt{3 + \sqrt{6}}$, the three-term polynomial y_3 is a worse approximation than the two-term polynomial y_2.

The addition of two terms to y_2 produces y_4, which is a better approximation for cos x, but only if $|x| < \sqrt{30}$. The graph of y_4 intersects that of y_2 at $(\pm \sqrt{30}, -14)$. For $|x| > \sqrt{30}$ it can be shown that $y_4 < y_2$, the graph of y_4 lies below that of y_2, and y_4 is a worse approximation than y_2.

BIBLIOGRAPHIC ENTRIES: TAYLOR POLYNOMIALS

The references below are to the AMERICAN MATHEMATICAL MONTHLY.

1. Blumenthal, L. M., Concerning the remainder term in Taylor's formula, vol. 33, p. 424.

2. Kammerer, H. M., Sine and cosine approximation curves, vol. 43, p. 293.

   Graphs showing Taylor polynomials for sin x and cos x up to degree 19.

3. Moritz, R. E., A note on Taylor's theorem, vol. 44, p. 31.

   An attempt to motivate the auxiliary function in the usual proof of Lagrange's version of Taylor's theorem.

4. Miller, N., The Taylor series approximation curves for the sine and cosine, vol. 44, p. 96.

   Comments about the way in which Taylor polynomials approximate the sine function. (See Kammerer, vol. 43, p. 293.) Discussion of zeros (real and complex) and some statements about asymptotic behavior.

5. Hummel, P. M., and Seebeck, C. L., Jr., A generalization of Taylor's expansion, vol. 56, p. 243.

   The authors develop a series that converges more rapidly than the Taylor series of f but requires calculation of the derivatives of f at two points instead of only one.

6. Rudinger, G., A theorem on the remainder of a Taylor series, vol. 57, p. 411.

   Concerns the sign of the remainder term.

7. Nicholas, C. P., Taylor's theorem in a first course, vol. 58, p. 559.

8. Hamilton, H. J., Cauchy's form of R_n from the iterated integral form, vol. 59, p. 320.

9. Ogilvy, C. S., A theorem on the Taylor expansion, vol. 62, p. 654.
10. Diananda, P. H., On Taylor's theorem with remainder, vol. 64, p. 492.
11. Goodner, D. B., An extension of Taylor's formula, vol. 70, p. 303.
12. Ballard, W. R., Livingston, A. E., and Myers, W. M., Jr., A variant of Taylor's theorem, vol. 70, p. 865.
13. Beesack, P. R., A general form of the remainder in Taylor's theorem, vol. 73, p. 64.
14. Maria, Alfred J., Taylor's formula with derivative remainder, vol. 73, p. 67.

(b)

OTHER POLYNOMIALS

A PROOF OF WEIERSTRASS'S THEOREM*†

DUNHAM JACKSON, University of Minnesota

Weierstrass's theorem with regard to polynomial approximation can be stated as follows:

If $f(x)$ is a given continuous function for $a \leq x \leq b$, and if ϵ is an arbitrary positive quantity, it is possible to construct an approximating polynomial $P(x)$ such that

$$|f(x) - P(x)| < \epsilon$$

for $a \leq x \leq b$.

This theorem has been proved in a great variety of different ways. No particular proof can be designated once for all as the simplest, because simplicity depends in part on the preparation of the reader to whom the proof is addressed. A demonstration which follows directly from known facts about power series or Fourier series, for example, is not so immediate if a derivation of those facts has to be gone through first. A proof commonly regarded as among the simplest and neatest is the one due to Landau‡ in which an approximating polynomial is given explicitly by means of a certain type of "singular" integral. The purpose of this note is to present a modification or modified formulation of Landau's proof which is believed to possess further advantages of simplicity, at least from some points of view.

Let $f(x)$ be a given continuous function for $a \leq x \leq b$. Without essential loss of generality it can be supposed that $0 < a < b < 1$, since any finite interval whatever can be carried over into an interval contained in (0, 1) by a linear transformation, under which any continuous function will go into a continuous function and any polynomial into a polynomial of the same degree. For convenience in the writing of the formulas which enter into the proof, let the function $f(x)$, supposed given in the first instance only for $a \leq x \leq b$, be defined outside the

---

* From AMERICAN MATHEMATICAL MONTHLY, vol. 41 (1934), pp. 309–312.

† Presented to the Minnesota Section of the Association, May 13, 1933.

‡ E. Landau, *Über die Approximation einer stetigen Funktion durch eine ganze rationale Funktion*, Rendiconti del Circolo Matematico di Palermo, vol. 25 (1908), pp. 337–345; see also Courant-Hilbert, *Methoden der Mathematischen Physik*, vol. I, second edition, Berlin, 1931, pp 55–57.

interval (a, b) as follows:

$$f(x) = 0 \qquad \text{for} \qquad x \leq 0,$$

$$f(x) = \frac{x}{a} f(a) \qquad \text{for } 0 < x < a,$$

$$f(x) = \frac{1-x}{1-b} f(b) \qquad \text{for } b < x < 1,$$

$$f(x) = 0 \qquad \text{for} \qquad x \geq 1.$$

Then $f(x)$ is defined and continuous for all values of x. The question at issue is that of approximating it by means of a polynomial for values of x belonging to the interval (a, b).

Let J_n denote the constant

$$J_n = \int_{-1}^{1} (1 - u^2)^n du,$$

and let

$$P_n(x) = \frac{1}{J_n} \int_0^1 f(t) [1 - (t - x)^2]^n dt.$$

The integrand in the latter integral is a polynomial of degree $2n$ in x with coefficients which are continuous functions of t, and the integral is for each value of n a polynomial in x of degree $2n$ (at most) with constant coefficients.

If $0 \leq x \leq 1$, the value of the integral is not changed if the limits are replaced by $-1+x$ and $1+x$, since $f(t)$ vanishes everywhere outside the interval $(0, 1)$, and vanishes in particular from $-1+x$ to 0 and from 1 to $1+x$:

$$P_n(x) = \frac{1}{J_n} \int_{-1+x}^{1+x} f(t) [1 - (t - x)^2]^n dt.$$

By the substitution $t - x = u$ this becomes

(1) $$P_n(x) = \frac{1}{J_n} \int_{-1}^{1} f(x + u)(1 - u^2)^n du.$$

If the equation

$$1 = \frac{1}{J_n} \int_{-1}^{1} (1 - u^2)^n du$$

is multiplied by $f(x)$, this factor, being independent of u, may be written under the integral sign:

(2) $$f(x) = \frac{1}{J_n} \int_{-1}^{1} f(x)(1 - u^2)^n du.$$

Hence, by subtraction of (2) from (1),

(3) $$P_n(x) - f(x) = \frac{1}{J_n}\int_{-1}^{1}[f(x+u) - f(x)](1-u^2)^n du.$$

The problem now is to show that the value of this expression approaches zero as n becomes infinite.

Let ϵ be any positive quantity. Since $f(x)$ is (uniformly) continuous there is a $\delta > 0$ (independent of x) such that $|f(x+u) - f(x)| \leq \epsilon/2$ for $|u| \leq \delta$. Let M be the maximum of $|f(x)|$. Then $|f(x+u) - f(x)| \leq 2M$ for all values of u. For $|u| \geq \delta$, $1 \leq u^2/\delta^2$, and

$$|f(x+u) - f(x)| \leq 2Mu^2/\delta^2.$$

For any value of u, one or the other of the quantities $\epsilon/2$, $2Mu^2/\delta^2$ is greater than or equal to $|f(x+u) - f(x)|$, and their sum therefore is certainly greater than or equal to $|f(x+u) - f(x)|$:

$$|f(x+u) - f(x)| \leq \epsilon/2 + 2Mu^2/\delta^2$$

for all values of u. Consequently, for $0 \leq x \leq 1$,

$$|P_n(x) - f(x)| \leq \frac{1}{J_n}\int_{-1}^{1}(\epsilon/2)(1-u^2)^n du + \frac{1}{J_n}\int_{-1}^{1}\frac{2Mu^2}{\delta^2}(1-u^2)^n du$$

$$= \epsilon/2 + \frac{2M}{\delta^2 J_n}\int_{-1}^{1}u^2(1-u^2)^n du.$$

Let the last integral be denoted by J'_n. By integration by parts,

$$J'_n = \int_{-1}^{1} u \cdot u(1-u^2)^n du = \left[-u \cdot \frac{(1-u^2)^{n+1}}{2(n+1)}\right]_{-1}^{1}$$

$$+ \int_{-1}^{1}\frac{(1-u^2)^{n+1}}{2(n+1)}du = \frac{J_{n+1}}{2(n+1)}.$$

But $J_{n+1} < J_n$, since $1 - u^2 < 1$ throughout the interior of the interval of integration and hence $(1-u^2)^{n+1} = (1-u^2)(1-u^2)^n < (1-u^2)^n$. So

$$J'_n < \frac{J_n}{2(n+1)}, \qquad \frac{J'_n}{J_n} < \frac{1}{2(n+1)}.$$

It follows that as soon as n is sufficiently large

$$\frac{2MJ'_n}{\delta^2 J_n} < \epsilon/2$$

and consequently

$$|P_n(x) - f(x)| < \epsilon,$$

for $0 \leq x \leq 1$ and in particular for $a \leq x \leq b$. This is the substance of the conclusion to be proved.

The above proof has something in common with that of S. Bernstein,* the fundamental difference being that Landau's integral is used here instead of the algebraic formula for a binomial frequency distribution. It was in fact suggested to the writer, not by consideration of Bernstein's proof as such, but by a conversation with Professor W. L. Hart on the subject of Bernoulli's theorem. A noteworthy characteristic of Bernstein's proof is that it makes Weierstrass's theorem in effect a corollary of that of Bernoulli.

An alternative organization of the present method of proof is as follows. Let $f(x)$ at first be not merely continuous, but subject to the Lipschitz condition

(4) $$|f(x_2) - f(x_1)| \leq \lambda |x_2 - x_1|.$$

If this condition is satisfied by $f(x)$ as originally defined for $a \leq x \leq b$, it will be satisfied, possibly with a different value of λ, when the definition is extended to cover all values of x. Then, in (3), $|f(x+u) - f(x)| \leq \lambda |u|$, and

$$|P_n(x) - f(x)| \leq \frac{\lambda}{J_n} \int_{-1}^{1} |u|(1-u^2)^n du = \frac{2\lambda}{J_n} \int_0^1 u(1-u^2)^n du.$$

Let $\delta = 1/n^{1/2}$, and let

$$I_1 = \int_0^\delta u(1-u^2)^n du, \qquad I_2 = \int_\delta^1 u(1-u^2)^n du,$$

so that $|P_n(x) - f(x)| \leq 2\lambda(I_1 + I_2)/J_n$. In I_1, $u \leq 1/n^{1/2}$, and

$$I_1 \leq \int_0^\delta n^{-1/2}(1-u^2)^n du \leq \int_0^1 n^{-1/2}(1-u^2)^n du = \tfrac{1}{2} n^{-1/2} J_n, \quad 2I_1/J_n \leq 1/n^{1/2}.$$

In I_2, $1/u \leq n^{1/2}$ and $u = u^2/u \leq n^{1/2} u^2$, so that by application of the previous reckoning with J_n'

$$I_2 \leq n^{1/2} \int_\delta^1 u^2(1-u^2)^n du \leq n^{1/2} \int_0^1 u^2(1-u^2)^n du = \tfrac{1}{2} n^{1/2} J_n' \leq \tfrac{1}{4} n^{1/2} J_n/(n+1),$$

$$2I_2/J_n \leq \tfrac{1}{2} n^{1/2}/(n+1) < 1/n^{1/2}.$$

So it appears not merely that $|P_n(x) - f(x)|$ is less than a quantity independent

---

* See e.g. Pólya and Szegö, *Aufgaben und Lehrsätze aus der Analysis*, vol. I, Berlin, 1925, pp. 63–65.

of x which approaches zero as n becomes infinite, but also, more specifically,* that it is less than a quantity of the order of $1/n^{1/2}$. This is proved, to be sure, only for a function satisfying (4). But (4) is satisfied in (a, b) by any continuous function whose graph is a broken line made up of a finite number of straight line segments with finite slope, and as any function whatever that is continuous for $a \leq x \leq b$ can be approximated with any desired accuracy by a function of this special type, the general conclusion of Weierstrass's theorem follows immediately.

The method can be applied equally well to the proof of Weierstrass's theorem on the trigonometric approximation of a periodic continuous function, by the use of de la Vallée Poussin's integral†

$$\frac{1}{H_n}\int_{-\pi}^{\pi} f(t) \cos^{2n}\left(\frac{t-x}{2}\right) dt, \qquad H_n = \int_{-\pi}^{\pi} \cos^{2n}\frac{u}{2} du,$$

and to the proof of corresponding theorems on the approximate representation of continuous functions of more than one variable.

BIBLIOGRAPHIC ENTRIES: OTHER POLYNOMIALS

The references below are to the AMERICAN MATHEMATICAL MONTHLY.

1. Hedrick, E. R., The significance of Weierstrass's theorem, vol. 20, p. 211.

    Discusses the contrast between approximating a function by Taylor polynomials and approximating by the polynomials in Weierstrass's theorem.

2. Olds, C. D., The best polynomial approximation of functions, vol. 57, p. 617.

---

* Cf. C. de la Vallée Poussin, *Sur l' approximation des fonctions d' une variable réelle et de leurs dérivées par des polynômes et des suites limitées de Fourier*, Bulletins de la Classe des Sciences, Académie Royale de Belgique, 1908, pp. 193–254; pp. 221–224.

† See de la Vallée Poussin, loc. cit., pp. 228–230.

8

MAXIMA AND MINIMA

MAXIMUM PARCELS UNDER THE NEW PARCEL POST LAW*

W. H. BUSSEY

The size of a parcel that can be mailed under the rules of the Parcel Post Law is limited in three ways. It must be not more than three feet and six inches in length. It must not weigh more than eleven pounds. The sum of its length and girth must not exceed six feet. The girth is measured around a cross section perpendicular to the length; and since the greatest girth is the one measured, all such cross sections should be equal to the largest of them if it is desired that the volume be a maximum. Furthermore, since the circle has a larger area for a given perimeter than any other closed curve, the parcel should be in the form of a right circular cylinder if the volume is to be a maximum. Let R denote the radius of the base of the cylinder, V its volume and x its altitude or length. Then $V = \pi R^2 x$, and $2\pi R + x = 6$. From these two equations it is easily found by differential calculus that for a maximum value of V the length must be two feet and the girth four feet. The maximum value of V is 2.546 cubic feet. This is the largest mailable volume.

If the parcel to be mailed has its cross sections all equal and of a specified shape which is not circular, it is still true that the length should be two feet and the girth four feet for maximum volume. To prove this, let P denote the perimeter of a cross section and A its area. As P changes in value, the shape of the cross section remaining constant, A varies as P^2. That is, $A = kP^2$. Then if L denotes the length of the parcel, and V its volume, $V = kP^2L$ and $L + P = 6$. From these two equations the value of P that makes V a maximum is found to be 4. The corresponding value of L is 2.

Since a square has a larger area than any other rectangle having the same perimeter, the largest mailable box in the form of a rectangular parallelopiped is therefore one whose dimensions are $1 \times 1 \times 2$ feet. Its volume is 2 cubic feet. The largest mailable cube is one whose edge is 14.4 inches long. Its volume is 1.728 cubic feet. The largest mailable sphere has a diameter of 17.38 inches. Its volume is 1.592 cubic feet.

A cube whose edge is 15 inches is not mailable because the sum of its length and girth is 75 inches. But a right circular cylinder, whose length is 24 inches and whose diameter is 15.27 inches, is mailable, because the sum of its length and girth is only 72 inches, in spite of the fact that its volume exceeds that of the cube

---

* From AMERICAN MATHEMATICAL MONTHLY, vol. 20 (1913), pp. 58–59.

by more than 1,000 cubic inches, and the fact that it would take up more room in a mail car. Thus the use of the sum of length and girth as a measure of size leads to the absurd result that sometimes the smaller of two parcels will be rejected by the postal authorities because it is too large, while the larger of the two will be accepted as being small enough. In spite of this absurd result, the rule can be amply justified on the ground of the ease with which it can be applied. With a six-foot tape, the postal clerk can tell in a moment whether or not a parcel is too large to be mailed. The man who is waiting his turn at the post office window will be glad that the clerk does not have to compute the cubical content of every parcel presented.

THE SHORTEST CIRCULAR PATH FROM A POINT TO A LINE*

ARNOLD DRESDEN, University of Wisconsin

Introduction. In connection with the study of a certain type of problem in the calculus of variations, I was led to consider the following question:

"Let there be given a line $x=b$ and a system of circles tangent at the origin to a line $y=mx\,(m>0)$ (see Fig. 1). We take on each of these circles the arc from the origin to the nearer of the two points of intersection with the line $x=b$ and inquire which among the circles of the system furnishes for this arc the minimum value; that is, we inquire which among the circular arcs passing through the origin with slope m will furnish the shortest path from the origin to the line $x=b$."

The intuitive reply to this question was that the shortest circular path is furnished by the circle which cuts the line $x=b$ at right angles. This reply proved to be incorrect. It seemed therefore worth while to publish a discussion of this question which incidentally yielded other points of interest.

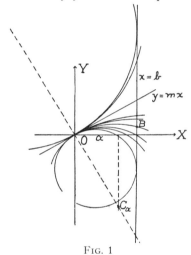

FIG. 1

1. The circles of the system are given by the equation

$$(x - \alpha)^2 + \left(y + \frac{\alpha}{m}\right)^2 = \frac{\alpha^2(m^2 + 1)}{m^2}, \tag{1}$$

the parameter α being the x-coordinate of the center. The points of intersection of this circle with the line $x=b$ are real if $\alpha \geq b/(1+\mu)$ or $\alpha \leq b/(1-\mu)$, where

$$\mu = \frac{\sqrt{1+m^2}}{m} = \operatorname{cosec} \phi,$$

---

* From AMERICAN MATHEMATICAL MONTHLY, vol. 28 (1921), pp. 437–439.

MAXIMA AND MINIMA

($\phi < \pi/2$) being the inclination of the line $y = mx$. The length of the arc OB then becomes

$$L(\alpha) = \int_0^b \sqrt{1+y'^2}\, dx = \alpha\mu \int_0^b \frac{dx}{\sqrt{\alpha^2\mu^2 - (x-\alpha)^2}} = \alpha\mu \left\{ \arcsin \frac{b-\alpha}{\alpha\mu} + \arcsin \frac{1}{\mu} \right\},$$

this function being defined only for values of α greater than $b/(1+\mu)$ or less than $b/(1-\mu)$. The question we raised is that of determining the value of α for which $L(\alpha)$ is a minimum.

2. We introduce as a new variable the angle θ, defined by the relation

$$\theta = \arcsin \frac{b-a}{\alpha\mu} \quad \text{or} \quad \alpha = \frac{b}{1 + \mu \sin \theta},$$

which transforms* the ranges $\alpha \geq b/(1+\mu)$ and $\alpha \leq b/(1-\mu)$ into the single continuous range $-(\pi/2) \leq \theta \leq (\pi/2)$. It is shown without difficulty that this angle θ represents the supplement of, or the negative, the inclination of the tangent to the circle C_a at its point of intersection with the line $x = b$.

We find then

$$L(\alpha) = L\left(\frac{b}{1+\mu \sin \theta}\right) = L_1(\theta) = \frac{b(\theta + \phi)}{\sin \theta + \sin \phi},$$

and hence

$$L'(\alpha) = L_1'(\theta) \frac{d\theta}{d\alpha} = -\frac{\sin \theta + \sin \phi - (\theta + \phi)\cos \theta}{\sin \phi \cos \theta}.$$

The maximum and minimum values of the function $L_1(\theta)$ will be found among the zeros of the function

$$F(\theta) \equiv \sin \theta + \sin \phi - (\theta + \phi)\cos \theta.$$

While for our purpose we are concerned only with the values of $F(\theta)$ on the range $(-\pi/2, \pi/2)$, we discuss the function in its entirety.

We find $F'(\theta) = (\theta + \phi)\sin \theta$, so that

$$F'(\theta) = 0, \quad \text{for} \quad \theta = -\phi, \quad \text{and for} \quad \theta = n\pi,$$

while $F'(\theta) > 0$ on the intervals

$$\cdots (-5\pi, -4\pi), (-3\pi, -2\pi), (-\pi, -\phi), (0, \pi), (2\pi, 3\pi), (4\pi, 5\pi) \cdots,$$

and $F'(\theta) < 0$ on the intervals

$$\cdots (-4\pi, -3\pi), (-2\pi, -\pi), (-\phi, 0), (\pi, 2\pi), (3\pi, 4\pi) \cdots.$$

---

* No value of α corresponds to $\theta = -\arcsin 1/\mu$; but $\lim_{\alpha \to \infty} \theta = -\arcsin 1/\mu$.

Moreover since

$$F(0) = \sin\phi - \phi < 0, \quad F(-\phi) = 0, \quad F(\phi) = 2(\sin\phi - \phi\cos\phi) > 0,$$

we obtain for the graph of $F(\theta)$ a curve as sketched in Fig. 2. We conclude that there exists a value θ_1 such that $0 < \theta_1 < \phi$ and such that $F(\theta_1) = 0$.

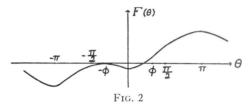

Fig. 2

Corresponding to the values $\theta = -\pi/2, -\phi, 0, \theta_1, \phi, \pi/2$, we have the values $\alpha = b/(1-\mu), \pm\infty, b, a_1, b/2, b/(1+\mu)$, so that we can now obtain the graph of the function $L(a)$ as in Fig. 3. The values $\alpha = b/(1-\mu)$ and $\alpha = b/(1+\mu)$ yield the circles of the system that are tangent to the line $x = b$ while $\alpha = \pm\infty$ yields the straight line $y = mx$.

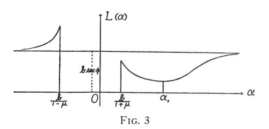

Fig. 3

3. It is readily verified that

$$L\left(\frac{b}{1-\mu}\right) > L(\infty)$$

and

$$L\left(\frac{b}{1-\mu}\right) > L\left(\frac{b}{1+\mu}\right);$$

but that $L[b/(1+\mu)]$ may be greater or less than $L(\infty)$. There is exactly one value of α, viz., α_1, for which $L(\alpha)$ has a minimum; this value α_1 lies between b and $b/2$, so that the shortest circular path from the origin to the line $x = b$ which departs from the origin with slope m lies between the circle which cuts this line perpendicularly and the one which passes through the point $(b, 0)$.

We have seen that $\alpha_1 = b/(1+\mu \sin \theta_1)$ while θ_1 is the unique root of the equation $\sin \theta + \sin \phi = (\theta+\phi) \cos \theta = 0$ on the interval $(0, \pi/2)$. Hence, θ_1 is determined as a function of ϕ by the relation

$$\frac{\sin \theta_1 + \sin \phi}{\theta_1 + \phi} = \cos \theta_1.$$

From this it follows that $\lim_{\phi \to 0} \theta_1 = 0$; that is, among the circles tangent to the X-axis at the origin, the straight line furnishes the shortest path from the origin to the line $x = b$, which shows that the result obtained above is in accord with known facts.

AN EXAMPLE IN MAXIMA AND MINIMA*

ELIJAH SWIFT, University of Vermont

It sometimes puzzles the beginning student to find functions of two variables, x and y, which possess a minimum in x alone and also in y alone without possessing one in both variables. Thus, if we set $x=0$, the expression $x^2-3xy+y^2$ (which then becomes a function of y alone, namely y^2) takes on a smaller value, namely 0, for $y=0$ than for any other value of y and the same thing is true for x when we set $y=0$; but the original expression may take on negative values for pairs of values of x and y arbitrarily near the origin, e. g. for $x=\epsilon$, $y=\epsilon$, and consequently does not possess a minimum at $(0,0)$.

In this connection it may be interesting to exhibit a function of n variables which has the property that it does not possess a minimum in all these, but is such that if we set any one of them equal to zero it will possess a minimum in the remaining $n-1$. Such a function is

$$(2x_1 + x_2)^2 + (2x_1 + x_3)^2 + \cdots + (2x_1 + x_n)^2 - x_1^2.$$

A moment's inspection shows that if we set any variable, say x_n, equal to zero, the above expression will have a positive value for all values of the remaining variables not all zero, while it vanishes if all the variables vanish. On the other hand if we set $x_1=\epsilon$, $x_2=x_3=\cdots=x_n=-2\epsilon$, the function is negative and consequently does not have a minimum for the values $x_1=x_2\cdots=x_n=0$.

---

* From AMERICAN MATHEMATICAL MONTHLY, vol. 34 (1927), p. 263.

A PROBLEM IN MAXIMA AND MINIMA*

ROGER A. JOHNSON, Hunter College of the City of New York

PROBLEM. *A frustum of a right circular cone with a fixed altitude h and a fixed lower base of radius r has a variable upper base of radius x. Consider the area of the lateral surface as a function of x in the interval from $x=0$ (cone) to $x=r$ (cylinder). In particular, for what values of x in this interval is the area a maximum or minimum?*

This problem is noteworthy for several reasons. It is suggested by a standard problem of the calculus of variations, that of the surface of revolution of minimum area.† It is a typical problem in maxima and minima, of the class ordinarily solved in a first course in the calculus, but the author has not seen it proposed in any calculus text. Finally, the nature of the solution is by no means intuitionally evident.

The reader is invited at this point to consider the problem, and before turning the page to formulate intuitively his own conclusions as to its solution.

The straightforward solution of the problem by the usual methods of the calculus should not tax the powers of the average undergraduate. The area in question is

$$(1) \qquad S = \pi(r + x)[h^2 + (r - x)^2]^{1/2},$$

the last factor representing the slant height g. In particular, the areas of the cylinder $(x=r)$ and of the cone $(x=0)$ are, respectively,

$$S_c = 2\pi rh, \qquad S_0 = \pi r[h^2 + r^2]^{1/2}.$$

Differentiating (1), we have

$$(2) \qquad \frac{dS}{dx} = \pi\left[g - \frac{(r+x)(r-x)}{g}\right] = \pi\frac{g^2 - (r^2 - x^2)}{g} = \frac{\pi(x^2 - rx + \tfrac{1}{2}h^2)}{2g}.$$

Now this derivative is positive when $x=0$, and also when $x=r$, thus tempting the unwary to the inference that S increases monotonically from $x=0$ to $x=r$, and that within this interval there is no maximum nor minimum. A second thought, however, to the effect that when h is small as compared with r the area of the cylinder is less than that of the cone, impresses on us that the situation is not so simple. We therefore proceed to transform the derivative,

$$(3) \qquad \frac{dS}{dx} = \frac{\pi}{2g}\left[\left(x - \frac{r}{2}\right)^2 + \frac{1}{4}(2h^2 - r^2)\right]$$

---

* From AMERICAN MATHEMATICAL MONTHLY, vol. 35 (1928), pp. 187–188.

† For if we consider the surfaces generated by revolving about an axis the various curves joining two fixed points A and B, where AB is parallel to the axis, we naturally think first of the cylinder generated by AB itself, and secondly of the double frustum generated by a broken line ACB, with $AC=CB$.

from which it appears that if $2h^2 > r^2$, the derivative is always positive and S does indeed increase monotonically as the frustum changes from conical to cylindrical form. But if $r^2 > 2h^2$, we have the surprising result that the derivative vanishes twice between $x = 0$ and $x = r$; in other words, under these conditions, the area has both a maximum and a minimum within the interval. As the upper base of the frustum increases, the lateral area first increases to a maximum, then decreases to a minimum, then increases to that of the cylinder. The maximum and the minimum are given by values of x equidistant from $r/2$:

$$x_1 = \tfrac{1}{2}(r - (r^2 - 2h^2)^{1/2}), \quad x_2 = \tfrac{1}{2}(r + (r^2 - 2h^2)^{1/2}).$$

It is of interest to inquire further whether the relative maximum at x_1, or the cylinder S_c, is greater; and similarly to compare the relative minimum at x_2 with the cone S_0. Without much difficulty the following results are established:

I. If $r < h\sqrt{2}$, the area increases monotonically as x increases from 0 to r. If $r = h\sqrt{2}$, the area is stationary when $x = \tfrac{1}{2}r$.

II. If $r > h\sqrt{2}$, the area increases from S_0 to a maximum S_1 at x_1, then decreases to a minimum S_2 at x_2, then increases to the final value S_c when $x = r$.

(a) The area of the cone S_0 is less than that of the relative minimum frustum at x_2, provided that $r/h < \tfrac{1}{2}\sqrt{(5 + 3\sqrt{3})} = 1.596$. If the ratio of r to h exceeds this value, then x_2 gives the minimum area in the whole interval.

(b) Similarly the cylinder S_c is greater than the relative maximum at x_1, provided $r/h < \tfrac{1}{8}\sqrt{(11 + 5\sqrt{5})} = 1.665$; but when the ratio exceeds this value, x_1 furnishes an absolute maximum.

(c) When r/h exceeds $\sqrt{3}$, the area of the cone S_0 exceeds that of the cylinder S_c.

TWO RECTANGLES IN A QUARTER-CIRCLE*

B. M. STEWART, Michigan State College

1. The problem. Consider the rectangle A, bounded by $x=0$, $x=\cos a$, $y=0$, $y=\sin a$, and of area $A=\cos a \sin a$; and the rectangle B, bounded by $x=0$, $x=\cos b$, $y=\sin a$, $y=\sin b$, and of area $B=\cos b (\sin b - \sin a)$; the angles, a and b, being subject to the restriction $0<a<b<90°$. (See Figure 1.) The problem is to maximize the function $z=A+B$, subject to the auxiliary condition $A=B$.

2. The interest of the problem. A substitution shows $z=2A=\sin 2a$, so that at first sight the solution seems to be $a=45°$, but the restriction on the companion angle b—which must also be in the first quadrant—makes this solution not acceptable. Here then is a situation of interest to students of the calculus, for the ordinary method of finding a maximum has failed because of a restriction on the range of the variable.

3. The source of the problem. Furthermore, the restriction is not artificial, for this problem arises in electrical engineering in the design of a winding-core with the figure described above representing one-fourth of the total cross section. (See Figure 2.)

4. General methods of solution. The student with broad interests will want to investigate the general methods for solving problems of the type here encountered—namely, to maximize the function $z=f(a, b)$ subject to the auxiliary condition $g(a, b)=0$. These general methods are presented in advanced calculus texts under the topics of partial differentiation, or differential geometry, or Lagrange's method of multipliers. Even with these methods caution must be used if there are restrictions imposed on the range of the variables.

5. A solution by elementary calculus. But this problem is of a special character. It is clear from the figure that if angle a is between $45°$ and $90°$, then area B cannot be as great as area A. Since the function $z=\sin 2a$ is increasing for $0<a<45°$, the problem is to find the greatest a in the range $0<a<45°$ for which the equation $A=B$ has a solution b satisfying $0<a<b<90°$. Conceivably this may be solved by treating the relation $A=B$ as in elementary calculus for a maximum of a—namely, by setting $da/db=0$. The following equations are obtained:

(1) $\quad A = B; \quad \cos a \sin a = \cos b \sin b - \cos b \sin a;$

(2) $\quad da/db = 0; \quad \cos^2 b - \sin^2 b + \sin a \sin b = 0.$

By squaring and by using simple trigonometric identities it is possible to elimi-

---

* From AMERICAN MATHEMATICAL MONTHLY, vol. 52 (1945), pp. 92–94.

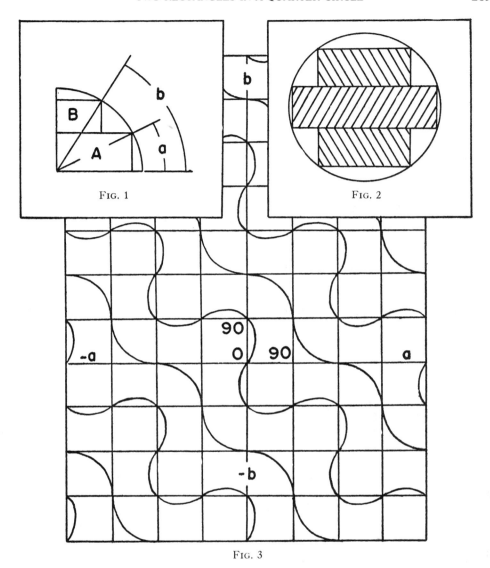

FIG. 1

FIG. 2

FIG. 3

nate b from equations (1) and (2) and to obtain the following equation in which $Q = \sin^2 a$:

(3) $$(Q - 1)(15Q^3 + 3Q^2 + 10Q - 1) = 0.$$

From equation (3) one and only one value of a, with an acceptable companion b,

can be obtained. The approximate solution of the problem is as follows:
$$a = 18° \ 2' \ 27'', \qquad b = 52° \ 4' \ 10''.$$

The meaning of this solution is clarified by constructing the graph of the relation $A = B$ (see Figure 3), a task which is much simplified by noticing the symmetry in the points (0, 0) and (90°, 90°) and the symmetry in the line $b = a + 90°$.

6. Another solution by elementary calculus. The following method is indirect, yet worthy of a brief description. For each fixed choice of angle a there can be found by solving $dB/db = 0$ just one value of b such that $0 < a < b < 90°$ and such that B has a maximum value, say $M(a)$, so designated because it depends upon a. It is evident from the figure that for small values of a the value of $M(a)$ exceeds A, hence it is possible to solve the equation $A = B$; whereas for large values of a the value of $M(a)$ is less than A, hence it is impossible to solve $A = B$. The equation $M(a) = A$ determines the boundary between these cases and determines the largest value of a for which the equation $A = B$ has a solution with acceptable values of a and b. The equation $M(a) = A$ is equivalent to equation (3). Since the solution $a = 18° \ 2' \ 27''$ is in the range $0 < a < 45°$ where z is increasing, this maximum a serves to maximize z under the auxiliary condition $A = B$.

7. A solution without calculus. The maximum a with a companion b satisfying $A = B$ and $0 < a < b < 90°$ may also be determined by using the theory of equations.

Consider the rectangles in the figure. For small angles a there are two choices of b such that $A = B$. For large angles a there are no choices of b. Between these cases is a certain angle a for which there is one choice of b. This situation may be studied algebraically by replacing the equation (1) after some rearrangement, squaring, and use of identities by the following equation which may be described as a reduced quartic in the variable $\cos b$ with coefficients involving $\cos a$:

(4) $\cos^4 b - \cos^2 a \cos^2 b + 2(\cos a - \cos^3 a) \cos b + (\cos^2 a - \cos^4 a) = 0.$

First, Descartes' rule of signs shows there are two or no real positive roots, confirming the geometric intuition used above. Secondly, the inbetween case of two real and positive but identical roots can be determined by the vanishing of the discriminant of the quartic. The computation of the discriminant for equation (4) is not an easy matter, but after considerable simplification the discriminant may be written in the following way with $P = \cos^2 a$:

$$- 16P^2(P - 1)(15P^3 - 48P^2 + 61P - 27).$$

Under the substitution $P = \cos^2 a = 1 - \sin^2 a = 1 - Q$ the discriminant takes the form which follows:

$$- 16Q(1 - Q)^2(15Q^3 + 3Q^2 + 10Q - 1).$$

Setting the discriminant equal to zero leads to but one acceptable solution, in agreement with that obtained from equation (3).

8. A trigonometric oddity. In the equation $A = B$ one of the two solutions for b, when $a = 18°$, is $b = 54°$, exactly! The proof of this fact is an interesting exercise in trigonometry. The second solution of $A = B$, when $a = 18°$, is given by $b = 50° \ 7'$, approximately.

MINIMAL TANGENTS

IRVING KAPLANSKY, New York City

A popular problem in elementary calculus texts is to find the shortest tangent to a curve cut off between the axes, or the tangent that cuts off the triangle of smallest area, *etc.* The algebra is usually found to be fairly heavy, unlike the gratifyingly simple answers. It is perhaps interesting to solve once for all a general problem of this kind.

At the point (x, y) of a curve let the tangent be drawn; its x- and y-intercepts, say u and v, are given by

$$u = (xy' - y)/y', \qquad v = y - xy'.$$

We have $u' = yy''/(y')^2$, $v' = -xy''$. Suppose $z = f(u, v)$ is to be minimized. We set

$$z' = f_u u' + f_v v' = y''[f_u y/(y')^2 - f_v x]$$

equal to zero. The root $y'' = 0$ corresponds to a point of inflection which yields an uninteresting extremum. The desired solution is therefore given by

(1) $$(y')^2 f_v/f_u = y/x.$$

Examples. 1. If $z = u^n + v^n$, then $f_v/f_u = v^{n-1}/u^{n-1} = (-y')^{n-1}$ and (1) becomes

(2) $$(-y')^{n+1} = y/x.$$

The case $n = 2$ corresponds to minimizing the length of the tangent, the case $n = 1$ to minimizing the sum of its intercepts.

2. If $z = uv$, then $f_v/f_u = u/v = -1/y'$, yielding $-y' = y/x$. This is the problem of minimizing the area of the triangle cut off by the tangent, and fits in as the case $n = 0$ of (2). The answer has the geometrical interpretation that the triangle formed by the origin, the point of tangency, and the x-intercept is isosceles.

3. A further variant is to seek a curve for which z is constant. This is given by solving (1) as a differential equation. The solution of (2) is

$$xy =. c \qquad (n = 0),$$
$$x^{n/(n+1)} + y^{n/(n+1)} = a^{n/(n+1)} \qquad (n > 0).$$

We thus establish well known properties of the rectangular hyperbola, the parabola ($n = 1$), and the 4-cusped hypocycloid ($n = 2$).

---

* From AMERICAN MATHEMATICAL MONTHLY, vol. 52 (1945), p. 439.

A RIGOROUS TREATMENT OF THE FIRST MAXIMUM PROBLEM IN THE CALCULUS*

J. L. WALSH, Harvard University

Even some of the best calculus texts fail to set forth adequately the logic underlying the usual maximum and minimum problems, studied subsequently to curve-plotting. The present note outlines the treatment the writer has used for a number of years, in a course primarily for freshmen and sophomores.

PROBLEM. *A piece of sheet tin three feet square is to be made into a rectangular box open at the top, by cutting out equal squares from the corners and bending up the sides of the resulting piece parallel with the edges. Among all such boxes, to find the box of greatest volume.*

We consider an arbitrary box that can be made in this way. Let x denote the length in feet of the side of the square corners cut out, so that x is also the height of the box. The volume V of the box is then (in cubic feet)

$$(1) \qquad V = V(x) = x(3 - 2x)^2.$$

We study V as a function of x in the *closed* interval $0 \leq x \leq 3/2$. When x is very small and positive, so also is the volume V, as appears both by (1) and by considering the shallow box formed; we have $V(0) = 0$. When x is smaller than but near $3/2$, the volume V is also very small and positive as is seen both by (1) and by considering the small area of the base of the corresponding box, we have $V(3/2) = 0$. Between these two values $x = 0$ and $x = 3/2$, the continuous function V is always positive and hence possesses a maximum; the maximum V is characterized by the vanishing of $D_x V$ for the corresponding value of x. We find

$$D_x V = x \cdot 2(3 - 2x)(-2) + (3 - 2x)^2 = (3 - 2x)(3 - 6x).$$

Between the two values 0 and $3/2$ the only value of x for which $D_x V$ vanishes is $x = 1/2$, which therefore yields the box of maximum volume.

The logic underlying this formal work can be analyzed as follows:

1. The function $V(x)$ is continuous in the closed interval $0 \leq x \leq 3/2$, hence possesses a maximum† there. [Any function continuous in a closed finite interval possesses a maximum there.]

2. The function V vanishes for $x = 0$ and $x = 3/2$ and is positive between those values, so the maximum occurs at some point or points in the *interior* $0 < x < 3/2$ of the original interval.

3. At this maximum of $V(x)$ we must have $D_x V = 0$. [If $D_x V > 0$, the function V increases as x increases; if $D_x V < 0$, the function V decreases as x increases and increases as x decreases; in either case V has no maximum.]

---

* From AMERICAN MATHEMATICAL MONTHLY, vol. 54 (1947), pp. 35–36.

† To be more explicit, $V(x)$ possesses a (weak) absolute maximum, which is necessarily also a (weak) relative maximum.

4. Among the values $0 < x < 3/2$ we have $D_x V = 0$ if and only if $x = 1/2$.
5. Hence $x = 1/2$ yields the maximum value of V.

It will be noticed that $x = 1/2$ cannot yield a minimum of V, nor a point of inflection. Moreover, the function $V(x)$ possesses a *unique* maximum in the interval $0 \leq x \leq 3/2$; there is no choice here among several relative maxima.

The logic as given is complete (although the theorem quoted in 1 can hardly be proved at this stage), and will satisfy the most critical student even at any stage of his later career. This logical analysis, with suitable modifications, applies to all maximum-minimum problems in one variable that the student is likely to meet.

END-POINT MAXIMA AND MINIMA*

C. O. OAKLEY, Haverford College

It sometimes happens that a very elementary maximum-minimum problem will lead to seemingly impossible or even foolish results. The following examples are of this category and belong to the type commonly known as *end-point maxima*. They are of such simple nature that they may be readily assigned to a beginning class in the calculus.† Although not original, these two particular problems do not seem to be widely known and should prove to be both interesting and instructive to teachers and students alike.

PROBLEM I. *Find the position of the point $P(x, y)$, on an ellipse, such that the distance to a focus F is a maximum (minimum).*

Let the ellipse have the equation $b^2x^2 + a^2y^2 = a^2b^2$. Then the distance PF is a continuous function of x in the interval $(-a, a)$ and hence it has both a maximum and a minimum value. (Any function which is continuous in a closed finite interval possesses both a maximum value and a minimum value in that interval.) Now the fortunate student, bright or dull, might begin to solve this problem as follows:

SOLUTION 1. Let the ellipse be represented parametrically by $x = a \cos \theta$, $y = b \sin \theta$ and let F be the right hand focus $(c, 0)$. Then

$$L \equiv \overline{PF}^2 = (a \cos \theta - c)^2 + b^2 \sin^2 \theta,$$

and

$$\frac{dL}{d\theta} = 2(a \cos \theta - c)(-a \sin \theta) + 2b^2 \sin \theta \cos \theta = 0,$$

from which it follows that $\sin \theta = 0$, or $\cos \theta = a/c$. The solutions of $\sin \theta = 0$ yield the answers while $\cos \theta$ cannot equal a/c since $a/c > 1$.

But even an able student might attack the problem differently—and come to grief.

SOLUTION 2. Now $PF = a - ex$. It's derivative is $-e \neq 0$. And yet $a - ex$ is obviously least when $x = a$ and greatest when $x = -a$, if the condition $-a \leq x \leq a$ is kept in mind. Even though the square of the distance is used, $L \equiv PF^2 = (a - ex)^2$, the unwary student is still in trouble for $dL/dx = -2e(a - ex) = 0$ yields $x = a/e$, a point outside of the ellipse (on the directrix, indeed). The function L *does* have a (relative) minimum at $x = a/e$ (what about the maximum?) but there is no (real) y-coordinate for the corresponding point P on the ellipse. Whether PF or $\overline{PF}^2$ is used, the end-points of the interval determine the (absolute) maximum and the (absolute) minimum values wanted.

---

*From AMERICAN MATHEMATICAL MONTHLY, vol. 54 (1947), pp. 407–409.

†See J. L. Walsh, A Rigorous Treatment of the First Maximum Problem in the Calculus, this MONTHLY, vol. 54, (1947), pp. 35–36. [*Ed. note:* Reprinted on pp. 299–301, this book.]

Even the method that most beginning students are likely to employ will result in the same difficulty:

SOLUTION 3.
$$L \equiv \overline{PF}^2 = (x-c)^2 + y^2,$$
$$= (x-c)^2 + \frac{b^2}{a^2}(a^2 - x^2);$$
$$\frac{dL}{dx} = 2(x-c) - 2\frac{b^2}{a^2}x = 0,$$

or again,
$$x = a^2/c = a/e.$$

PROBLEM II. *A man is in a boat at P one mile from the nearest point A on shore. He wishes to go to B which is farther down the shore M miles from A. If he can row r miles an hour and walk w miles an hour, toward what point C should he row in order to reach B in least time?*

Usually this problem is given with simple numerical values that present no difficulty. The shore line is assumed to be straight; let $x = AC$. Then the total time t that it takes to go from A to B is readily determined as

$$t = \frac{1}{r}(1+x^2)^{1/2} + \frac{1}{w}(M-x),$$

whence

$$\frac{dt}{dx} = \frac{x}{r}(1+x^2)^{-1/2} - \frac{1}{w} = 0,$$

and

$$x = \frac{r}{\sqrt{w^2 - r^2}}.$$

The first astonishing thing about this "answer" is that it is independent of the distance M. If $r < w$ and if $r < M\sqrt{w^2 - r^2}$, routine testing of the derivative will show that a minimum time is attained. But if $r < w$ and $r > M\sqrt{w^2 - r^2}$, then the derivative vanishes for a value of x that lies outside of the interval AB and the problem has an end-point minimum, that is, the man should row directly toward B. Finally if $r \geq w$, the man should again row directly to B.

Plotting the graphs of the functions involved in these problems is a great aid in clarifying the points at issue.

DISCUSSION. Problems involving end-point maxima and minima usually arise only when the independent variable is restricted to a certain interval in order to insure reality of the function or for certain physical reasons such as the avoidance of negative values of distance or time. In Problem I we found that the answers were either relative maxima and minima or end-point maxima and minima according to the particular choice of the independent variable. This is indeed a general situation: extreme values of the one type can be converted into those of the other type by an appropriate choice of a parameter or, what amounts to the same thing, by an appropriate transformation on the independent variable. Indeed let the function to be maximized be $y=f(x)$ and let the range of x be (a, b). Then put $x = a \cos^2 \theta + b \sin^2 \theta$. This transformation restricts x to the interval (a, b) but permits the new independent variable θ to vary from $-\infty$ to $+\infty$. It is readily seen that now $dy/d\theta = (dy/dx)2(b-a) \sin \theta \cos \theta$ which vanishes at $\theta = 0$ and $\pi/2$, namely, at the points $x = a$ and $x = b$, as well as at any relative maxima and minima inside the interval (a, b). Applying this transformation to the above exercises will convert them into the usual relative maxima and minima problems. For example in Problem II if we set $x = M \sin^2 \theta$, we find $dt/d\theta$ vanishes when $\sin \theta = 0$, $\cos \theta = 0$ and when $\sin^4 \theta = r^2/M^2(w^2 - r^2)$. An analysis of these values gives the results stated above.

If the interval is an infinite one, say (a, ∞), the transformation $x = a + \tan^2 \theta$ will change an end-point maximum (minimum) at $x = a$ into a relative one.

Editorial Note. Another example of an end-point maximum has been submitted by V. L. Klee, Jr. of the University of Viriginia. It is as follows:

PROBLEM. We are given a straight fence 100 feet long, and wish by adding 200 feet more to form a rectangular enclosure whose boundary contains the original fence. How shall this be done so as to enclose the greatest possible area?

Let x denote the length of new fence which is aligned with the original 100 feet of fence. Then the dimensions of the enclosure are $100 + x$ and $50 - x$, and its area $A(x)$ is $-x^2 - 50x + 5000$. The student may reason as follows: $D_x A = -2(x+25)$; hence we would choose $x = -25$ in order to enclose the maximum area. He will then be perplexed to note that this solution obviously does not satisfy the conditions of the problem. The correct solution, of course, is $x = 0$ which is an end-point maximum.

C. B. ALLENDOERFER

ONE-SIDED MAXIMA AND MINIMA*

J. D. MANCILL, University of Alabama

1. Introduction. Repeated inquiries from students and colleagues concerning the subject of this discussion, doubtless due to the incomplete treatment of it in the literature, have prompted this attempt to give a complete analysis of the problem of one-sided (unilateral) extremes for functions of a single variable. Almost without exception, the text books on the calculus restrict their discussion of maxima and minima of functions to the interior points of the range of definition of the function. The author knows of no text which treats fully sufficient conditions for an end-point of the range to be a maximum or a minimum. When the end-point extremes are mentioned they are frequently referred to as absolute extremes, since the usual definition of relative extreme excludes the end-points. But this is undesirable since the end-point extreme may not be absolute. It is better to state the definition of relative extreme so as to include the end-points, and develop necessary conditions and sufficient conditions that an end-point be a relative maximum or minimum. Then the procedure, even in elementary calculus, would be to test the end-points along with the zeros of the first derivative for relative extremes, and then determine the absolute extremes of the function from all the extremes on the closed range.

In a recent note Oakley* notes that the end-point extremes may be transformed into interior point extremes by a suitable transformation on the independent variable. Although this explains why a point may be an end-point maximum in one formulation of a problem and an interior maximum in another, such a transformation is not a practical way of handling end-point maxima. For first, it renders even the simple problems very complicated, especially when it comes to checking the sufficient conditions for a maximum or a minimum. Second, it is entirely unnecessary. This is due to the fact that an end-point is in general an extreme and one only needs to determine whether it is a relative maximum or a relative minimum. Some examples will illustrate these remarks.

2. Functions of one variable. We shall consider a single-valued function

$$y = f(x), \qquad a \leq x \leq b,$$

defined on the closed interval (a, b). The following definitions will be needed:

DEFINITION 1. $f(x)$ belongs to the class C of continuous functions at $x = x_1$, $a < x_1 < b$ if and only if $\lim_{x \to x_1} f(x) = f(x_1)$.

DEFINITION 2. $f(x)$ belongs to the class of functions C at the end-point a if and only if $\lim_{x \to a^+} f(x) = f(a)$ and at the end-point b if and only if $\lim_{x \to b^-} f(x) = f(b)$.

---

*From AMERICAN MATHEMATICAL MONTHLY, vol. 55 (1948), p. 311–315.

*End-point maxima and minima, this MONTHLY, vol. 54 (1947), pp. 407–409. [*Ed. note:* Reprinted on pp. 303–305, this book.]

DEFINITION 3. $f(x)$ belongs to the class of functions C^n if and only if $f^{(n)}(x)$ belongs to C.

DEFINITION 4. $f'_+(x_1) = \lim_{h \to 0^+}[f(x_1+h) - f(x_1)]/h$, and similarly for $f'_-(x_1)$.

DEFINITION 5. $f(x)$ belongs to the class of functions D' on (a, b) if and only if $f(x)$ belongs to C on (a, b) and $f'(x)$ belongs to C on (a, b) except at a finite number of points x_1 at which the derivatives $f'(x_1+)$ and $f'(x_1-)$, on the right and left of x_1 respectively, exist, where $f'(x_1+) = \lim_{x \to x_1^+} f'(x)$ and similarly for $f'(x_1-)$.

DEFINITION 6. $f(x)$ has an *absolute maximum* at a point x_1 of the interval (a, b) if and only if $f(x_1) \geq f(x)$ for all values of x in (a, b). A function $f(x)$ has a *relative maximum* at a point x_1 of (a, b) if and only if $f(x_1) \geq f(x)$ for all values of x on (a, b) in a certain neighborhood of x_1.

Obvious modifications of the inequalities are necessary for the definition of absolute and relative minima. A well-known property of continuous functions states that a function $f(x)$ which is continuous on the closed interval (a, b) takes its absolute maximum and minimum values at least once in that interval. Therefore, one must consider the end-points of the closed interval in determining the absolute extremes of the function on that interval.

The reader should distinguish between $f'_+(x_1)$ and $f'(x_1+)$. This is easily done by considering the function $x^2 \sin(1/x)$. In this case $f'_+(0) = 0$ but $f'(0+)$ does not exist.

The geometric meaning of Definition 5 is that the curve defined by the function $y = f(x)$ on the range (a, b) has a finite number of corners, or in other words, the curve is composed of a finite number of sub-arcs each of class C'.

THEOREM 1. *If a continuous function $f(x)$ assumes a relative maximum value at $x = x_1$ of the interval (a, b); and if $f'_-(x_1)$ exists, then $f'_-(x_1) \geq 0$; and if $f'_+(x_1)$ exists, then $f'_+(x_1) \leq 0$. The inequalities are reversed if x_1 is a minimum.*

This theorem follows immediately from Definition 3 and the inequalities

$$[f(x_1 + h) - f(x_1)]/h \geq 0 \qquad \text{for } h < 0,$$
$$[f(x_1 + h) - f(x_1)]/h \leq 0 \qquad \text{for } h > 0,$$

for the case when x_1 is a maximum.

COROLLARY (FERMAT'S THEOREM). *If a function $f(x)$ assumes a relative maximum or a minimum at $x = x_1$, $a < x_1 < b$ at which $f(x)$ is differentiable, then $f'(x_1) = 0$.*

The corollary follows from the fact that $f'(x_1)$ exists if and only if $f'_-(x_1)$ and $f'_+(x_1)$ exist and are equal. The end-points are excluded from the corollary since the left-hand derivative at $x = a$ and the right-hand derivative at $x = b$ are not considered in the theorem.

THEOREM 2. *A continuous function $f(x)$ assumes a relative maximum at $x = x_1$; x_1 interior to (a, b) if $f'_-(x_1) > 0$ and $f'_+(x_1) < 0$: at $x = a$ if $f'_+(a) < 0$; and at $x = b$ if*

$f'_-(b) > 0$. *Similar conditions with inequalities reversed are sufficient for a minimum.*

These conclusions follow immediately from Definition 4. The function $f(x) = 1 - |x|$, $-1 \leq x \leq 1$ affords a simple illustration of these results. In this example $f'_-(0) = 1$ and $f'_+(0) = -1$. Therefore, $f(x)$ has a relative maximum at $x = 0$ which is also the absolute maximum on that range.

THEOREM 3. *If $f(x)$ is of class C^n within (a, b), $f^{(n)}(a+)$ exists, and*

$$f^{(k)}(a) = 0, \qquad k = 1, \cdots, n-1,$$
$$f^{(n)}(a+) < 0,$$

then $f(x)$ has a relative maximum at $x = a$. Similar conditions with the inequalities reversed are sufficient for a minimum.

From Taylor's formula and the hypotheses of the Theorem, we have

$$f(x) - f(a) = (x - a)^n f^{(n)}(t)/n!, \qquad a < t < x.$$

The sign of the right member of this equation depends only upon the sign of $f^{(n)}(t)$, since $x > a$. Also, t approaches a with x and therefore, if x is sufficiently near a the sign of the right member is negative.

As an example consider the function $f(x) = -x^{3/2}$, $0 \leq x \leq 1$. It is easily seen that $f'(0) = 0$ and $f''(0+) = -\infty$. Therefore $f(x)$ has a relative maximum at $x = 0$ which is also the absolute maximum on that range.

THEOREM 4. *If $f(x)$ is of class C^n within (a, b), $f^{(n)}(b-)$ exists and*

$$f^{(k)}(b) = 0, \qquad k = 1, \cdots, n-1,$$
$$f^{(n)}(b-) < 0, \qquad n \text{ even},$$
$$f^{(n)}(b-) > 0, \qquad n \text{ odd},$$

then $f(x)$ has a relative maximum at $x = b$. Similar conditions with the inequalities reversed are sufficient for a minimum.

It follows from the hypotheses of the theorem and from Taylor's formula that

$$f(x) - f(b) = (x - b)^n f^{(n)}(t)/n!, \qquad x < t < b.$$

In this case the sign of the right member of this equation depends upon the sign of $f^{(n)}(t)$ and whether n is even or odd since $x < b$. Therefore, the conclusions of the theorem follow as in the proof of Theorem 3.

As an example consider the function $f(x) = -(-x)^{5/2}$, $-1 \leq x \leq 0$. It is easily seen that $f'(0) = f''(0) = 0$, and $f^{(3)}(0-) = +\infty$. Therefore, $f(x)$ has a relative maximum at $x = 0$ which is also the absolute maximum on that range.

We may say that $f(x)$ on (a, b) has a *left-hand maximum* at $x = x_1$ if $f(x_1) \geq f(x)$ for all values of x in a certain right hand neighborhood of x_1, for example in Theorem 3. Likewise, we may say $f(x)$ has a *right-hand maximum* at $x = x_1$ if

$f(x_1) \geq f(x)$ for all values of x in a certain left-hand neighborhood of x_1, for example in Theorem 4. Obviously then, if x_1 is interior to the interval (a, b) and $f(x)$ has a left-hand and a right-hand maximum at x_1, at which $f(x)$ is continuous, then $f(x)$ has a maximum at $x = x_1$. For example, the function

$$f(x) = -(-x)^{5/2}, \qquad -1 \leq x \leq 0,$$
$$= -x^{3/2}, \qquad 0 \leq x \leq 1,$$

has a maximum at $x = 0$. The following well-known theorem follows easily:

THEOREM 5. *If $f(x)$ is of class C^{2n} on (a, b), x_1 is interior to (a, b), and*

$$f^{(k)}(x_1) = 0, \qquad k = 1, \cdots, 2n - 1,$$
$$f^{(2n)}(x_1) < 0,$$

then $f(x)$ has a maximum at $x = x_1$. Similar conditions with inequalities reversed are sufficient for a minimum.

It may also be remarked here that if a continuous function $f(x)$ has a left-hand maximum and a right-hand minimum (or vice-versa) at x_1 interior to (a, b), then $f(x)$ has an inflection at x_1. If $f'_-(x_1) \neq f'_+(x_1)$, then $f(x)$ has an abrupt inflection at x_1. For example the function

$$f(x) = -(-x)^{5/2}, \qquad -1 \leq x \leq 0,$$
$$= x^{3/2}, \qquad 0 \leq x \leq 1,$$

has an inflection at $x = 0$.

3. **An application.** As a physical application, let us consider one of the problems treated by Oakley in his note already referred to:

A man is in a boat at P one mile from the nearest point A on shore. He wishes to go to B which is farther down the shore M miles from A. If he can row r miles an hour and walk w miles an hour, toward what point C should he row in order to reach B in least time?

Assume the shore line to be straight with the point B to the right of A and let $x = AC$. Then the time $t(x)$ that it takes to go from P to B is

$$t(x) = (1 + x^2)^{1/2}/r + (M - x)/w, \qquad 0 \leq x \leq M,$$

whence

$$t'(x) = x/r(1 + x^2)^{1/2} - 1/w,$$

and $t'(0) < 0$. Therefore, $t(x)$ has a relative maximum at $x = 0$ for all positive values of r, w, and M, and consequently the man would not row towards A. If $r \geq w$, then $t'(M) < 0$ and $t(x)$ has a relative minimum at $x = M$. This is the absolute minimum since in this case $t'(x) \neq 0$ on $0 \leq x \leq M$. Therefore, the man should row directly to B. If $r < w$, then $t'(w) = 0$ has the unique solution

$$x_C = r/(w^2 - r^2)^{1/2}.$$

It is easily shown that in this case $t(x_C) < t(M)$ if $0 < x_C < M$, and consequently the man should row to x_C.

There is still one other case which was not considered by Oakley, that is when $x_C = M$, or when $r = wM/(1+M^2)^{1/2} < w$. In this case $t'(M) = 0$ and $t''(M) > 0$. It follows from Theorem 4 that $t(x)$ has a relative minimum at $x = M$ and this is the absolute minimum since $t'(x)$ vanishes only at x_C. Therefore, the man should row directly to B in this case.

It is the author's firm conviction that if maxima and minima problems, even in elementary calculus, are analyzed as exemplified in these pages, the student will meet no difficult or astonishing "answers." This is due in part to the fact that the student will not be led to the erroneous conclusion that maxima and minima occur only at the vanishing of the first derivative.

FERMAT'S PRINCIPLE AND CERTAIN MINIMUM PROBLEMS*

A. V. BAEZ, Stanford University

The object of this note is to show the relation between Fermat's Principle and the end-point minimum problems discussed by Oakley.†

It is known that of all possible paths connecting two points A and B in space a light ray leaving A chooses the path which enables it to reach B in the least time. This is Fermat's Principle from which the laws of reflection and refraction follow. They can be used to solve certain minimum problems without differentiation. Before discussing Oakley's problems consider a simpler one.

A farmer has to walk from his house to a straight river, fill his pail and carry it to his barn in the least time. At what point should he fill his pail? The solution by geometrical optics is obvious. His path to and from the river must make equal angles with the normal to the river.

Suppose he always walks to the river at a speed v_1 and away from it at a speed v_2. Where then should he dip his pail? Snell's Law immediately suggests that his path of approach to the river should make an angle θ_1 and that away from the river an angle θ_2 with the normal so that

$$\frac{\sin \theta_1}{\sin \theta_2} = \frac{v_1}{v_2}.$$

The problem which follows was discussed by Oakley, and his notation is used, together with the above equation which expresses Snell's Law.

PROBLEM. *A man is in a boat at P one mile from the nearest point A on shore. He wishes to go to B which is farther down the shore M miles from A. If he can row r miles an hour and walk w miles an hour, toward what point C should he row in order to reach B in least time?*

Let $AC = x$. Snell's Law indicates that

$$\frac{\sin \theta_1}{\sin \theta_2} = \frac{r}{w}.$$

Use $\theta_2 = 90°$ and geometry to obtain

$$\frac{x}{\sqrt{1+x^2}} = \frac{r}{w}.$$

Solving for x we have

$$x = \frac{r}{\sqrt{w^2 - r^2}}$$

which Oakley obtains by calculus. The optical analysis makes clear why the result is independent of M (θ_1 is the "critical" angle). Optically it is also apparent that if $r < w$ and $r \geq M\sqrt{w^2 - r^2}$ or if $r \geq w$ (giving $\sin \theta_1 \geq 1$) the man should go straight to B.

---

*From AMERICAN MATHEMATICAL MONTHLY, vol. 55 (1948), p. 316.
†End-point maxima and minima, this MONTHLY, vol. 54 (1947), pp. 407–409. [*Ed. note:* Reprinted on pp. 303–305, this book.]

THE PROBLEM OF A NON-VANISHING GIRDER ROUNDING A CORNER*

NORMAN MILLER, Queen's University

In the usual problem of a girder being carried around a corner from one passageway to another, the girder is etherialized into a line segment (fig. 1). The problem gains in interest as well as in reality if we suppose the girder to have positive width. Moreover, the geometry to which it gives rise extends beyond the "practical" problem of the girder.

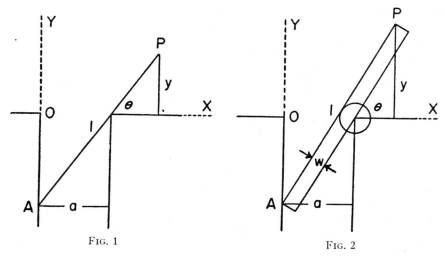

Fig. 1   Fig. 2

Denote by w and l the width and length of the girder, supposed square, and by a the width of the passage from which the girder is being moved. We take $w < a < l$. Then, with the notation of fig. 2, the problem is to find the maximum of y where

$$y = l \sin \theta - (a - w \csc \theta) \tan \theta$$
$$= l \sin \theta - a \tan \theta + w \sec \theta.$$

Since

$$dy/d\theta = l \cos \theta - a \sec^2 \theta + w \sec \theta \tan \theta,$$

the maximizing value of θ must satisfy

(1) $$l \cos^3 \theta - a + w \sin \theta = 0.$$

Attention to the graphs of the separate terms of this equation shows that it has exactly one root between 0 and $\pi/2$ and one between $-\pi/2$ and 0. If $w=0$, these two roots are equal in absolute value, but otherwise unequal.

---

* From AMERICAN MATHEMATICAL MONTHLY, vol. 56 (1949), pp. 177–179.

The root of (1) which lies between 0 and $\pi/2$ gives the maximum value of y and hence the minimum width of the horizontal passageway which will allow the girder to turn the corner.

The root of (1) between $-\pi/2$ and 0 has also some interest. The further discussion will be concerned not with the problem of a physical girder but with the curve which is the complete locus of the point P. This locus will be defined as follows: A variable straight line touches a fixed circle of radius w and intersects in a variable point A a fixed straight line which is at a distance a from the centre of the circle. The locus in question is that of a point on the moving line which is at a distance l from A, where $w < a < l$. In case $w = 0$ the locus is a Conchoid of Nicomedes which (in fig. 1) is symmetrical about OX. In case $w > 0$ the locus is a sort of distorted Conchoid, having two points of contact with the fixed circle and containing an unsymmetrical loop. The parametric equations of this locus, with the axes indicated in fig. 2, are

$$x = l \cos \theta, \qquad y = l \sin \theta - a \tan \theta + w \sec \theta,$$

which give the single quartic equation

$$(xy - wl)^2 = (x - a)^2(l^2 - x^2).$$

This equation becomes that of the Conchoid when $w = 0$.

When the point A (fig. 2) is on the negative y-axis, the angle θ is positive and when A moves up to the positive y-axis θ becomes negative. Thus all values of θ between $-\pi/2$ and $\pi/2$ apply to points on the curve and a critical point occurs on the loop for one positive value of θ and for one negative value (fig. 3). For $w > 0$ the abscissas of these points are different as well as the absolute values of their ordinates.

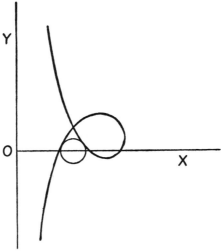

FIG. 3

Example. In the solution of the following numerical problem the maximizing root of (1) is easily identified: A $1\frac{1}{2}$ foot square beam 50 feet long is carried through a passageway 12 feet wide into a lane at right angles to the passageway. Find the minimum width of the lane which will allow the beam to turn the corner while remaining horizontal.

Here $w = 1.5$, $a = 12$, $l = 50$, and (1) reduces to

(2) $$50 \cos^3 \theta - 12 + 1.5 \sin \theta = 0.$$

This equation has the positive root given by $\cos \theta = 3/5$, $\sin \theta = 4/5$, for which we verify that $d^2y/d\theta^2 < 0$. This root gives the maximum value of y, which is found to be 26.5. The required minimum width of the lane is then 26.5 feet. The minimizing root of (2) is the negative value of θ given by $\cos \theta = 0.64$, $\sin \theta = -0.79$, each correct to two figures. For this root it is easily verified that $d^2y/d\theta^2 > 0$. The corresponding point on the curve, the lowest point on the loop, has the approximate coordinates $(32, -23)$.

ON THE HARMONIC MEAN OF TWO NUMBERS*

G. POLYA, Stanford University

We know that x satisfies the double inequality

$$9 \leq x \leq 11,$$

but we know nothing else about the value of x. In such a case, we should take for x the approximate value 9.9; then we can be certain that the relative error cannot exceed 10% and this is the best we can get. The following lines explain this example and prove the underlying theorem.

About an otherwise unknown quantity x we have just one piece of definite information: we know that x is contained between two given positive bounds a and b,

(1) $$a \leq x \leq b,$$

where

(2) $$0 < a < b.$$

In choosing a proximate value p for x we risk a certain *error* $p-x$ and a certain *relative error* $(p-x)/x$. We may wish to reduce the risk of an extreme error, or the risk of an extreme relative error, to a minimum; and so two different problems arise.

PROBLEM I. *Find*

$$\min_{p} \left(\max_{x} |p - x| \right)$$

and the value of p for which it is attained under the conditions (1) *and* (2).

PROBLEM II. *Find*

$$\min_{p} \left(\max_{x} |p - x|/x \right)$$

and the value of p for which it is attained under the conditions (1) *and* (2).

The solution of problem I is obvious and certainly known to many persons although I am not able to give a definite quotation:

$$p = \frac{a+b}{2}, \qquad \min_{p} \max_{x} |p - x| = \frac{b-a}{2}.$$

---

* From AMERICAN MATHEMATICAL MONTHLY, vol. 57 (1950), pp. 26–28.

The solution of problem II is a little less obvious and may be new, but is certainly very little known:

$$p = \frac{2ab}{a+b}, \qquad \min_{p} \max_{x} \frac{|p-x|}{x} = \frac{b-a}{b+a}.$$

We can state the second result in words as follows: *The approximation that yields the minimum for the greatest possible absolute value of the relative error, committed in approximating an unknown quantity contained between two known positive bounds, is the harmonic mean of these bounds.*

In both problems we can find the desired p by the same prescription: Choose p so that the errors (relative errors) committed in the two extreme admissible cases, $x=a$ and $x=b$, are equal in amount, but opposite in sign. This yields

(3) $\qquad p - a = -(p-b), \qquad p = (a+b)/2, \qquad p - a = (a+b)/2,$

(4) $\qquad \dfrac{p-a}{a} = -\dfrac{p-b}{b}, \qquad p = \dfrac{2ab}{a+b}, \qquad \dfrac{p-a}{a} = \dfrac{b-a}{b+a},$

in problems I and II, respectively. Following this prescription, we are bound to obtain the best possible values, by virtue of a classical theorem of Tchebischeff.* Our problems are so simple, however, that the result can be proved independently of any special knowledge. I give the proof for the solution of the less obvious problem II.

We regard the approximate value p as chosen. We focus our attention on the more interesting case in which $a \leq p \leq b$. The absolute value of the relative error depends on x:

(5) $\qquad \dfrac{|p-x|}{x} = \begin{cases} \dfrac{p}{x} - 1 & \text{when } a \leq x \leq p \\[4pt] 1 - \dfrac{p}{x} & \text{when } p \leq x \leq b. \end{cases}$

When x varies from a to b, (5) first decreases from $(p-a)/a$ to 0 then increases from 0 to $(b-p)/b$. Therefore, the maximum of (5) must be attained at one of the two end-points, at $x=a$ or at $x=b$. Now,

(6) $\qquad \dfrac{a}{a+b} \dfrac{p-a}{a} + \dfrac{b}{a+b} \dfrac{b-p}{b} = \dfrac{b-a}{b+a}.$

The right-hand side of (6) is a weighted mean of $(p-a)/a$ and $(b-p)/b$. If these quantities are *different*, one of them is *greater* than the right-hand side of (6),

---

*See e.g. Serge Bernstein, Leçons sur les propriétés extrémales des fonctions analytiques d'une variable réelle, Paris, 1926, pp. 2–4.

that is

(7) $$\max_x \frac{|p-x|}{x} > \frac{b-a}{b+a}.$$

It happens only in the case (4) that (7) goes over into an equation. This proves our theorem in the case in which the value of p lies between a and b. The two remaining (*a priori* unplausible) cases, $p<a$ and $p>b$, can be discussed even more simply.

A SIMPLE ENDPOINT MINIMUM*

FRANK HAWTHORNE, Hofstra College

C. O. Oakley (this MONTHLY, vol. 54, p. 407†) has discussed endpoint maxima and minima with examples. A particularly simple example which may be used to illustrate this point is:

Find that point on the circle $x^2+y^2=1$ which is nearest to $(2, 0)$.

The unsuspecting student sets up this problem using x as independent variable and obtains, for the square of the distance from (x, y) to $(2, 0)$,

$$L^2 = (2 - x)^2 + 1 - x^2 = 5 - 4x.$$

Differentiating and equating the result to zero he obtains the disconcerting expression $-4=0$. This certainly does not lead to the obvious solution $(1, 0)$.

If, however, the student uses the central angle as independent variable this difficulty is avoided.

---

*From AMERICAN MATHEMATICAL MONTHLY, vol. 58 (1951), p. 188.
†Ed. note: Reprinted on pp. 303–305, this book.

RELATIVE MAXIMA AND MINIMA OF FUNCTIONS OF TWO OR MORE VARIABLES*

A. S. HENDLER, Rensselaer Polytechnic Institute

Of our current calculus texts that include a treatment of partial derivatives, a number fail to include any discussion of relative extremes of functions of two variables. Some discuss only necessary conditions. Others add to necessary conditions a statement without proof of sufficient conditions. And a few to be sure do present sufficient conditions with a proof usually based on Taylor's theorem. Every text and every first course, however, certainly includes both necessary and sufficient conditions for relative extremes of functions of one variable. The purpose of this paper is to present an elementary derivation, that should have wide appeal among beginning students, of the usual sufficient conditions for relative extremes of functions of two variables, and to generalize this derivation to functions of more than two variables.

We take as our basis

THEOREM 1. *If*:

1. $f(x)$ belongs to C^2, $a \leq x \leq b$,
2. $f'(c) = 0$, $a < c < b$,
3. $f''(c) > 0$,

then $f(c)$ is a relative minimum value of $f(x)$.

DEFINITION 1. *$f(a, b)$ is a relative minimum value of $f(x, y)$ in a region R if and only if $f(a, b) < f(x, y)$ for all (x, y) in R satisfying the inequality $0 < (x-a)^2 + (y-b)^2 < \delta$ for some positive δ.*

If $f(x, y)$ belongs to C^2 in some region R and (a, b) is an interior point of R then

$$F(s) = f(a + s \cos \alpha, b + s \sin \alpha)$$

belongs to C^2, $-d \leq s \leq d$.

$$F'(s) = f_1 \cos \alpha + f_2 \sin \alpha.$$
$$F''(s) = f_{11} \cos^2 \alpha + 2f_{12} \cos \alpha \sin \alpha + f_{22} \sin^2 \alpha.$$

Here the subscripts denote partial derivatives; in particular, since $f(x, y)$ belongs to C^2, $f_{12} = f_{21}$. If $f_1(a, b) = f_2(a, b) = 0$, then $F'(0) = 0$. $F''(0) > 0$ if

$$f_{11}(a, b) \cos^2 \alpha + 2f_{12}(a, b) \cos \alpha \sin \alpha + f_{22}(a, b) \sin^2 \alpha > 0.$$

For $\sin \alpha = 0$ this inequality can only be satisfied if $f_{11}(a, b) > 0$; for $\sin \alpha \neq 0$ we must have in addition

$$[f_{12}(a, b) \sin \alpha]^2 - f_{11}(a, b) \cdot f_{22}(a, b) \sin^2 \alpha < 0.$$

---

* From AMERICAN MATHEMATICAL MONTHLY, vol. 61 (1954), pp. 418–420.

Hence we have

THEOREM 2. *If:*

1. $f(x, y)$ belongs to C^2 *in some region R,*
2. $f_1(a, b) = f_2(a, b) = 0$, (a, b) *an interior point of R,*
3. $f_{11}(a, b) > 0$,
4. $f_{11}(a, b)f_{22}(a, b) - f_{12}^2(a, b) > 0$,

then $f(a, b)$ is a relative minimum value of $f(x, y)$ in R.

For appropriate functions $f(x_1, x_2, x_3, \cdots, x_n)$ we make the transformation

$$x_i = a_i + sh_i \qquad \text{where } \sum_{i=1}^{n} h_i^2 = 1.$$

Also, we shall use the notation f_i and f_{ij} for the various derivatives of f evaluated at $x_i = a_i$. Thus:

$$F(s) = f(a_1 + sh_1, a_2 + sh_2, \cdots, a_n + sh_n)$$

$$F(0) = f(a_1, a_2, a_3, \cdots, a_n)$$

$$F'(0) = \sum_{i=1}^{n} f_i h_i$$

$$F''(0) = \sum_{i,j=1}^{n} f_{ij} h_i h_j.$$

$F'(0) = 0$ for all sets of values h_i if $f_i = 0$, $(i = 1, 2, 3, \cdots, n)$.

For $F''(0) = \sum_{i,j=1}^{n} f_{ij} h_i h_j$ to be greater than zero we must have:
1. If $h_i = 0$, $(i = 2, 3, 4, \cdots, n)$, $f_{11} > 0$;
2. If $h_i \neq 0$, $(i = 1, 2, 3, \cdots, n)$ then, since

$$\sum_{i,j=1}^{n} f_{ij} h_i h_j = f_{11} h_1^2 + 2 h_1 \sum_{j=2}^{n} f_{1j} h_j + \sum_{i,j=2}^{n} f_{ij} h_i h_j,$$

we must have in addition to $f_{11} > 0$,

$$f_{11} \sum_{i,j=2}^{n} f_{ij} h_i h_j - \left[\sum_{j=2}^{n} f_{1j} h_j\right]^2 = [f_{11} f_{22} - f_{12}^2] h_2^2$$

$$+ 2\left[f_{11} \sum_{j=3}^{n} f_{2j} h_j - \sum_{k=3}^{n} f_{12} f_{1k} h_k\right] h_2 + f_{11} \sum_{i,j=3}^{n} f_{ij} h_i h_j - \sum_{j,k=3}^{n} f_{1j} f_{1k} h_j h_k > 0.$$

If $h_i = 0$, $(i = 3, 4, \cdots, n)$, or if $n = 2$, we must have then

$$f_{11} f_{22} - f_{12}^2 > 0.$$

3. If $h_i \neq 0$, $(i = 1, 2, 3, 4, \cdots, n)$, and $n > 2$, we must have in addition

$$[f_{11}f_{22} - f_{12}^2]\left[f_{11}\sum_{i,j=3}^{n} f_{ij}h_ih_j - \sum_{j,k=3}^{n} f_{1j}f_{1k}h_jh_k\right] - \left[f_{11}\sum_{j=3}^{n} f_{2j}h_j - \sum_{k=3}^{n} f_{12}f_{1k}h_k\right]^2 > 0.$$

If $h_i = 0$, $(i = 4, 5, 6, \cdots, n)$, or if $n = 3$, we must have

$$[f_{11}f_{22} - f_{12}^2][f_{11}f_{33} - f_{13}^2] - [f_{11}f_{23} - f_{12}f_{13}]^2 > 0.$$

We have thus established

THEOREM 3. *If*:

1. $f(x, y, z)$ belongs to C^2 *in some region R,*
2. $f_1 = f_2 = f_3 = 0$ *at (a_1, a_2, a_3) an interior point of R,*
3. $f_{11} > 0$ *at (a_1, a_2, a_3),*
4. $\begin{vmatrix} f_{11} & f_{21} \\ f_{12} & f_{22} \end{vmatrix} > 0$ *at (a_1, a_2, a_3),*
5. $\begin{vmatrix} f_{11} & f_{21} & f_{31} \\ f_{12} & f_{22} & f_{32} \\ f_{13} & f_{23} & f_{33} \end{vmatrix} > 0$ *at (a_1, a_2, a_3),*

then $f(a_1, a_2, a_3)$ is a relative minimum value of $f(x, y, z)$.

We are now in position to derive the corresponding theorems for functions of four or more variables.

The above can easily be adapted to maximum values by the reversal of appropriate inequality signs.

Editorial Note: The argument on p. 291 does not imply Theorem 2 as claimed. It is based on the false assumption that if a function of two variables has a local minimum at a point when restricted to every straight line through this point then it also has a local minimum at this point. Counterexamples are found in several texts. Apostol's *Calculus*, Vol. 2, 2nd ed. (Wiley, 1969) gives the following example as Exercise 16 on p. 313:

$$f(x, y) = (y - x^2)(y - 3x^2).$$

This function takes both positive and negative values in every two-dimensional neighborhood of $(0, 0)$ so it has no local minimum there although it has a local minimum at $(0, 0)$ on every line $y = mx$.

What the argument on p. 291 actually proves is that if f satisfies conditions (1) through (4) of Theorem 2, and if for every fixed angle α we let $F(s) = d(a + s \cos \alpha, b + s \sin \alpha)$, then $F(0)$ is a relative minimum of $F(s)$. The error in this paper can serve to stimulate valuable class discussion.

A CALCULUS PROBLEM WITH OVERTONES IN RELATED FIELDS*

C. S. OGILVY, Hamilton College

Several elementary extremum problems in the calculus which seem at first glance to be quite distinct from one another can be linked together through considerations not usually presented in the texts.

I. One ship, A, is anchored 9 miles offshore. Opposite a point 6 miles down the coast a second ship, B, is anchored 3 miles offshore. A boat from A is to put a passenger ashore and then proceed to B (Fig. 1).

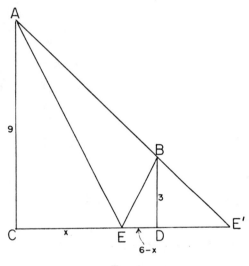

Fig. 1

(a) What is the shortest course for the boat?

(b) Suppose the passenger pays each member of the boat's crew \$1 per mile for every mile they carry him; but each member of the crew must forfeit \$1 for every mile travelled from shore to B. What is the course which will yield them the maximum profit, under the restriction that it must consist of two straight-line segments?

II. A man can row 4 miles per hour and run 5 miles per hour. He leaves ship B of problem I wishing to go to point Q on the shore 5 miles from D (Fig. 2).

(a) Where should he land in order to reach Q most quickly?

These do not look to the student like connected problems. I(a) seeks a minimum distance, I(b) a maximum profit, and II(a) a minimum time. We observe first that the two (a) parts are both solved by applying Fermat's principle of

---

* From AMERICAN MATHEMATICAL MONTHLY, vol. 65 (1958), pp. 765–767.

optics, where in I the shore is a reflector for light travelling from A to B and in II the shore is a refractor with light travelling from the slower medium into the faster. From this point of view, both problems do minimize the same variable, time. In the first instance, the physicist seeks two similar triangles so that the angle of incidence will equal the angle of reflection: $9/x = 3/(6-x)$, or $x = CE = 4.5$ miles. In II(a) the light must travel in conformance with Snell's Law: $\sin \alpha / \sin \beta = 4/5$. But this is the limiting case where $\beta = \pi/2$, which means $\sin \alpha = 4/5$, or $DS = 4$ miles.

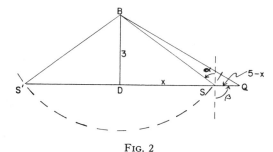

FIG. 2

We turn now to the calculus method of handling these two problems.

I(a). The distance function to be minimized is $\sqrt{81+x^2}+\sqrt{9+(6-x)^2}$. Setting the first-derivative equal to zero gives immediately the equation

(1) $$\frac{x}{\sqrt{81+x^2}} = \frac{6-x}{\sqrt{9+(6-x)^2}}.$$

The "solutions" of (1) are $x = 4.5, 9$. The student throws away the second value because he feels pretty certain that $x < 6$. This is the wrong reason for throwing it away. The function is not stationary at $x = 9$ and indeed 9 does not satisfy (1). It is purely extraneous, introduced by squaring in the course of solving (1). But although "extraneous" roots may have no place in algebra, there is usually some reason for their appearance in the calculus. We now discover the meaning of this one.

I(b). The profit function to be maximized is $\sqrt{81+x^2}-\sqrt{9+(6-x)^2}$, which leads, on differentiation, to

(2) $$\frac{x}{\sqrt{81+x^2}} = \frac{x-6}{\sqrt{9+(6-x)^2}}.$$

Observe that when we square both sides of (2) it is identical with the equation resulting from squaring both sides of (1). Hence solutions again appear to be 4.5 and 9; but this time 9 satisfies (2) and 4.5 does not. The crew should land the passenger at E'.

At this stage we can formulate a General Remark:

If two problems requiring different solutions lead at some stage to the same equation, then this equation must deliver at least two solutions, one of which will be extraneous to each problem.

II(a). The time function to be minimized is $\sqrt{9+x^2}/4+(5-x)/5$. Setting the first derivative equal to zero we get

$$(3) \qquad \frac{x}{\sqrt{9+x^2}} = \frac{4}{5},$$

or $x = \pm 4$. As before, only one of the two values of x is admissible. It is $+4$ which satisfies (3). And now our General Remark *predicts* the existence of a companion problem:

II(b). Suppose the man has to pay the owner of the boat $1 for each hour he rows, but some altruistic soul pays him $1 per hour for each hour he runs. What now is his most profitable course, assuming still that he runs only straight along the shore after landing?

The profit function to be maximized is $-\sqrt{9+x^2}/4+(5-x)/5$, which yields

$$(4) \qquad \frac{x}{\sqrt{9+x^2}} = -\frac{4}{5},$$

and only -4 satisfies (4). He should row to S'.

These results have interesting geometric interpretations. Equations (1) and (2) are proportions calling for similar right triangles. There are *two* pairs of similar right triangles formed by the ships and the shoreline, and E and E' give them both. Equations (3) and (4) ask for $\alpha = \sin^{-1}(\pm 4/5) = \cos^{-1}(3/5)$. This is the "ambiguous case" of high school trigonometry: triangles BQS and BQS' both satisfy the required conditions. Student reaction to these unexpected cross-connections with previously studied topics has been favorable.

We leave to the reader the statement of the more difficult problems III(a) and III(b) where the point Q is inland. Snell's Law this time provides no shortcut, and the solution is technically burdensome because the terms in x^3 and x^4 do not drop out in either the physics or the calculus method. It is believed that only two of the roots of the resulting equation are ever real, namely, the answers to III(a) and III(b) respectively.

I am indebted to Professor H. M. Gehman of the University of Buffalo for the statement of problem I(b).

EXCEPTIONAL EXTREMUM PROBLEMS*

C. S. OGILVY, Hamilton College

The elementary calculus method of locating the maximum or minimum points of a function of one variable may fail for either of two reasons. (1) The function is stationary outside the range applicable to the problem, resulting in the well known "end point" maximum or minimum. This situation is usually, but not always, easily recognizable [1]. (2) The failure is caused by the use of what Widder calls an unsuitable independent variable [2]. It is often difficult to predict when this will happen.

The method of Lagrange multipliers may or may not get around the difficulty. In the case of an end point maximum or minimum it will frequently be of no help. Consider the following problem. Given a straight fence 100 feet long, we wish by adding 200 feet more to form a rectangular enclosure whose boundary contains all of the original fence. How shall this be done so as to enclose the greatest possible area? [3] If we let x be the length of the new fence which is to be aligned with the original 100 feet and proceed in the usual fashion to maximize the resulting expression for area, x comes out to be -25, which is not permitted by the conditions of the problem. If we use a Lagrange multiplier to maximize the function $f(x, y) = (100+x)y$ subject to the restraint $g(x, y) = 100 + 2x + 2y - 200 = 0$, we get the same result. The method of Lagrange must lead to a point or points where one of the curves $f(x, y) = c$ is tangent to $g(x, y) = 0$ [4]. Figure 1 indicates what is happening. A member of the family of hyperbolas $(100+x)y = c$ is tangent to the line $x+y-50=0$ at the point $P: (-25, 75)$, the correct values of (x, y) for absolute maximum area, a square enclosure 75 feet on a side. But in order to bring the solution within the stipulation of the original problem we need $x=0$, an end point maximum. The function is not stationary there, of course; and from this point of view the failure of both the elementary and the Lagrange method is essentially trivial. They could not be expected to succeed.

Consider next Widder's example: to find the shortest distance from the focus $(1, 0)$ to the parabola $y^2 = 4x$. Here $f(x, y) = (x-1)^2 + y^2$ must be minimized subject to the restraint $g(x, y) = y^2 - 4x = 0$. An attempt to eliminate y from $f(x, y)$ and set $f' = 0$ yields $x = -1$, a point not on the parabola. Of course the method of Lagrange succeeds, because the circle $f(x, y) = c^2$, c being the distance from the focus, can become tangent to the parabola at the required point $(0, 0)$. We have $u = f(x, y) + \lambda g(x, y) = (x-1)^2 + y^2 + \lambda(y^2 - 4x)$,

$$\partial u/\partial x = 2(x-1) - 4\lambda = 0,$$
$$\partial u/\partial y = 2y + 2\lambda y = 0,$$
$$\partial u/\partial \lambda = y^2 - 4x = 0.$$

---

* From AMERICAN MATHEMATICAL MONTHLY, vol. 67 (1960), pp. 270–275.

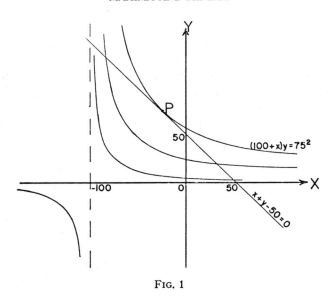

Fig. 1

In the simultaneous solution of these three equations in this particular example one "could not eliminate λ by solving the second equation thereof for λ. For, $\partial g/\partial y = 0$ at the very point which yields the minimum" [2]. But actually, the implication that the two conditions in quotations are generally interdependent is unjustified. We shall see that inability to eliminate λ in this way is not a consequence of the vanishing of $\partial g/\partial y$.

What we should like, ideally, is a criterion for judging in advance the "suitability" of an independent variable. Unfortunately this seems rather too much to ask. We might surmise, from the above example, that the difficulty occurs because x and the distance function, say s, behave in the same way at the minimum point. That is, ds/dx becomes meaningless at $(0, 0)$ because, as one moves along the parabola, x and s become stationary at the same time. But this circumstance is in fact neither necessary nor sufficient. If one seeks the extrema of the distance from the point $(2, 0)$ to the hypocycloid whose equation is $x^{2/3} + y^{2/3} = 1$, the attempt to maximize s as a function of x fails to yield the local maxima at $(0, \pm 1)$ even though s and x are not "behaving similarly" at these points. (Later we shall find the converse counter example.) This is also an illuminating illustration of the limitations of Lagrange multipliers, which yield only the local maxima at P_1, P_2, where the circle representing the distance function is tangent to the curve, and none of the cusps (Fig. 2).

We digress temporarily to investigate the apparently peculiar state of affairs near the focus in Widder's problem. In order to simplify the details of the necessary algebra, we consider instead a similar example: [5] to find that point on the circle $x^2 + y^2 = 1$ nearest to $(\frac{1}{2}, 0)$. If we let $L^2 = s$, the square of the distance

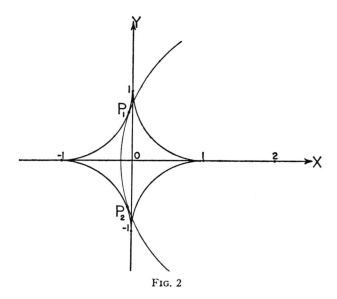

FIG. 2

from $(\frac{1}{2}, 0)$ to any point (x, y) on the circle, we have $s = (\frac{1}{2} - x)^2 + 1 - x^2$, and setting $ds/dx = 0$ yields the absurd result $-1 = 0$. To be slightly more general, we can say that if s is the squared distance from a point on the circle to any point (a, b), $a \neq 0$, the elementary method fails when $b = 0$. However, for any $b \neq 0$ but *arbitrarily close to zero*, the same method is successful. For instance, if P is a point which is allowed to assume various positions along the line $x = a \neq 0$, then for every position except $(a, 0)$ there is no trouble. What causes this sudden discontinuity?

The expression for s is

(1) $$s = (x - a)^2 + (\pm \sqrt{1 - x^2} - b)^2.$$

If $b = 0$

(2) $$s = a^2 + 1 - 2ax,$$

a straight line of slope $-2a$ in the (x, s)-plane. Admissible values of x lie in the interval $-1 \leq x \leq 1$, and if x equals 1 or -1, $L = |a - 1|$ or $|a + 1|$, the end point minimum or maximum.

If $b \neq 0$, the curve of (1) is an ellipse. For any fixed a (we take $a = \frac{1}{2}$ in Fig. 3), we have a family of ellipses with b as the parameter, whose envelope is the pair of parallel lines $x = \pm 1$. Each ellipse is tangent to both lines and hence is centered on the s-axis; its maximum and minimum points are the maximum and minimum s for that particular a and b. The major axis of each ellipse is inclined to the

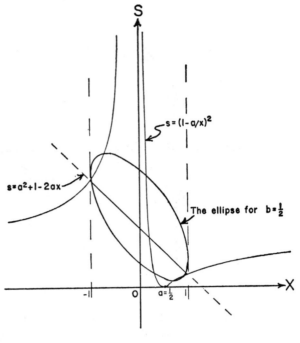

Fig. 3

horizontal at an angle ϕ whose tangent is negative and numerically larger than $2a$. As b tends toward ∞ or $-\infty$ for any fixed a, $\phi \to \frac{1}{2}\pi$ and the ellipse moves upward, both $s(\max)$ and $s(\min)$ tending toward $+\infty$. As $b \to 0$, $\tan \phi \to -2a$. The segment of the straight line of (2) lying in the interval $-1 \leq x \leq 1$ is a degenerate ellipse, the limiting position of the ellipses of the family. If we differentiate (1) with respect to x and set $ds/dx = 0$, and then eliminate the parameter b between this equation and (1), we obtain the locus of the maximum and minimum of s, namely

$$(3) \qquad s = (1 - a/x)^2.$$

As $b \to 0$, the ordinary maximum and minimum points of the nondegenerate ellipses approach the end point maximum and minimum points of the degenerate ellipse continuously along the two branches of the curve of (3), thus accounting for the apparent discontinuity mentioned at the outset. This kind of behavior is typically encountered in such problems.

We turn now to our final example: to find the cone of maximum lateral surface area with vertex at the origin and inscribed in the ellipsoid formed by rotating the curve $y^2/a^2 + x^2/b^2 = 1$ around the Y-axis (Fig. 4). If one expresses the lateral surface area S as a function of x and sets $dS/dx = 0$, the result yields the true maximum if $a^2 > 2b^2$. If $a^2 = 2b^2$, the maximum occurs at the point $(b, 0)$ where the cone has flattened out into a circular disc. The elementary method yields this result even though x and S are stationary at the same time, the converse counter example mentioned earlier. If $a^2 < 2b^2$, the maximum is again at $(b, 0)$, but this time it is an end point case and hence not revealed by the elementary attack.

If one uses the Lagrange method to maximize $f(x, y) = \pi x \sqrt{(x^2 + y^2)}$ subject to the restraint $g(x, y) = y^2/a^2 + x^2/b^2 - 1 = 0$, one finds: (1) If $a^2 > 2b^2$, the maximum cone is delivered and also a second solution, $y = 0$, overlooked by the first method. (2) If $a^2 = 2b^2$, the circle is the answer. This is the case where $\partial g/\partial y = 2y/a^2 = 0$ at the point where the maximum occurs, yet it is perfectly feasible to eliminate λ in the course of the solution. (3) If $a^2 < 2b^2$, $y = 0$ as in (1).

It should be noted that there is always another extremal at $(0, a)$, the limiting case where the lateral surface area tends to zero. As usual, the reason for the failure of the Lagrange method to produce this minimum is that the curves $f(x, y) = $ constant can never be tangent to $g(x, y) = 0$ there; they have the line $x = 0$ as a vertical asymptote. On the other hand, one of them can be tangent to $g(x, y) = 0$ at the point $(b, 0)$, which explains why the method picks up the end points in cases (1) and (3) of the previous paragraph. Figure 4 shows an

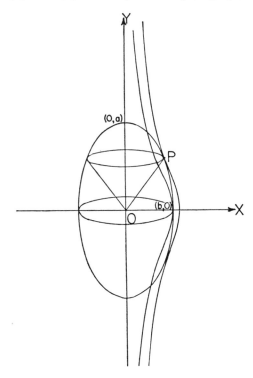

FIG. 4

ellipse for which $a^2 > 2b^2$, and two tangent curves $f(x, y) = c$. The lateral surface area of the inscribed cone has its absolute maximum at the point P and a local minimum at the point $(b, 0)$.

There exists a device which will yield *all* the extrema: a suitable parametrization of the original variables. This is not difficult to effect. The method of choosing the parametrization has been indicated by Oakley [6]. The interested reader can find all six of the extrema in the hypocycloid example by the use of the conventional parametrization, $x = \cos^3 \theta$, $y = \sin^3 \theta$.

References

1. *Cf.* Problem E 1276, A well-concealed endpoint maximum, this MONTHLY, vol. 65, 1958, pp. 205–206.
2. David V. Widder, Advanced Calculus, Englewood Cliffs, N. J., 1947, p. 115.
3. V. L. Klee, Jr., this MONTHLY, vol. 54, 1947, p. 409.
4. See, for instance, Angus E. Taylor, Advanced Calculus, Boston, 1955, Sec. 6.8.
5. Frank Hawthorne, A simple endpoint maximum, this MONTHLY, vol. 58, 1951, p. 188. (The problem is stated but not analyzed.) [*Ed. note:* Reprinted on p. 321, this book.]
6. C. O. Oakley, Endpoint maxima and minima, this MONTHLY, vol. 54, 1947, pp. 407–409. [*Ed. note:* Reprinted on pp. 303–305, this book.]

SO-CALLED "EXCEPTIONAL" EXTREMUM PROBLEMS*

HUGH A. THURSTON, University of British Columbia

Although the invitation "rejoinders to earlier notes are encouraged" is no longer printed in the MONTHLY, I should like to make a rejoinder to Ogilvy's Classroom Note [1]. I shall refer also to Thurston [2], Oakley [3], and Walsh [4].†

Ogilvy maintains that the "elementary calculus method" of finding maxima and minima may fail for two reasons, and he gives an example of each. The second failure he imputes to an "unsuitable independent variable" and remarks: "what we should like is a criterion for judging in advance the suitability of an independent variable. Unfortunately this seems rather too much to ask."

I maintain that the elementary method does not fail in either case and that therefore no criterion is needed. By the elementary method I mean the method suggested by common sense: we examine the stationary values of the function being investigated, the values where the derivative fails to exist, and the endpoint values. The greatest of these is the maximum and the least the minimum, provided of course the function does have a maximum and a minimum. A similar method occurs in Walsh's paper (the case where there is just one extremum in the range under consideration) and is sufficient to exemplify the principle that we should realize precisely what we are looking for, rather than mechanically apply a learnt test.

Ogilvy's first problem is essentially to maximize $(50-x)(100+x)$ under the conditions $0 < x < 50$. He writes that if we proceed "in the usual fashion" then "x comes out to be -25, which is not permitted by the conditions of the problem." But to get this result, the standard (and, I would have hoped, usual) method must be replaced by something like "set the derivative equal to zero and solve."

The second problem is essentially:—

Minimize $(x-1)^2+y^2$ under the condition $y^2=4x$.

This is clearly equivalent to: Minimize $(x-1)^2+4x$ under the condition $x \geq 0$. Because the derivative is always positive throughout the range considered, the minimum occurs when $x=0$. Ogilvy states that his method of solution "yields $x=-1$, a point not on the parabola."

In both cases, the incorrect results can be imputed to *neglect of the domain of the function under investigation*. (E.g., the minimum of $(x-1)^2+4x$ with no restriction on x does occur when $x=-1$; the maximum of $(50-x) \cdot (100+x)$ does occur when $x=-25$.) In my note, I said regretfully that the concept of domain is apt to be pushed into the background, and gave an example (from a well-

---
*From AMERICAN MATHEMATICAL MONTHLY, vol. 68 (1961), pp. 650–652.

†*Ed. note:* Ogilvy reprinted on pp. 297–302, Thurston on pp. 80–82, Oakley on pp. 275–277, and Walsh on pp. 273–274, this book.

known textbook) of a problem whose solution goes wrong if domains are neglected. There are no such things as "unsuitable" variables: the use of this term is an attempt to put on the variable the blame which rightly belongs to an illogical method of solution.

Why are both common sense and Walsh's thirteen-year-old paper neglected? It must be because teaching is too mechanical: finding stationary points is a convenient way for finding certain local maxima and minima; and the *technique* of setting the derivative equal to zero is emphasized at the expense of *understanding* what the problem is and why the technique is being used. Why else would a student expect to find the least value of $a-ex$ as x varies from $-a$ to a by differentiation? And yet Oakley found that "even an able student" might do this.

The fact is that as one goes through life most maximum and minimum problems are endpoint problems. How fast do I run to get to the telephone as quickly as possible? Answer: as fast as I can—an endpoint solution. Of course, most of these problems are trivial; the interesting ones are where there is some balancing factor: in maximizing $x(1-x)$ the increase in x balances the decrease in $(1-x)$. However, we teachers of calculus seem to have so lost our sense of proportion that we regard these as normal and the endpoint solutions as wicked exceptions.

To show how far our sense of proportion has been lost, let us consider the following problem. Choose P between P_0 and the vertex P_m on the parabola $y=9x^2-28x+24$ (Fig. 1), complete the shaded region and rotate it about the y-axis. Where should P be taken to give a maximum volume? Because P is restricted to a finite closed interval, the endpoint maximum (at P_m) and minimum (at P_0) stare us in the face; it would certainly be possible to have extrema in between, but at any rate the endpoint values are the most obvious suspects.

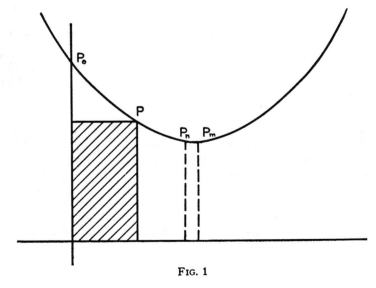

Fig. 1

Indeed, no one would doubt that the value given by P_m is greater than that given by P_n. And yet the problem is described [5] as "A *well-concealed* endpoint maximum"! (My italics.) However, we are not alone in making this mistake. I heard that, until recently, geographers, when asked for the highest point in Florida, would cite the highest of Florida's few hills. Now, however, someone has noticed that the highest point in Florida is on the Alabama border: an endpoint maximum.

References

1. C. S. Ogilvy, Exceptional extremum problems, this MONTHLY, vol. 67, 1960, pp. 270–275.
2. H. A. Thurston, Concerning domains of real functions, this MONTHLY, vol. 66, 1959, pp. 900–902.
3. C. O. Oakley, Endpoint maxima and minima, this MONTHLY, vol. 54, 1947, pp. 407–409.
4. J. L. Walsh, A rigorous treatment of the first maximum problem in the calculus, this MONTHLY, vol. 54, 1947, pp. 35–36.
5. Problem E 1276, this MONTHLY, vol. 65, 1958, pp. 205–206.

BIBLIOGRAPHIC ENTRIES: MAXIMA AND MINIMA

Except for the entry labeled MATHEMATICS MAGAZINE, the references below are to the AMERICAN MATHEMATICAL MONTHLY.

1. Fish, E., Curious process for maximum or minimum, vol. 1, p. 332.

    A curiosity that could serve as a source for classroom discussion.

2. Bliss, G. A., A note concerning maxima and minima of functions of several variables, vol. 14, p. 47.

    Sufficient conditions for a function of n variables to have an extremum at a point where all partial derivatives of order less than n vanish. One must decide whether a certain quadratic form in n variables is definite, semidefinite, or indefinite.

3. Hildebrandt, T. H., Existence of a minimum of a quadratic function, vol. 15, p. 57.

4. Lennes, N. J., Note on maxima and minima by algebraic methods, vol. 17, p. 9.

    Extrema of certain polynomials without calculus.

5. Jackson, D., Relating to a problem in minima, vol. 24, p. 42.

    A simple solution to the following problem: Find the point so situated that the sum of its distances from the three vertices of an acute-angled triangle is a minimum. The problem is also discussed by Wilson, vol. 24, p. 241, and in Johnsons' paper below.

6. Johnson, R. A., Relating to a problem in minima discussed by Professor Dunham Jackson in the January, 1917, number (p. 46) of this Monthly, vol. 24, p. 243.

7. DeCou, E. E., A practical printer's problem in maxima and minima, vol. 27, p. 415.

    Requires minimizing $Ax+B/x$.

8. Bennett, A. A., On the treatment of maxima and minima in calculus texts, vol. 31, p. 293.

Criticizes texts that ignore endpoint extrema.

9. Uspensky, J. F., A minimum problem, vol. 40, p. 5.

The problem is to choose λ and μ so that the maximum of

$$\left| (x^2+y^2)^{1/2} - \lambda x - \mu y \right|$$

over the rectangle $0 \leq x \leq a$, $0 \leq y \leq b$ will be as small as possible. The detailed solution is a possible source of examples or exercises.

10. Britton, J. R., A note on polynomial curves, vol. 42, p. 306.

Conditions under which n given points will serve as the extreme points of the graph of a polynomial of degree $n+1$.

11. Garver, R., The solution of problems in maxima and minima by algebra, vol. 42, p. 435.

Shows that many extremum problems can be reduced to the purely algebraic fact that if $\sum_{j=1}^{n} k_j x_j = c$, where the k_j, x_j and c are positive, then the product $x_1 \cdots x_n$ is largest when $k_1 x_1 = \cdots = k_n x_n$.

12. Brown, O. E., Maximum and minimum values of functions of several variables, vol. 44, p. 161.

13. Rinehart, R. F., On extrema of functions which satisfy certain symmetry conditions, vol. 47, p. 145.

A sequel to Garver, vol. 42, p. 435 (see entry 11).

14. Courant, R., Soap film experiments with minimal surfaces, vol. 47, p. 167.

15. Coe, C. J., Problems on maxima and minima, vol. 49, p. 33.

Ingenious but sometimes devious methods for finding extrema without calculus.

16. Stewart, B. M., A maximum problem, vol. 49, p. 454.

Treats the following question: Of all planes through a given point in a solid, which one cuts off the maximum volume?

17. Phipps, C. G., Maxima and minima under restraint, vol. 59, p. 230.

A discussion leading up to Lagrange multipliers.

18. Mancill, J. D., One-sided maxima and minima of functions of two or more variables, vol. 60, p. 80.

19. Ogilvy, C. S., and Gunderson, N. G., The razor's edge?, vol. 65, p. 769.

20. Saltzer, C., Lagrange's multipliers, vol. 66, p. 225.

21. Jacobson, R. A., and Yocom, K. L., Paths of minimal length within a cube, vol. 73, p. 634.

22. Hoggatt, Vern, Maximum area in a corner, MATHEMATICS MAGAZINE, vol. 26, p. 95.

9

INTEGRATION

(a)
THEORY

THE FUNDAMENTAL THEOREM OF THE CALCULUS*

LOUIS BRAND, University of Cincinnati

Both proof and content of the fundamental theorem of the integral calculus are foreshadowed by the

FUNDAMENTAL THEOREM OF THE SUM CALCULUS. *If $F(n)$ is a function of the integral variable n having $f(n)$ as difference, then*

(1) $$\sum_{n=a}^{b} f(n) = F(b+1) - F(a).$$

The proof is immediate; for the sum in (1) is

$$\sum_{n=a}^{b} f(n) = F(a+1) - F(a) + F(a+2) - F(a+1) + \cdots + F(b+1) - F(b)$$

$$= F(b+1) - F(a).$$

This theorem is readily and effectively illustrated by forming a short table of *anti-differences*. Thus to verify the brief table

$f(n)$	$F(n) = \Delta^{-1} f(n)$
r^n	$\dfrac{r^n}{r-1} \quad (r \neq 1)$
$n^{(k)}$	$\dfrac{n^{(k+1)}}{k+1}$
$\cos n\alpha$	$\dfrac{\sin(n - \tfrac{1}{2}\alpha)}{2 \sin \tfrac{1}{2}\alpha}$

---
* From AMERICAN MATHEMATICAL MONTHLY, vol. 62 (1955), pp. 440–441.

we need only show that $\Delta F(n) = f(n)$. We can now compute sums such as

$$\Sigma ar^n, \qquad \Sigma n^2 = \Sigma(n^{(2)} + n^{(1)}), \qquad \Sigma \cos n\alpha,$$

between integral limits by use of (1).

A direct analogue of (1) is the

FUNDAMENTAL THEOREM OF THE INTEGRAL CALCULUS. *If $f(x)$ is integrable in the interval (a, b) and $F(x)$ is a function having $f(x)$ as derivative, then*

(2) $$\int_a^b f(x)dx = F(b) - F(a).$$

The proof again depends upon "telescopic" cancellation. Divide (a, b) into subintervals δ_i by the intermediate points

$$a = x_0 < x_1 < x_2 < \cdots < x_n = b$$

and let $\delta = \max \delta_i$; then

(3) $$\int_a^b f(x)dx = \lim_{\delta \to 0} \sum_{i=1}^n f(t_i)\delta_i, \qquad x_{i-1} \leq t_i \leq x_i.$$

Since $F(x)$ is differentiable it is also continuous in (a, b); and from the mean value theorem

(4) $$F(x_i) - F(x_{i-1}) = f(\xi_i)\delta_i, \qquad x_{i-1} \leq \xi_i \leq x_i.$$

Since the points t_i in (3) are at our disposal we may take $t_i = \xi_i$. Then the sum in (3) becomes

$$F(x_1) - F(x_0) + F(x_2) - F(x_1) + \cdots + F(x_n) - F(x_{n-1}),$$

that is, $F(b) - F(a)$. Since this is true for any subdivision of (a, b), it must hold as $\delta \to 0$.

The analogy between the theorems is enhanced by writing them in the form:

$$\sum_{n=a}^b f(n)\,\Delta n = \Delta^{-1}f(n)\Big|_a^{b+1}; \qquad \int_a^b f(x)dx = D^{-1}f(x)\Big|_a^b.$$

The requirement in the fundamental theorem that $f(x)$ be integrable is essential, for the derivative of a function, $f(x) = F'(x)$, is not necessarily integrable. For example the function

$$F(x) = x^2 \sin \frac{1}{x^2} \ (x \neq 0), \qquad F(0) = 0$$

has the derivative

$$F'(x) = 2x \sin \frac{1}{x^2} - \frac{2}{x} \cos \frac{1}{x^2} \quad (x \neq 0), \qquad F'(0) = 0;$$

but since $F'(x)$ is unbounded in any neighborhood of the origin, $F'(x)$ is not integrable over $0 \leq x \leq 1$.

INTEGRATION*

D. G. MEAD, Pratt Institute

Although we frequently think of indefinite integration as being more complex than the determination of roots of algebraic equations, we shall see that in some respects this is not the case. It is well known that solutions of some algebraic equations cannot be written in terms of radicals, and also that certain functions do not possess elementary integrals. However, due to the length and depth of the proofs of these facts, few texts go beyond merely exhibiting some examples. Confusing a sufficient condition with one which is both necessary and sufficient, many a student finds security in knowing that all equations of degree ≤ 4 have been solved. With respect to integration, however, he is well aware of his rather unhappy position, that of knowing that some functions cannot be integrated (in finite terms) but of possessing no way of determining whether a particular function is in this class or not. This latter situation need not be accepted, for Liouville obtained a test [1] which is both necessary and sufficient for the integrability of any one of a rather large class of functions. We shall recall this test and show both how it can be applied by the college freshman and its utility as an integration technique.

The foundation for the test is the following theorem which is sufficiently "natural" to be readily accepted and easily remembered.

LIOUVILLE'S THEOREM. *If $\int f e^g dx$ is an elementary function, f and g are rational functions of x, and the degree of $g > 0$, then*

$$(1) \qquad \int f e^g dx = R e^g$$

where R is a rational function of x.

This is actually a special case of Liouville's original theorem (see [1] page 114 or [2] page 47), but it is sufficiently general for our purposes.

In this note, by rational function we mean the quotient of two polynomials with coefficients in any field of characteristic zero, for example, the complex numbers. The term elementary function is more difficult to define (indeed, Ritt takes 12 pages to do so). However, the student is willing to accept as meaningful such statements as "the general algebraic equation of degree 5 cannot be solved in terms of radicals," although we know that "in terms of radicals" requires some preliminary discussion. Therefore, he finds no difficulty with the following definition: An elementary function is one which can be constructed by means of any finite combination of the operations addition, subtraction, multiplication, division, raising to powers, taking roots, forming trigonometric functions and their inverses, taking exponentials and logarithms. In short, no matter how complicated the function, if we can write down all of its terms, the function is

---

* From AMERICAN MATHEMATICAL MONTHLY, vol. 68 (1961), pp. 152–156.

elementary. (Actually, the construction of elementary functions includes the forming of algebraic functions, but it seems advisable to omit this generality for the beginning calculus student.)

We return to the test. If we wish to determine whether fe^g can be integrated (*i.e.* has an integral which is an elementary function, or, as we shall also say, $\int fe^g dx$ is elementary), we know, by Liouville's Theorem the form of the integral. Differentiating equation (1) and cancelling the nonzero e^g we find $f = R' + Rg'$ or, letting $R = P/Q$, where P and Q are relatively prime polynomials in x,

(2) $$fQ^2 = P'Q - PQ' + PQg'.$$

Thus $\int fe^g dx$ is elementary if and only if there exist polynomials P and Q satisfying the differential equation (2).

Besides Liouville's theorem, the test requires only one further fact; namely, the following

LEMMA. *If the polynomial $f(x)$ has an r-fold zero at $x = \alpha$ and $r > 0$, then $f'(x)$ has an $(r-1)$-fold zero at $x = \alpha$; in other words, if $f(x) = (x-\alpha)^r h(x)$ where $r > 0$, $h(x)$ is a polynomial and $h(\alpha) \neq 0$, then $f'(x) = (x-\alpha)^{r-1} k(x)$ where $k(\alpha) \neq 0$.*

The proof of the lemma is a simple differentiation exercise which the student can supply.

It will make things easier if we define the term multiplicity. The number α is called a zero of the polynomial $f(x)$ of multiplicity r (or a root of $f(x) = 0$ of multiplicity r) if $f(x) = (x-\alpha)^r h(x)$, where the polynomial $h(\alpha) \neq 0$. In terms of multiplicity the lemma reads: If α is a zero of the polynomial $f(x)$ of multiplicity $r > 0$, then α is a zero of $f'(x)$ of multiplicity $r - 1$.

By examining some of the examples most frequently quoted in texts, we shall show how easily the analysis of equation (2) can be carried out, even by the student who has not previously encountered differential equations.

Example 1: e^{-x^2}. If $\int e^{-x^2} dx$ is elementary, then $\int e^{-x^2} dx = Re^{-x^2}$ or $1 = R' - 2xR$. Letting $R = P/Q$, where P and Q are relatively prime polynomials and $Q \neq 0$, we find:

(1.2) $$Q^2 = QP' - PQ' - 2xPQ$$

which is equation (2). Rearranging, we obtain

(1.3) $$Q(Q - P' + 2xP) = -PQ'.$$

Let us assume that the degree of Q is positive. Then $Q = 0$ has a root; let α be such a root and call its multiplicity $r(r > 0)$. Since P and Q are relatively prime, $P(\alpha) \neq 0$. Now, α is a zero of the left side of (1.3) of multiplicity $\geq r$ but α is a zero of the right side of multiplicity $r - 1$. This is a contradiction, hence our assumption that the degree of Q is positive must be false. Q is a constant ($\neq 0$) which we can assume is unity.

From (1.3) we obtain

(1.4) $$P' - 2xP = 1.$$

Since P is a polynomial in x, it is clear that the degree of $-2xP>$ degree of P', and the degree of $-2xP>0$. The degree of the left side of (1.4) is always greater than the degree of the right side, which is a contradiction. We have proved that there is no polynomial P satisfying (1.4), hence no rational function satisfying (1.2). Consequently $\int e^{-x^2}dx$ is not elementary.

Example 2: e^{bx}/x, with b a nonzero constant. If $\int(e^{bx}/x)dx$ is elementary, then $\int(e^{bx}/x)dx = Re^{bx}$ or $(1/x) = R' + bR$. Letting $R = P/Q$, where P and Q are relatively prime polynomials, $Q \neq 0$, we find:

(2.2) $$Q^2 = xQP' - xPQ' + xbPQ$$

(2.3) $$Q(Q - xP' - bxP) = -xPQ'.$$

If we assume that Q has positive degree, then $Q=0$ has a root. Let α be such a root and call its multiplicity r. If $\alpha \neq 0$, we encounter the same contradiction met in the first example, that α is a zero of the left side of (2.3) of multiplicity $\geq r$, while α is a zero of the right side of multiplicity $r-1$. Thus α must be zero, and $Q = cx^r$, for some $c \neq 0$. Putting this expression for Q in (2.2) we have $cx^{r+1}(cx^{r-1} - P' - bP) = -crx^rP$. Again there is a contradiction, for the number 0 is a zero of the left side of multiplicity $\geq r+1$, while it is a zero of the right side of multiplicity r. Our assumption that Q has positive degree is no longer tenable; hence Q is a constant, which we can assume is unity.

From (2.3) we obtain

(2.4) $$xP' + bxP = 1.$$

As before, since P is a polynomial in x, the degree of the left side = degree of $(bxP) > 0 =$ degree of the right side. We have proved that there is no polynomial P satisfying (2.4), hence no rational function satisfying (2.2). Consequently $\int(e^{bx}/x)dx$ with $b \neq 0$ is not elementary.

Example 3: $(\sin x)/x$. It is clear that if $f(x) = u(x) + iv(x)$, where $u(x)$ and $v(x)$ are real valued functions, then

$$\mathcal{R}\int f(x)dx = \int \mathcal{R}f(x)dx = \int u(x)dx,$$

$$\mathcal{I}\int f(x)dx = \int \mathcal{I}f(x)dx = \int v(x)dx,$$

and if $\int f(x)dx$ is elementary, both $\int u(x)dx$ and $\int v(x)dx$ are elementary. ($\mathcal{R}$ and $\mathcal{I}$ stand for "the real and imaginary parts of," respectively.)

Although $(\sin x)/x$ is not in the form of Liouville's Theorem, by Euler's relation ($e^{ix} = \cos x + i \sin x$), we have $(\sin x)/x = \mathcal{I}(e^{ix}/x)$. Since e^{ix}/x does not possess an elementary integral, by example 2, neither does $\mathcal{I}(e^{ix}/x) = (\sin x)/x$.

Example 4: $1/\log x$. Again Liouville's theorem is not immediately applicable. If $y = \log x$, then $\int (1/\log x) dx = \int (e^y/y) dy$. By Example 2, the latter integral is not elementary, hence the same is true of the former.

Example 5: $(x^2 + ax + b)e^x/(x-1)^2$, with a and b constants. Not only do the usual integration techniques require a considerable amount of skill, but there is no a priori assurance that we could find the integral, if it exists. Let us apply the test.

Assume $\int [(x^2 + ax + b)e^x/(x-1)^2] dx = Re^x = Pe^x/Q$. Then

(5.2) $$(x^2 + ax + b)Q^2 = (P'Q - Q'P + PQ)(x-1)^2$$

(5.3) $$Q(Q(x^2 + ax + b) - (x-1)^2 P' - (x-1)^2 P) = - Q'P(x-1)^2.$$

Assume Q has positive degree, and let α be a zero of Q of multiplicity r. If $\alpha \neq 1$, α is a zero of the left side of multiplicity $\geq r$, but a zero of the right side of multiplicity $r-1$. This is a contradiction, hence $\alpha = 1$ and $Q = (x-1)^r$. Substituting this into (5.3) we find

$$(x-1)^r[(x-1)^r(x^2 + ax + b) - (x-1)^2 P' - (x-1)^2 P] = - r(x-1)^{r+1} P.$$

Using the fact that the multiplicities of 1 as a zero of the left and right sides must be the same, we see that $r = 1$; i.e. $Q = (x-1)$. In the last equation we can cancel a common factor of $(x-1)^2$ from both sides, giving

$$(x^2 + ax + b) - (x-1)P' - (x-1)P = -P,$$

$$(x-1)P' + (x-2)P = x^2 + ax + b.$$

P is clearly linear, $P = cx + d$, which means

$$cx - c + cx^2 + dx - 2cx - 2d = x^2 + ax + b.$$

Since these two polynomials are identical, the coefficients of like powers are the same, $c = 1$; $d - c = a$, $-c - 2d = b$, whence $b = -2a - 3$.

Consequently $\int [(x^2 + ax + b)e^x/(x-1)^2] dx$ is elementary if and only if $b = -2a - 3$, in which case the integral is $e^x(x + a + 1)/(x - 1) + C$.

By using one unproved theorem, we have seen how it is possible for even the beginning calculus student to test the integrability of certain transcendental functions, and the logical structure of the test, though not trivial, is sufficiently similar to others he has seen for it to be easily grasped (e.g., the test for the rationality of say $\sqrt{2}$). Although it is obvious that a table of functions which cannot be integrated could be constructed by a careful analysis of examples, to do so for the student would be no better than is done at present. Rather we feel that the student, by applying the test to a few functions, will have a fruitful introduction to differential equations and will gain well-founded confidence in his ability to follow and reconstruct proofs requiring more than one or two steps. Also, the last example illustrates the usefulness of the test as an integration technique.

References

1. J. Liouville, Memoire sur l'integration d'une classe de fonctions transcendantes, J. Reine Angew. Math., vol. 13, 1835, pp. 93–118.
2. J. Ritt, Integration in Finite Terms, New York, 1948.

THE TWO FUNDAMENTAL THEOREMS OF CALCULUS

F. CUNNINGHAM, Jr., Bryn Mawr College

This note is purely polemical, rather than offering anything new. The Fundamental Theorems of calculus are those which state the quasi-inverse relationship between differentiation and integration. There are really two of them, and they are independent of each other.

THEOREM 1. *If F is differentiable on $[a, b]$ and F' is (Riemann) integrable there, then $\int_a^b F' = F(b) - F(a)$.*

Proof. For any partition $a = x_0 < x_1 < \cdots < x_n = b$ we have

$$F(b) - F(a) = \sum_{i=1}^n [F(x_i) - F(x_{i-1})]$$

$$= \sum_{i=1}^n F'(c_i)(x_i - x_{i-1}) \qquad c_i \in (x_{i-1}, x_i)$$

by the Mean Value Theorem for derivatives. As the partition gets finer, the Riemann sums in the last line converge by definition to $\int_a^b F'$, because by hypothesis this limit exists.

THEOREM 2. *If f is continuous on $[a, b]$ and $F(x) = \int_a^x f$ for $x \in [a, b]$ then $F'(x) = f(x)$.*

The well-known proof consists in estimating

$$[F(x + \Delta x) - F(x)]/\Delta x = \int_x^{x + \Delta x} f \Delta x,$$

and then using the continuity of f at x to get $f(x)$ for the limit as $\Delta x \to 0$.

The procedure followed by so many current calculus texts, that it has to be regarded as standard, is to prove Theorem 2 first, and then base the proof of Theorem 1 on it. (One compares F with G defined by $G(x) = \int_a^x F'$. Since F and G have the same derivative by Theorem 2, they differ by a constant. Hence $F(b) - F(a) = G(b) - G(a) = \int_a^b F'$.) I submit that this procedure suffers from the following defects:

(a) It proves Theorem 1 only with the additional and unnecessary hypothesis that F' be continuous.

(b) Making Theorem 1 depend on Theorem 2 obscures the fact that the two theorems are saying different things, having different applications, and may give the student the impression that Theorem 2 is *the* fundamental theorem. This also requires that Theorem 2 be presented first, whereas actually Theorem 1 most naturally precedes Theorem 2, both in motivation (i.e. usefulness) and in

---

* From AMERICAN MATHEMATICAL MONTHLY, vol. 72 (1965), pp. 406–407.

ease of understanding. (To understand Theorem 2, you have to envisage the indefinite integral defined by replacing one limit of integration by x.)

The need for Theorem 1 is immediate. No sooner has one motivated the definite integral by applications such as area and work and then produced the definition itself, than one wants to evaluate definite integrals. Theorem 1 tells how to do this in those numerous cases where the integrand can be recognized as the derivative of some function F. The proof given above is simple, direct and rigorous, and requires no machinery which is not available at this point. I feel further that it connects with the motivating applications of the definite integral in a way which encourages understanding. One annoying detail which the student will of course not understand is that to apply Theorem 1, you have to know that the integral exists. How much to insist on this will depend on taste and circumstances.

The more sophisticated problem to which Theorem 2 is applicable is the creation of transcendental functions, for instance the logarithm, having a given derivative. You need first an existence theorem to make $F(x) = \int_a^x f$ defined (f being continuous), and then Theorem 2 to conclude that $F' = f$ as desired.

BIBLIOGRAPHIC ENTRIES: THEORY OF INTEGRATION

Except for the entries labeled MATHEMATICS MAGAZINE, the references below refer to the AMERICAN MATHEMATICAL MONTHLY.

1. Dean, G. R., Integration as a summation, vol. 10, p. 34.

   Proof that $\int_a^{a+h} f(x)dx = \lim_{n \to \infty} \sum_{k=0}^n f(a+kh/n)h/n$ when f has a convergent Taylor expansion.

2. Huntington, E. V., On setting up a definite integral without the use of Duhamel's theorem, vol. 24, p. 271.
3. Slobin, H. L., A theorem on improper definite integrals, vol. 34, p. 265.

   By treating even and odd parts of f separately, the author proves that $\int_{-\infty}^{\infty} f(x)dx = \int_{-\infty}^{\infty} f(x - x^{-1})dx$, provided both integrals exist.

4. Brown, A. B., A proof of the Lebesgue condition for Riemann integrability, vol. 43, p. 396.
5. Agnew, R. P., Convergence in mean and Lebesgue integration, vol. 44, p. 4.
6. Walsh, J. L., and Sewell, W. E., Note on degree of approximations to an integral by Riemann sums, vol. 44, p. 155.
7. Hildebrandt, T. H., Definitions of Stieltjes integrals of the Riemann type, vol. 45, p. 265.
8. Shohat, J. A., Definite integrals and Riemann sums, vol. 46, p. 538.
9. Robbins, H. E., A note on the Riemann integral, vol. 50, p. 617.
10. Rosenthal, A., On Bliss' substitute for Duhamel's principle, vol. 60, p. 409.
11. Rademacher, H., On the condition of Riemann integrability, vol. 61, p. 1.
12. Livingston, A. E., A necessary condition for the convergence of $\int_a^\infty f(x)\, dx$, vol. 61, p. 250.

13. Hoyt, J. P., A natural approach to the fundamental theorem of the integral calculus, vol. 61, p. 413.

14. Schillo, Paul, On the applications of the fundamental theorem of integral calculus, vol. 63, p. 340.

15. Potts, D. H., Elementary integrals, vol. 63, p. 545.

> Discusses the commonly asked question, "When is the integral of an elementary function again an elementary function?" See also Mead, vol. 68, p. 152, reproduced on pp. 274–278, this book.

16. Sklar, Abe, On the definition of the Riemann integral, vol. 67, p. 897.

> The Riemann integral as an ordinary limit of an ordinary sequence.

17. Kelman, R. B., and Rivlin, T. J., Conditions for the integrand of an improper integral to be bounded or tend to zero, vol. 67, p. 1019.

18. Marcus, S., On the Riemann integral in two dimensions, vol. 71, p. 544.

19. Morduchow, M., Integrals and equal division sums, MATHEMATICS MAGAZINE, vol. 27, p. 65.

20. Manheim, J., A classroom presentation of the definite integral, MATHEMATICS MAGAZINE, vol. 34, p. 157.

21. Pursell, L. E., Riemann integrals as mappings in elementary applications, MATHEMATICS MAGAZINE, vol. 37, p. 311.

(b)

TECHNIQUES OF INTEGRATION

THE DIRICHLET FORMULA, AND INTEGRATION BY PARTS*

L. M. GRAVES, University of Chicago

It seems not to have been generally noticed that the calculus formula for integration by parts is a special case of the Dirichlet formula for interchange of order of integration in an iterated integral, namely,

$$\int_a^b \int_a^x F(x, s)ds\, dx = \int_a^b \int_s^b F(x, s)dx\, ds.$$

This relation holds true for either Riemann or Lebesgue integrals. The essential idea can be indicated by specializing the indefinite integrals as follows. Let

$$f(x) = \int_a^x f'(s)ds, \quad \int_b^x g(s)ds = h(x).$$

Then

$$\int_a^b f(x)g(x)dx = \int_a^b \int_a^x f'(s)g(x)ds\, dx$$

$$= \int_a^b \int_s^b f'(s)g(x)dx\, ds = -\int_a^b f'(s)h(s)ds.$$

Professor C. J. Coe has used the proof of this relation as an exercise for some of his classes at the University of Michigan. The relation first came to my attention through a calculus of variations problem in which the first variation could be transformed by the use of Dirichlet's formula, but not by the usual integration by parts.

---

* From AMERICAN MATHEMATICAL MONTHLY, vol. 38 (1931), pp. 277–278.

A REMARK ON INTEGRATION BY PARTS*

J. L. BORMAN, Purdue University

In employing the formula $\int u\,dv = uv - \int v\,du$, one makes a skillful choice for u and dv so as to have a product $v\,du$ which is readily integrable. Once the choice for dv is made, the function v is obtained by integration. The constant of integration is invariably chosen as zero, since it is felt that this is the simplest choice to make. The point of this note is that the popular choice is, in some cases, the worst rather than the best.

For the sake of completeness let us examine the theory. Suppose it is required of us to integrate

$$\int f(x)g(x)dx$$

and we let

$$u = f(x), \quad dv = g(x)dx.$$

If $G(x)$ is a function such that $G'(x) = g(x)$, then

$$\int f(x)g(x)dx = f(x)G(x) - \int f'(x)G(x)dx.$$

But it is equally true that

$$\int f(x)g(x)dx = f(x)[G(x) + c] - \int f'(x)[G(x) + c]dx$$

for any constant c.

The object is to choose c so that the last integration is as simple as possible. In most cases the choice $c=0$ is best; indeed, the choice is made automatically, but this is not a rule without exceptions as the following example shows.

In the integral

$$\int x \arctan x \, dx$$

let $u = \arctan x$ and $dv = x\,dx$, then

$$du = dx/(1 + x^2) \quad \text{and} \quad v = x^2/2 + c.$$

Now if we let $c = 0$,

$$\int x \arctan x \, dx = \frac{x^2}{2} \arctan x - \int \frac{x^2 dx}{2(1 + x^2)}.$$

---

* From AMERICAN MATHEMATICAL MONTHLY, vol. 51 (1944), pp. 32–33.

While the last integral is not difficult to evaluate we submit that a tremendous simplification can be produced by choosing $c=1/2$. In this case

$$\int x \arctan x \, dx = \frac{(x^2+1)}{2} \arctan x - \frac{1}{2}\int dx.$$

And now the last integration is obvious.

The reader might be interested in constructing other examples of his own. The method seems specially suitable in cases where one is asked to integrate such functions as $\arctan \sqrt{x+k}$ and $\log(x+k)$.

UNDETERMINED COEFFICIENTS IN INTEGRATION*

J. B. REYNOLDS, Lehigh University

Teachers of calculus usually find difficulty in getting students to be able to integrate readily expressions in which the integrand is a product of an exponential term and a binomial factor each term of which contains a sine or a cosine of an angle. The integration is usually accomplished by integrating by parts twice. Sometimes a table of integrals is resorted to and the student never really understands how the integral was found. The integration of the forms cited can be performed readily by the use of undetermined coefficients. Suppose we want to integrate the following product

$$e^{3x}(2 \sin 5x - 4 \cos 5x)dx.$$

A little thought shows that the only terms the integral can contain are given in

$$\int e^{3x}(2 \sin 5x - 4 \cos 5x)dx = e^{3x}(A \sin 5x + B \cos 5x) + C$$

in which A and B are coefficients to be determined. Upon taking the derivative of both sides with respect to x we have

$$e^{3x}(2 \sin 5x - 4 \cos 5x) = e^{3x}([3A - 5B] \sin 5x + [3B + 5A] \cos 5x).$$

Upon cancelling e^{3x} from both sides and equating coefficients of $\sin 5x$ and $\cos 5x$ we have

$$3A - 5B = 2 \quad \text{and} \quad 3B + 5A = -4$$

whence $A = -7/17$ and $B = -11/17$ and, therefore,

$$\int e^{3x}(2 \sin 5x - 4 \cos 5x)dx = -e^{3x}(7 \sin 5x + 11 \cos 5x)/17 + C.$$

It is equally easy to apply this method to the integral of the product of a sine and a cosine of two angles. Thus, if we asume

$$\int \sin 3x \cos x \, dx = A \sin 3x \sin x + B \cos 3x \cos x + C$$

we find after differentiating and equating coefficients of like terms $A - 3B = 1$ and $3A - B = 0$, giving $A = -1/8$ and $B = -3/8$.

Of course the method can be used to derive general formulas for the integration of these and other forms.

---

* From AMERICAN MATHEMATICAL MONTHLY, vol. 54 (1947), pp. 37–38.

INTEGRATION BY PARTS*

K. W. FOLLEY, Wayne University

A schematic method for integrating by parts the product of two functions of x is illustrated in the following examples.

EXAMPLE 1. The product of a power of x by a function which can be integrated successively.

$$I_1 = \int x^5 \cos x\,dx.$$

Let $f = x^5$ and $g = \cos x$. Write the derivatives of f with respect to x in one column and the integrals of g with respect to x in a second column, continuing to the row in which $f^{(n)} = 0$.

$$
\begin{array}{ll}
x^5 & \cos x \\
5x^4 & \sin x \\
20x^3 & -\cos x \\
60x^2 & -\sin x \\
120x & \cos x \\
120 & \sin x \\
0 & -\cos x \\
\end{array}
$$

$I_1 = x^5 \sin x + 5x^4 \cos x - 20x^3 \sin x - 60x^2 \cos x + 120x \sin x + 120 \cos x + C.$

EXAMPLE 2. The product of an exponential function by a trigonometric function.

$$I_2 = \int e^{ax} \sin bx\,dx.$$

Let $f = e^{ax}$ and $g = \sin bx$. Proceed as in Example 1, but continue to the row in which the product $f^{(n)}g_n$ is a constant multiple of I_2.

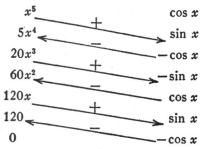

$$I_2 = e^{ax}\left(-\frac{1}{b}\cos bx + \frac{a}{b^2}\sin bx\right) - \frac{a^2}{b^2} I_2$$

$$I_2 = \frac{e^{ax}(a \sin bx - b \cos bx)}{a^2 + b^2} + C.$$

---

* From AMERICAN MATHEMATICAL MONTHLY, vol. 54 (1947), pp. 542–543.

Example 3.

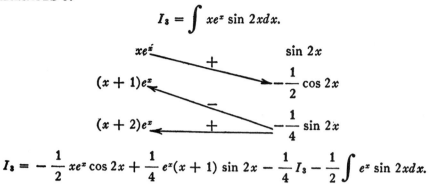

$$I_3 = -\frac{1}{2}xe^x\cos 2x + \frac{1}{4}e^x(x+1)\sin 2x - \frac{1}{4}I_3 - \frac{1}{2}\int e^x \sin 2x\,dx.$$

The final integral is a special case of Example 2. If we substitute its value and solve for I_3, we obtain

$$I_3 = \frac{e^x}{25}\left[(3+5x)\sin 2x + (4-10x)\cos 2x\right] + C.$$

The justification for the preceding method follows from the continued application of the formula for integration by parts.

$$\int fg\,dx = fg_1 - \int f'g_1\,dx = fg_1 - f'g_2 + \int f''g_2\,dx = \cdots$$

$$= fg_1 - f'g_2 + \cdots + (-1)^{n-1}f^{(n-1)}g_n + (-1)^n\int f^{(n)}g_n\,dx,$$

where

$$f^{(i+1)} = \frac{d}{dx}f^{(i)} \quad \text{and} \quad g_{i+1} = \int g_i\,dx.$$

REMARKS ON INTEGRATION BY PARTS*

C. D. OLDS, San Jose State College

This note originated in an attempt to show the calculus student that the formula for integration by parts is not merely a useful trick for finding antiderivatives, but is somewhat deeper than that, and is, moreover, tied up intimately with the early history of elliptic integrals and functions, starting with the work of Fagnano (1682–1766) and leading up to the great discoveries of Abel. We also have here an example of a formula which leads to new results when in a certain situation it seems to be practically useless.

The formula for integration by parts shows that there is a mutual relation between the two integrals $\int y dx$ and $\int x dy$ by virtue of the equation

$$(1) \qquad \int x dy + \int y dx = xy + c.$$

The advantage of this, of course, is that one integration may be substituted for the other. The student soon discovers that there are cases in which no alternation is produced by this substitution and that is when $x = f(y)$ and $y = f(x)$, that is, when x is the same function of y that y is of x. But even in this case the equation (1) is not entirely useless, for the equation

$$(2) \qquad \int f(x) dx + \int f(y) dy = xy + c$$

still holds whether $\int f(x) dx$ can be integrated in finite terms or not, Clearly, the equations $x = f(y)$ and $y = f(x)$, manifestly imply that a symmetrical equation exists between x and y. Fagnano clearly perceived the importance of this simple remark and he deduced important consequences from it.

For example, let x be the abscissa of an hyperbola whose principal semi-axis is unity. Then its arc is given by

$$s = \int \left(\frac{e^2 x^2 - 1}{x^2 - 1} \right)^{1/2} dx,$$

where e is the eccentricity. Let y be another abscissa, so related to the abscissa x that

$$ey = \left(\frac{e^2 x^2 - 1}{x^2 - 1} \right)^{1/2},$$

or

$$e^2 x^2 y^2 = e^2(x^2 + y^2) - 1.$$

---

* From AMERICAN MATHEMATICAL MONTHLY, vol. 56 (1949), pp. 29–30.

Since this equation is symmetrical with respect to x and y, it follows that

$$ex = \left(\frac{e^2y^2 - 1}{y^2 - 1}\right)^{1/2}$$

and hence from (2) that

$$\int \left(\frac{e^2x^2 - 1}{x^2 - 1}\right)^{1/2} dx + \int \left(\frac{e^2y^2 - 1}{y^2 - 1}\right)^{1/2} dy = exy + c.$$

The two integrals in this equation are elliptic and cannot be expressed in terms of a finite number of the elementary functions. Nevertheless the algebraic sum of two such integrals can be so expressed provided a symmetrical relation exists between x and y. Fagnano, about 1718 discovered many theorems of this type in his attempts to rectify the ellipse and the hyperbola. His methods are the starting point of the theory of elliptic functions, and are suggestive of those used by Euler in his proof (1761) of the addition theorem for elliptic integrals, this theorem in turn being a special case of Abel's famous theorem (1826) on the sum of Abelian integrals. Of course, how much of this can be made intelligible to the calculus student depends upon the knowledge and skill of the teacher. The writer has often been surprised at how much of this story can be told and he is convinced that opportunities like this one should not be overlooked. There is always the hope that by such discussions some of the teacher's enthusiasm for mathematics can be imparted to the student along with a little of the spirit for independent investigation. A few students, at least, will be encouraged to read more about the history of mathematics.

Equation (2) can be generalized so that three or more integrals are simultaneously combined. Thus, if two symmetrical equations exist between the variables x, y, z so that $yz = f(x)$, $xz = f(y)$, $xy = f(z)$, then

(3) $$\int f(x)dx + \int f(y)dy + \int f(z)dz = xyz + c.$$

For example, if the two symmetrical equations are

$$x + y + z = a, \qquad x^2 + y^2 + z^2 = b^2,$$

then $f(x) = yz = 1/2(a^2 - b^2) - ax + x^2$, and (3) holds. It is an instructive exercise to verify this directly. Such a generalization is by no means as trivial as it seems, for it was in this manner that W. H. F. Talbot, the English pioneer of photography, was able to show that Abel's theorem is deducible from symmetric functions of the roots of equations and partial fractions. He was in possession of these results by 1825, about three years before Abel published his first theorem on the subject (1828).

References

1. W. H. F. Talbot, Researches in the Integral Calculus, Phil. Trans. Part 1, 1836, pp. 177–215.

2. One of Fagnanos' theorems is proved in Joseph Edwards, the Integral Calculus, vol. 1, Macmillan 1930, p. 582. See also G. N. Watson: The Marquis and the Land-Agent, Math. Gazette, vol. 17, 1933, pp. 5–17.

PARTIAL FRACTIONS WITH REPEATED LINEAR OR QUADRATIC FACTORS*

M. R. SPIEGEL, Rensselaer Polytechnic Institute

The great tedium involved in separating into partial fractions a rational algebraic function in which the denominator has repeated linear or quadratic factors has long been recognized. It is the purpose of this article to indicate a method which the author has found convenient in working with Laplace transforms where such separation into partial fractions is important from the practical viewpoint.

The best way to present the ideas involved in this labor saving device is by means of an example. We pose the problem:

Separate

$$\frac{3x+1}{(x-2)(x^2+x+1)^3}$$

into partial fractions.

Normally one would write

$$\frac{3x+1}{(x-2)(x^2+x+1)^3} \equiv \frac{A}{x-2} + \frac{Bx+C}{(x^2+x+1)^3} + \frac{Dx+E}{(x^2+x+1)^2} + \frac{Fx+G}{x^2+x+1}.$$

Since A can be calculated immediately (by the usual method of multiplying through in this identity by $(x-2)$ and then letting $x \to 2$) we have 6 coefficients to compute, *i.e.*, we have to solve 6 equations in 6 unknowns. Anyone who has used this method must realize how much labor is involved.

Suppose, however, we consider

$$\frac{3x+1}{(x-2)(x^2+x+a)} \equiv \frac{A}{x-2} + \frac{Bx+C}{x^2+x+a} \tag{1}$$

where a is a parameter and A, B, and C are of course functions of a.

Then

$$A = \frac{7}{a+6}.$$

Multiplying (1) through by x and letting $x \to \infty$ we observe that

$$0 = A + B \quad \text{or} \quad B = \frac{-7}{a+6}.$$

---

* From AMERICAN MATHEMATICAL MONTHLY, vol. 57 (1950), pp. 180–181.

Finally letting $x=0$ in (1) we find

$$-\frac{1}{2a} = -\frac{1}{2}\left(\frac{7}{a+6}\right) + \frac{C}{a}$$

or

$$C = -\frac{1}{2} + \frac{1}{2}\left(\frac{7a}{a+6}\right) = -\frac{1}{2} + \frac{1}{2}\left(\frac{7a+42-42}{a+6}\right)$$

$$= -\frac{1}{2} + \frac{7}{2} - \frac{21}{a+6} = 3 - \frac{21}{a+6}.$$

We now practically have the problem solved.
Differentiating (1) twice with respect to a we get

$$\frac{2(3x+1)}{(x-2)(x^2+x+a)^3} \equiv \frac{A''}{x-2} + \frac{B''x+C''}{x^2+x+a} - \frac{2(B'x+C')}{(x^2+x+a)^2} \qquad (2)$$

$$+ \frac{2(Bx+C)}{(x^2+x+a)^3}$$

where the primes denote differentiations with respect to a. In view of the fact that

$$A' = \frac{-7}{(a+6)^2} \qquad B' = \frac{7}{(a+6)^2} \qquad C' = \frac{21}{(a+6)^2},$$

$$A'' = \frac{14}{(a+6)^3} \qquad B'' = \frac{-14}{(a+6)^3} \qquad C'' = \frac{-42}{(a+6)^3}.$$

Putting $a=1$ and using the values thus obtained in (2) we have the result desired.

Further extensions of this and applications to linear repeated factors are obvious.

TRANSFORMATION OF STANDARD INTEGRALS*

V. PUNGA, Rensselaer Polytechnic Institute

In every textbook of integral calculus we can find the tables of standard integrals. The problem of integration consists in reducing the given integral to one of the standard type.

Probably not everybody is aware of the fact that there exist transformations transforming any one of the standard integrals into any other one. The transformation formulas can be found in the following way:

If $\int f_1(v)dv = \phi_1(v)$ and $\int f_2(u)du = \phi_2(u)$, then, in order to transform $\int f_1(v)dv$ into $\int f_2(u)du$, we have to solve $\phi_1(v) = \phi_2(u)$ for v in terms of u, say $v = \psi(u)$, which is the required transformation. The transformation is impossible if we can not solve $\phi_1(v) = \phi_2(u)$ for v in terms of u (in elementary functions).

For example, in order to transform $\int \sec v \, dv (= \ln(\sec v + \tan v))$ into $\int e^u du (= e^u)$, setting $C = 0$, we set up: $\ln(\sec v + \tan v) = e^u$ or $\sqrt{1+\tan^2 v} + \tan v = e^{(e^u)}$. Solving this equation for v, we obtain

$$v = \arctan \tfrac{1}{2}(e^{e^u} - e^{-e^u}).$$

Hence,

$$dv = \frac{2e^u du}{e^{e^u} + e^{-e^u}}, \qquad \sec v = \frac{e^{e^u} + e^{-e^u}}{2}$$

and therefore:

$$\int \sec v \, dv = \int \frac{e^{e^u} + e^{-e^u}}{2} \cdot \frac{2e^u du}{e^{e^u} + e^{-e^u}} = \int e^u du.$$

As a curiosity I suggest the following method of integration (without taking it too seriously). Let the students memorize only one standard integral, say $\int e^u du = e^u + C$, and teach them how to reduce any other integral to this one. In this way we can reduce the table of standard integrals to only one formula.

---

* From AMERICAN MATHEMATICAL MONTHLY, vol. 62 (1955), pp. 363-364.

INTEGRALS OF INVERSE FUNCTIONS

F. D. PARKER, Clarkson College

If $y=f(x)$ and its inverse $x=g(y)$ are single-valued and continuously differentiable, the area under the curve $y=f(x)$ and between the ordinates $x=a$ and $x=b$ may be calculated by using the inverse function. In particular

$$\int_a^b f(x)dx = bf(b) - af(a) - \int_{f(a)}^{f(b)} g(y)dy.$$

This leads to an interesting formula for indefinite integrals

$$\int^x f(x)dx = xf(x) - \int^{f(x)} g(y)dy.$$

For the undergraduate student, this formula has two uses. First, it gives a method of formal integration for the logarithmic function, the inverse circular and hyperbolic functions, and many others. For example,

$$\int \ln x\, dx = x \ln x - \int^{\ln x} e^y dy$$
$$= x \ln x - x.$$

Second, it will sometimes determine whether an integral is an elementary function or not. For example, $\int (\ln x)^{1/2} dx$ is not elementary, since $\int e^{y^2} dy$ is not.

The method of deriving this formula suggests a comparison between the "ring" and "shell" methods of finding a volume of revolution, and the two methods of calculating the first moment of area. These lead to the result

$$\int^x f^2(x)dx = xf^2(x) - 2\int^{f(x)} yg(y)dy.$$

This appears to be less useful, but should suggest to the student the two methods of calculating moments of inertia, which lead to the result

$$\int^x f^3(x)dx = xf^3(x) - 3\int^{f(x)} y^2 g(y)dy.$$

This suggests the generalization

$$\int^x f^n(x)dx = xf^n(x) - n\int^{f(x)} y^{n-1}g(y)dy.$$

This relation is indeed valid and provides at the undergraduate level an interesting problem in differentiation.

The form of these results suggests that they are closely connected to the formulas obtained by integration by parts. In fact, the last relation may be quickly obtained by making the substitution $x=g(y)$ in the left hand side, then integrating by parts taking $u=y^n$ and $dv=g'(y)dy$.

---

* From AMERICAN MATHEMATICAL MONTHLY, vol. 62 (1955), pp. 439–440.

NOTE ON INTEGRATION BY OPERATORS

ROGER OSBORN, University of Texas

Certain convenient devices for performing indefinite integrations may be found in the consideration of differential operators. The validity of the results, and of intermediate steps, may be established by differentiation procedures. Such devices as these can be presented to students of integral calculus, and they are especially useful to students of differential equations. The three indefinite integrations (particular integrals) which are shown here are illustrative of several which can be performed easily by operator methods.

The first is

$$\int e^{ax} \sin bx \, dx = \frac{1}{D} e^{ax} \sin bx = e^{ax} \frac{1}{D+a} \sin bx$$

$$= e^{ax} \frac{D-a}{D^2-a^2} \sin bx = \frac{e^{ax}}{-b^2-a^2}(D-a)\sin bx$$

$$= \frac{e^{ax}}{-b^2-a^2}(b \cos bx - a \sin bx).$$

Similarly,

$$\int e^{ax} \cos bx \, dx = \frac{e^{ax}}{-b^2-a^2}(-b \sin bx - a \cos bx).$$

A third is

$$\int x^n e^{ax} dx = \frac{1}{D} e^{ax} x^n = e^{ax} \frac{1}{D+a} x^n = e^{ax} \frac{1}{a}\left(1+\frac{D}{a}\right)^{-1} x^n$$

$$= \frac{e^{ax}}{a}\left(1 - \frac{D}{a} + \frac{D^2}{a^2} - \frac{D^3}{a^3} + \cdots\right) x^n$$

$$= \frac{e^{ax}}{a}\left[x^n - \frac{nx^{n-1}}{a} + \frac{n(n-1)x^{n-2}}{a^2} - \cdots + (-1)^{n-1}\frac{n!}{a^n}\right].$$

---

* From AMERICAN MATHEMATICAL MONTHLY, vol. 64 (1957), p. 431.

INTEGRATION BY MATRIX INVERSION*

WILLIAM SWARTZ, Montana State College

The integration of several functions using differential operators was considered by Osborn.† The integration of these and certain other functions by matrix inversion can furnish an application of several aspects of matrix theory of interest to the student of matrix algebra.

Let V be the vector space of differentiable functions. Let the n-tuple f be a basis spanning a subspace S of V which is closed under differentiation. Then differentiation comprises a linear transformation T of S into itself. If the matrix A represents T relative to f, then when A is nonsingular the elements of fA^{-1} yield antiderivatives of the elements of f.

To integrate $e^{ax} \sin bx$ and $e^{ax} \cos bx$ consider $f = (e^{ax} \sin bx, e^{ax} \cos bx)$.

Then
$$fT = (ae^{ax} \sin bx + be^{ax} \cos bx, -be^{ax} \sin bx + ae^{ax} \cos bx)$$

and
$$A = \begin{pmatrix} a & -b \\ b & a \end{pmatrix}.$$

Furthermore
$$A^{-1} = \frac{1}{a^2 + b^2} \begin{pmatrix} a & b \\ -b & a \end{pmatrix}$$

and then
$$fA^{-1} = \left(\frac{e^{ax}}{a^2 + b^2} (a \sin bx - b \cos bx), \frac{e^{ax}}{a^2 + b^2} (b \sin bx + a \cos bx) \right)$$

yields antiderivatives of the elements of f.

To derive the formula

(1) $$\int x^n e^x dx = e^x [x^n - nx^{n-1} + n(n-1)x^{n-2} - \cdots + (-1)^n n!]$$

for positive integers n consider $f = (e^x, xe^x, x^2 e^x, \cdots, x^n e^x)$. Then
$$fT = (e^x, e^x + xe^x, \cdots, nx^{n-1}e^x + x^n e^x)$$

---

*From AMERICAN MATHEMATICAL MONTHLY, vol. 65 (1958), pp. 282–283.
†Roger Osborn, Note on integration by operators, this MONTHLY, vol. 64, 1957, p. 431. [*Ed. note:* Reprinted on p. 331, this book.]

and there follows the interesting matrix

$$A = \begin{pmatrix} 1 & 1 & 0 & \cdot & \cdot & \cdot & 0 & 0 & 0 \\ 0 & 1 & 2 & \cdot & \cdot & \cdot & 0 & 0 & 0 \\ 0 & 0 & 1 & \cdot & \cdot & \cdot & 0 & 0 & 0 \\ \cdot & \cdot & \cdot & \cdot & \cdot & \cdot & \cdot & \cdot & \cdot \\ \cdot & \cdot & \cdot & \cdot & \cdot & \cdot & \cdot & \cdot & \cdot \\ \cdot & \cdot & \cdot & \cdot & \cdot & \cdot & \cdot & \cdot & \cdot \\ 0 & 0 & 0 & \cdot & \cdot & \cdot & 1 & n-1 & 0 \\ 0 & 0 & 0 & \cdot & \cdot & \cdot & 0 & 1 & n \\ 0 & 0 & 0 & \cdot & \cdot & \cdot & 0 & 0 & 1 \end{pmatrix}$$

Since only an antiderivative of the last element of f is required, one is only interested in the last column of A^{-1}. Due to the peculiar form of A the inverse is easily deduced. One surmises that the last column of A^{-1} is the transpose of the row

(2) $\left((-1)^n n!, \ (-1)^{n-1} n!, \ (-1)^{n-2} \dfrac{n!}{2!}, \ \cdots, \ n(n-1), \ -n, \ 1 \right).$

That this supposition is correct may be verified by induction on n. In this connection it is useful to consider the $(n+2)$ rowed matrix corresponding to A as a partitioned matrix containing A as a principal submatrix. Finally one notes that multiplication of (2) by f yields the required formula (1).

A USEFUL INTEGRAL FORMULA*

DONALD K. PEASE, University of Connecticut, Hartford

Integrals of the type $\int e^{ax} \sin bx\, dx$ are bothersome because the formulas are confusing to remember and direct integration is tedious to perform. The formula developed here is relatively easy to remember and can be used in a variety of situations.

Consider $\int f(x)g(x)dx$ where $f''(x) = hf(x)$ and $g''(x) = kg(x)$. Thus f (and of course g) is circular or hyperbolic sine or cosine, or else exponential. Note that $k\int g(x)dx = g'(x)$. Integration by parts gives

$$k \int f(x)g(x)dx = f(x)g'(x) - \int f'(x)g'(x)dx.$$

Again integrating by parts, we have

$$k \int f(x)g(x)dx = f(x)g'(x) - f'(x)g(x) + h \int f(x)g(x)dx.$$

Solving algebraically for the integral we have the formula

$$\int f(x)g(x)dx = \frac{1}{h-k}[f'(x)g(x) - f(x)g'(x)] + C \qquad h \neq k.$$

Examples are:

$$\int e^{ax} \sin bx\, dx = \frac{1}{a^2 + b^2}[ae^{ax} \sin bx - be^{ax} \cos bx] + C.$$

$$\int \cosh ax \sin bx\, dx = \frac{1}{a^2 + b^2}[a \sinh ax \sin bx - b \cosh ax \cos bx] + C.$$

---

* From AMERICAN MATHEMATICAL MONTHLY, vol. 66 (1959), p. 908.

AN EXTENSION OF INTEGRATION BY PARTS*

JAMES W. BROWN, Stanford University

The undergraduate mathematics student spends much of his time learning basic formulas which become familiar tools in applications. An example is the method of Integration by Parts which he must often apply a number of times in the same problem. There is a useful extension of this method which can be simply stated and easily remembered.

The basic integration formula is given in the operational notation as

$$D^{-1}(fg') = fg - D^{-1}(gf').$$

If $g = D^{-1}F$, an indefinite integral of the function fF is given by

(1) $$D^{-1}(fF) = fD^{-1}(F) - D^{-1}[D(f)D^{-1}(F)].$$

It follows by mathematical induction that

(2) $$D^{-1}(fF) = \sum_{i=0}^{n}(-1)^i D^i(f)D^{-(i+1)}(F) - (-1)^n D^{-1}[D^{n+1}(f)D^{-(n+1)}(F)].$$

Equation (1) is the case for $n=0$, equation (2) is the inductive hypothesis, and the conclusion is that

$$D^{-1}(fF) = \sum_{i=0}^{n}(-1)^i D^i(f)D^{-(i+1)}(F)$$
$$- (-1)^n\{D^{n+1}(f)D^{-(n+2)}(F) - D^{-1}[D^{n+2}(f)D^{-(n+2)}(F)]\}$$
$$= \sum_{i=0}^{n+1}(-1)^i D^i(f)D^{-(i+1)}(F)$$
$$- (-1)^{n+1}D^{-1}[D^{n+2}(f)D^{-(n+2)}(F)].$$

If f is a polynomial of degree n, $D^{n+1}(f) = 0$ and (2) reduces to the integration formula:

(3) $$D^{-1}(fF) = \sum_{i=0}^{n}(-1)^i D^i(f)D^{-(i+1)}(F).$$

This result may be thought of as an algorithm where the polynomial f is successively differentiated, the function F is successively integrated, alternating signs are affixed to the terms, and an arbitrary constant is added for generality. For example,

$$\int (x^2 + x + 1)e^{2x}dx = D^{-1}[(x^2 + x + 1)e^{2x}]$$
$$= (x^2 + x + 1)(\tfrac{1}{2}e^{2x}) - (2x + 1)(\tfrac{1}{4}e^{2x}) + 2(\tfrac{1}{8}e^{2x}) + C.$$

---

* From AMERICAN MATHEMATICAL MONTHLY, vol. 67 (1960), p. 372.

USE OF HYPERBOLIC SUBSTITUTION FOR CERTAIN TRIGONOMETRIC INTEGRALS

WILLIAM K. VIERTEL, State University Agricultural and Technical College, Canton, N. Y.

In the "good old days," that is, the first half of this century, there was taught in at least one of the better engineering colleges in the United States, an ingenious method of integration of certain trigonometric expressions, which is now on the way to becoming a lost art. Not only is it ingenious, but it includes some very good basic mathematics.

As taught to this writer, the method was theoretically faulty, but it worked—that is, it gave results of integration problems which could be verified to be correct by the reverse process of differentiation. The mathematics teachers at the above-mentioned college at the time were engineers rather than mathematicians, and so did not concern themselves with what they considered to be mathematical technicalities; they were interested only in results.

The method was called by its author "imaginary substitution," since it involved complex numbers. A more accurate term would be hyperbolic substitution. It is applicable to integrals of the form

$$\int \cos^m x \sin^n x \, dx$$

where m and n are integers and $m+n$ is odd negative. For example,

	m	n	$m+n$
$\int \tan^2 x \sec x \, dx = \int \cos^{-3} x \sin^2 x \, dx$	-3	2	-1
$\int \csc^3 x \, dx = \int \sin^{-3} x \, dx$	0	-3	-3
$\int \dfrac{dx}{\cos x \sin^2 x}$	-1	-2	-3

These integrations done by any other method may be extremely tedious.

The method illustrated here is a mathematically correct version of the original method.

The substitution set used is the appropriate one of the following two:

Substitution set I: Let $\tan x = \sinh \theta$. Then $\sec^2 x \, dx = d(\tan x) = \cosh \theta \, d\theta$ and

$$\sec x = \sqrt{(1 + \tan^2 x)} = \sqrt{(1 + \sinh^2 \theta)} = \cosh \theta \qquad (-\pi/2 < x < \pi/2).$$

---

* From MATHEMATICS MAGAZINE, vol. 38 (1965), pp. 141–144.

Substitution set II: Let $\cot x = \sinh \theta$. Then $-\csc^2 x \, dx = d(\cot x) = \cosh \theta \, d\theta$ and
$$\csc x = \sqrt{(1 + \cot^2 x)} = \sqrt{(1 + \sinh^2 \theta)} = \cosh \theta \qquad (0 < x < \pi/2).$$

Both of these transformations are 1-1 and monotone, and, therefore, acceptable.

In using this method, it will need to be recalled that $e^\theta = \cosh \theta + \sinh \theta$ from which it follows that $\theta = \ln(\cosh \theta + \sinh \theta)$.

How these substitutions are used is shown in the following examples.

Example 1. $\int \sec x \, dx = \int (\cos x)^{-1} dx$ \qquad $(m = -1, n = 0, m+n = -1)$.

$$\int \sec x \, dx = \int \frac{\sec^2 x \, dx}{\sec x} = \int \frac{d(\tan x)}{\sec x}.$$

Using substitution set I, we obtain

$$\int \frac{\cosh \theta \, d\theta}{\cosh \theta} = \int d\theta = \theta = \ln(\cosh \theta + \sinh \theta) = \ln(\sec x + \tan x).$$

Example 2. $\int \csc^3 x \, dx = \int \sin^{-3} x \, dx$ \qquad $(m = 0, n = -3, m+n = -3)$,

$$\int \csc^3 x \, dx = -\int \csc x(-\csc^2 x \, dx) = -\int \csc x \, d(\cot x).$$

Using substitution set II, we obtain

$$-\int (\cosh \theta)(\cosh \theta \, d\theta) = -\int \cosh^2 \theta \, d\theta = -\frac{1}{2}\int (1 + \cosh 2\theta) d\theta$$

$$= -\frac{1}{2}\int (d\theta + \cosh 2\theta \, d\theta) = -\frac{1}{2}\theta - \frac{1}{4}\sinh 2\theta$$

$$= -\frac{1}{2}\theta - \frac{1}{2}\sinh \theta \cosh \theta$$

$$= -\frac{1}{2}\ln(\cosh \theta + \sinh \theta) - \frac{1}{2}\sinh \theta \cosh \theta$$

$$= -\frac{1}{2}\ln(\csc x + \cot x) - \frac{1}{2}\cot x \csc x$$

$$= \frac{1}{2}\ln(\csc x - \cot x) - \frac{1}{2}\cot x \csc x.$$

Example 3. $\int \tan^2 x \sec x \, dx = \int \cos^{-3} x \sin^2 x \, dx$ $(m = -3, n = 2, m+n = -1)$.
This integral may be rewritten as

$$\int \frac{\tan^2 x \sec^2 x \, dx}{\sec x} = \int \frac{\tan^2 x \, d(\tan x)}{\sec x}.$$

Using substitution set I, we obtain

$$\int \frac{\sinh^2 \theta \cosh \theta \, d\theta}{\cosh \theta} = \int \sinh^2 \theta \, d\theta = \frac{1}{2} \int (\cosh 2\theta - 1) d\theta$$

$$= \frac{1}{4} \sinh 2\theta - \frac{1}{2} \theta = \frac{1}{2} \sinh \theta \cosh \theta - \frac{1}{2} \ln(\cosh \theta + \sinh \theta)$$

$$= \frac{1}{2} \tan x \sec x - \frac{1}{2} \ln(\sec x + \tan x)$$

$$= \frac{1}{2} \tan x \sec x + \frac{1}{2} \ln(\sec x - \tan x).$$

Example 4. $\int \dfrac{dx}{\sin^2 x \cos x}$ $(m = -1, n = -2, m+n = -3)$.

This integral may be rewritten as

$$\int \frac{\sec^3 x \, dx}{\tan^2 x} = \int \frac{(\sec x)(\sec^2 x \, dx)}{\tan^2 x} = \int \frac{\sec x \, d(\tan x)}{\tan^2 x}.$$

Using substitution set I, we obtain

$$\int \frac{(\cosh \theta)(\cosh \theta \, d\theta)}{\sinh^2 \theta} = \int \coth^2 \theta \, d\theta = \int (\csch^2 \theta + 1) d\theta$$

$$= -\coth \theta + \theta = \frac{-\cosh \theta}{\sinh \theta} + \theta$$

$$= \frac{-\cosh \theta}{\sinh \theta} + \ln(\cosh \theta + \sinh \theta)$$

$$= \frac{-\sec x}{\tan x} + \ln(\sec x + \tan x)$$

$$= -\csc x + \ln(\sec x + \tan x).$$

(Constants of integration have been omitted in all cases in the interest of brevity.)

The present writer does not know who originated this very ingenious method. In its imperfect form, it was taught to several generations of engineering students at Stevens Institute of Technology by the late Professor Charles O. Gunther, who learned it from his predecessor, Professor J. Burkitt Webb. It was included in a text-book entitled "Integration by Trigonometric and Imaginary Substitution" by Professor Gunther, published by Van Nostrand in 1907. This book never had a large sale and has long been out of print.

Professor Gunther called the method "imaginary substitution," because he used the substitution

$$\tan x = i \sin \theta,$$

from which he obtained $d(\tan x) = i \cos \theta \, d\theta$, and

$$\sec x = \sqrt{(1 + \tan^2 x)} = \sqrt{(1 - \sin^2 \theta)} = \cos \theta.$$

As applied to example 1 above, these substitutions work out as follows:

$$\int \sec x \, dx = \int \frac{\sec^2 x \, dx}{\sec x} = \int \frac{d(\tan x)}{\sec x} = \int \frac{i \cos \theta \, d\theta}{\cos \theta} = \int i \, d\theta = i\theta.$$

Since $e^{i\theta} = \cos \theta + i \sin \theta$,

$$i\theta = \ln(\cos \theta + i \sin \theta) = \ln(\sec x + \tan x).$$

This illustrates that the mechanics and the results of the process, as used by Professor Gunther, are the same as described here.

THE INTEGRATION OF INVERSE FUNCTIONS

JOHN H. STAIB, Drexel Institute of Technology

Let f and g be inverse functions. It is often the case that an antiderivative for g is more accessible than one for f. Thus it would be nice to be able to replace integrals of f with integrals of g. The following formula does the trick:

$$\int_a^b f(x)dx = (bb^* - aa^*) - \int_{a^*}^{b^*} g(x)dx,$$

where $u^* = f(u)$. We make several observations about this identity.

1. *A pictorial derivation is available.* From the figure it is easy to see that

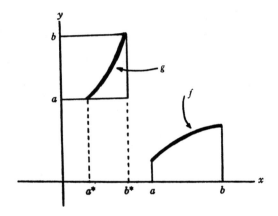

$$\int_a^b f(x)dx = b^*(b - a) - \int_{a^*}^{b^*} [g(x) - a]dx.$$

$$= \cdots$$

2. *Its proof provides a nice application for the integration-by-parts formula.* Let $F' = f$; then

$$\int_{a^*}^{b^*} [g(x) \cdot 1]dx = xg(x)\Big|_{a^*}^{b^*} - \int_{a^*}^{b^*} xg'(x)dx$$

$$= b^*b - a^*a - \left[F(g(x))\Big|_{a^*}^{b^*}\right]$$

$$= [bb^* - aa^*] - [F(b) - F(a)].$$

---

* From MATHEMATICS MAGAZINE, vol. 39 (1966), pp. 223–224.

3. It is instructive. Suppose that we know an antiderivative for g, say G. Then we may write

$$\int_a^x f(u)du = [xf(x) - af(a)] - \int_{f(a)}^{f(x)} g(u)du$$
$$= xf(x) - G(f(x)) + c_a.$$

Thus we have an analogue for the formula

$$f'(x) = \frac{1}{g'(f(x))}.$$

BIBLIOGRAPHIC ENTRIES: TECHNIQUES OF INTEGRATION

Except for the entry labeled MATHEMATICS MAGAZINE, the references below are to the AMERICAN MATHEMATICAL MONTHLY.

1. Dunkel, O., A method of handling certain elementary integrals, vol. 39, p. 163.

$$\int (x^2 + 2Ax + B)^{\pm 1/2} dx.$$

2. Doole, H. P., Integration of certain simple step functions, vol. 44, p. 222.

3. Baker, F. E., Comments on a paper by Dr. Doole (vol. 44, p. 222), vol. 45, p. 679.

4. MacNeish, H. F., Logarithmic integration, vol. 56, p. 25.

Because we have formulas for differentiating products and quotients, students often ask whether there are comparable formulas for integrating products and quotients. This paper gives a procedure with a sufficient condition that answers the question affirmatively for certain combinations of polynomials.

5. Phelps, C. R., "Integration by parts" as a method in the solution of exact differential equations, vol. 56, p. 335.

6. Sydnor, T. E., On integration by parts, vol. 57, p. 479.

7. Ballantine, J. P., Integration by long division, vol. 58, p. 104.

8. Wolinsky, A., An alternate derivation of a well-known integration formula, vol. 58, p. 630.

$$\int \tan x \, dx = \log \sec x.$$

9. Duncan, D. G., Some standard problems in integration simplified, vol. 61, p. 421.

Simplifications occur by the use of appropriate coordinates.

10. Ficken, F. A., Finding the change of variable carrying one finite interval onto another, vol. 67, p. 781.

11. Spiegel, M. R., Remarks on the integration of products of functions, vol. 28, p. 115.

Compare with Folley, vol. 54, p. 542, reproduced on pp. 283–285, this book.

(c)

SPECIAL INTEGRALS

NOTE ON THE QUADRATURE OF THE PARABOLA*

OTTO DUNKEL, Washington University

The article by Professor Moritz entitled "On the Quadrature of the Parabola" in the November issue of the MONTHLY† has suggested to the writer to present another derivation of the same result, since, in addition to being fairly simple, this second method follows directly the classic process of defining an area as the common limit of an inferior and a superior sum. This development might be found easy enough to serve as an illustrative example in the presentation of the summation formula in the integral calculus.

Let the equation of the curve be $y = x^m$, $m = $ a positive integer, and suppose that it is desired to obtain the area between the curve, the x-axis and the ordinates at $x=a$ and $x=b$, where b is greater than a and both are positive. Divide the interval from a to b on the x-axis into n subintervals, equal or unequal, and upon them as bases erect two sets of rectangles, the one inscribed and the other circumscribed. The sums of the areas of these rectangles are respectively,

$$(1) \qquad I_n = \sum x_i^m (x_{i+1} - x_i), \qquad S_n = \sum x_{i+1}^m (x_{i+1} - x_i).$$

It will be shown that as n becomes infinite so that the length of the longest subinterval, δ, approaches zero, each of these sums approaches the same limit, which by the usual definition is the area desired. This limit will also be determined in the process.

It may easily be seen from a figure, especially when all the subintervals are equal, that

$$(2) \qquad \text{Limit } (S_n - I_n) = 0.$$

This also follows algebraically, for

$$S_n - I_n = \sum (x_{i+1}^m - x_i^m)(x_{i+1} - x_i) \leq \delta \sum (x_{i+1}^m - x_i^m).$$

In the latter summation all the terms cancel except the first and last, and hence the difference, $S_n - I_n$, is less than or equal to $\delta(b^m - a^m)$. Thus it follows that (2) is true.

A quantity will now be determined which lies between I_n and S_n and is independent of n. It is clear that x_{i+1}^m, $x_{i+1}^{m-1}x_i$, $x_{i+1}^{m-2}x_i^2$, $\cdots$, x_i^m form a decreasing

---

* From AMERICAN MATHEMATICAL MONTHLY, vol. 27 (1920), pp. 116–117.
† *Ed. note*: Vol. 26 (1919), pp. 388–390.

sequence of $m+1$ positive terms and hence their arithmetic mean is greater than the smallest term x_i^m. Using this inequality it follows that

$$I_n < \sum (x_{i+1} - x_i) \frac{x_{i+1}^m + x_{i+1}^{m-1} x_i + x_{i+1}^{m-2} x_i^2 + \cdots + x_i^m}{m+1},$$

$$< \frac{1}{m+1} \sum (x_{i+1}^{m+1} - x_i^{m+1}) = \frac{b^{m+1}}{m+1} - \frac{a^{m+1}}{m+1}.$$

In a similar manner an inequality is found for S_n, and hence

(3) $$0 < I_n < \frac{b^{m+1}}{m+1} - \frac{a^{m+1}}{m+1} < S_n.$$

By representing the three quantities I_n, $b^{m+1}/(m+1)-a^{m+1}/(m+1)$, S_n by points on a straight line and by considering the meaning of (2) and (3) as applied to these points, it will be obvious that the common limit of I_n and S_n is $b^{m+1}/(m+1)-a^{m+1}/(m+1)$. This then is the expression for the desired area. If the equation of the curve is $y = px^m$, it will be readily seen that the above result must be multiplied by p.

The same method, with a slight amount of extra manipulation, may be used for negative values of m and also for fractional values, excepting, however, the special case $m = -1$.*

An elementary evaluation of the area of any segment of an ordinary parabola by means of special properties of the curve is given in the *Traité de Géométrie* by Rouché et Comberousse, 2d vol., p. 348 (8th ed., 1912). The properties here used are such as might be given in the ordinary text on analytics. A somewhat similar treatment occurs in the first volume of Goursat-Hedrick's *Mathematical Analysis*, p. 134, and is referred to as one of the processes used by Archimedes. Here the summation of a geometric series is employed, but this may be avoided and the proof simplified by comparing the areas of the interior triangles with certain corresponding exterior triangles. These two proofs are somewhat similar to the one employed by Professor Moritz.

THE PROBABILITY INTEGRAL[†]

C. P. NICHOLAS AND R. C. YATES, United States Military Academy

The following evaluation requires only a knowledge of elementary calculus and thus with reasonable assumptions may be presented at the sophomore level.

Assuming its existence, let

$$\int_0^\infty e^{-x^2}dx = A, \quad \text{where} \quad \int_0^\infty = \lim_{h\to\infty} \int_0^h.$$

If the curve $z = e^{-y^2}$ be revolved about the Z-axis, the surface generated is $z = e^{-(x^2+y^2)}$ with volume:

$$V = 4\int_0^\infty \int_0^\infty e^{-(x^2+y^2)}dxdy$$

$$= 4\int_0^\infty \left[\int_0^\infty e^{-x^2}dx\right] \cdot e^{-y^2}dy$$

$$= 4\int_0^\infty A \cdot e^{-y^2}dy = 4A^2.$$

But, using the method of hollow cylinders:

$$V = 2\pi \int_0^\infty ye^{-y^2}dy = \pi.$$

Thus

$$4A^2 = \pi, \quad A = \frac{\sqrt{\pi}}{2}.$$

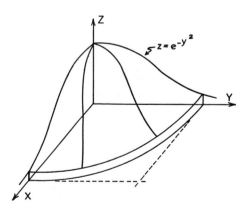

---

* An elementary treatment of this case was given by the writer under the title, A Geometric Treatment of the Exponential Function, in *Washington University Studies*, scientific series, vol. 6, no. 2, p. 33.

† From AMERICAN MATHEMATICAL MONTHLY, vol. 57 (1950), pp. 412–413.

THE CALCULUS OF ABSOLUTE VALUES*

KENNETH O. MAY, Carleton College

The customary way of dealing with functions involving absolute values is to consider separately the cases in which the expressions within absolute value symbols are positive and negative. This is clumsy, particularly where differentiation or integration is involved. It may be responsible for the neglect of such functions, even when they would be appropriate in applications. The fact is that the function defined by $y=|x|$ is a very well behaved continuous function with a continuous derivative everywhere except at the origin. There exist simple formulas by which it may be differentiated and integrated with no more attention to discontinuities or existence than is required for such functions as the negative powers of x.

A useful identity is

$$(1) \qquad |x| = x \operatorname{sg} x$$

where sg is the signum function defined by $\operatorname{sg} x = -1, 0, 1$ according as $x <, =, > 0$. Evidently $D \operatorname{sg} x = 0$ everywhere except at the origin, where $\operatorname{sg} x$ has a jump of 2. We note also that $\operatorname{sg} x = |x|/x = x/|x|$ for $x \neq 0$. Then it is easy to establish that

$$(2) \qquad D|x| = |x|/x.$$

We have also, for $x \neq 0$,

$$(3) \qquad D|x| = \operatorname{sg} x.$$

However, (2) has the advantage of calling attention to the non-existence of the derivative for $x=0$.

With the aid of (2) and the chain rule we can find derivatives of composite functions involving absolute values. By this means it is easy to show that

$$(4) \qquad D|x^n| = nx^{n-1}\operatorname{sg} x^n.$$

This holds for all x if $n>1$. If $n \leq 1$, the derivative fails to exist at the origin. More generally,

$$(5) \qquad D^n|f(x)| = f^{(n)}(x)\operatorname{sg} f(x)$$

wherever the derivative on the left exists.

The maxima and minima of $|f(x)|$ and $f(x)$ are related in a simple way. Obviously $|f(x)|$ has an absolute minimum at any zero of $f(x)$. Since in any interval where $f(x)>0$ the graphs of $|f(x)|$ and $f(x)$ coincide, they have relative maxima and minima at the same points in such intervals. In intervals where $f(x)<0$ the graphs of $|f(x)|$ and $f(x)$ are each other's mirror images in the x-axis, and $|f(x)|$ has a relative maximum (minimum) where $f(x)$ has a relative minimum (maximum). These remarks hold even if derivatives do not exist at the

---
* From AMERICAN MATHEMATICAL MONTHLY, vol. 62 (1955), pp. 651–653.

points in question. However, if the derivative tests are applicable, formula (5) gives the same results.

Integration by parts yields the following formula, and it can easily be verified by differentiating the right member.

(6) $$\int |x|\,dx = \tfrac{1}{2}x|x| = \tfrac{1}{2}x^2 \operatorname{sg} x.$$

Similarly one can find and verify

(7) $$\int |x^n|\,dx = \frac{1}{n+1}|x^n|x = \frac{1}{n+1}x^{n+1}\operatorname{sg} x^n \qquad (n \neq -1).$$

As an application of some of the above ideas, consider the problem of proving that a median of a probability distribution is a point from which the mean of the absolute deviations is a minimum. We consider only the case of a continuous distribution, though the discrete case or the general case may be treated by the same methods. The problem is to minimize the integral

(8) $$F(t) = \int_{-\infty}^{\infty} |x - t| f(x)\,dx$$

where $f(x)$ is non-negative and the integral of $f(x)$ from $-\infty$ to ∞ is 1. The usual proof is rather clumsy. However, using the above formulas, we have

(9) $$F'(t) = \int_{-\infty}^{\infty} \operatorname{sg}(t-x) f(x)\,dx$$

(10) $$= \int_{-\infty}^{t} f(x)\,dx - \int_{t}^{\infty} f(x)\,dx$$

(11) $$= 2\int_{-\infty}^{t} f(x)\,dx - 1.$$

But a median is defined as a value of t that makes the right member of (11) zero. Moreover, since

(12) $$F''(t) = 2f(t) \geq 0$$

for all values of t, $F'(t)$ is monotonically increasing, and equation (11) does yield an absolute minimum of $F(t)$.

As a second example, consider the following problem given in Kaplan's *Advanced Calculus* on page 515. Where is the function $|x^2 - y^2| + i|2xy|$ analytic? Since $u = |x^2 - y^2|$ and $v = |2xy|$ are continuous functions with continuous derivatives in any domain not including points on the axes or on the lines defined by $x^2 = y^2$, we need consider only the Cauchy-Riemann conditions. We have

$$\frac{\partial u}{\partial x} = 2x\operatorname{sg}(x^2 - y^2) \qquad \frac{\partial v}{\partial y} = 2x\operatorname{sg}(xy)$$

$$\frac{\partial u}{\partial y} = -2y\,\text{sg}(x^2 - y^2) \quad -\frac{\partial v}{\partial x} = -2y\,\text{sg}(xy).$$

The condition for analyticity is evidently $\text{sg}(x^2-y^2) = \text{sg}(xy)$, and hence the domains are defined by $xy(x^2-y^2) > 0$. The procedure is much simpler than considering the combinations of cases for different signs of x^2-y^2 and $2xy$.

ANOTHER LOOK AT THE PROBABILITY INTEGRAL*

C. P. NICHOLAS, U. S. Military Academy

A well-known method of showing that $\int_0^\infty e^{-x^2} dx = \sqrt{\pi}/2$ is first to demonstrate that the integral is convergent, and then by a transformation of coordinates to set up the equation

$$\int_0^\infty \int_0^\infty e^{-(x^2+y^2)} dx\, dy = \int_0^{\pi/2} \int_0^\infty e^{-\rho^2} \rho\, d\rho\, d\theta.$$

From this $[\int_0^\infty e^{-x^2} dx]^2 = \pi/4$, and we complete the solution by taking square roots.

An adaptation to instruction at the sophomore level appeared in this MONTHLY, vol. 57, 1950, pp. 412–413.† Simplicity was achieved by certain geometric devices, and by the *assumption* that the improper integral is convergent.

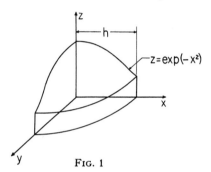

FIG. 1

The feeling of uneasiness that goes with this sweeping assumption can be avoided by the evaluation offered below. It is believed easy enough for a first course, and the final step establishes both the existence and the value of the integral.

Consider the surface generated by revolving about the Z-axis the curve $z = e^{-x^2}$, and let V_1 be the volume (Fig. 1) in the first octant bounded by this surface, the coordinate planes, and the cylinder $x^2 + y^2 = h^2$.

By the method of cylindrical shells

$$V_1 = \int_0^h \frac{\pi}{2} x e^{-x^2} dx = \frac{\pi}{4} [1 - e^{-h^2}].$$

Consider next the volume V_2, bounded as before except that we use the planes $x = h$ and $y = h$ instead of the cylinder $x^2 + y^2 = h^2$ (Fig. 2).

---
*From AMERICAN MATHEMATICAL MONTHLY, vol. 64 (1957), pp. 739–741.
†*Ed. note:* Reprinted on pp. 310–314, this book.

ANOTHER LOOK AT THE PROBABILITY INTEGRAL

By double integration

$$V_2 = \int_0^h \int_0^h e^{-(x^2+y^2)} dx dy = \left[\int_0^h e^{-x^2} dx \right]^2.$$

Consider finally the volume V_3, bounded as for V_1 except that we use the cylinder $x^2+y^2=2h^2$ in lieu of $x^2+y^2=h^2$ (Fig. 3).

Again by the method of cylindrical shells $V_3 = (\pi/4)[1-e^{-2h^2}]$.

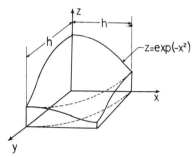

FIG. 2

Since by construction $V_1 < V_2 < V_3$, then for all positive values of h

$$\frac{\pi}{4}[1-e^{-h^2}] < \left[\int_0^h e^{-x^2} dx \right]^2 < \frac{\pi}{4}[1-e^{-2h^2}].$$

Now let h increase without bound, and seek limits of all three members. Evaluating the limits of the first and third members, we have

$$\frac{\pi}{4} \leq \left[\lim_{h \to \infty} \int_0^h e^{-x^2} dx \right]^2 \leq \frac{\pi}{4}.$$

Hence the middle limit must exist, and its value must be $\pi/4$. Therefore $\int_0^\infty e^{-x^2} dx = \frac{1}{2}\sqrt{\pi}$.

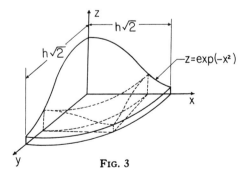

FIG. 3

$\int \sec \theta \, d\theta$*

NORMAN SCHAUMBERGER, Bronx Community College

The integral of $\sec \theta$, which is usually handled by multiplying numerator and denominator by $\sec \theta + \tan \theta$, can be treated in the following interesting manner (suggested by Jesse Douglas of The City College of New York).

The student is aware that the $\sec \theta$ and $\tan \theta$ are closely related. This might suggest that we take differentials of each,

$$d(\sec \theta) = \sec \theta \tan \theta \, d\theta, \qquad d(\tan \theta) = \sec^2 \theta \, d\theta.$$

If we now add the two expressions and factor the right side we obtain $d(\sec \theta + \tan \theta) = (\sec \theta + \tan \theta) \sec \theta \, d\theta$. Dividing by $\sec \theta + \tan \theta$ and integrating, we have the desired result.

---

* From AMERICAN MATHEMATICAL MONTHLY, vol. 68 (1961), p. 565.

NOTE ON $\int_a^x t^y$

M. J. PASCUAL, Watervliet Arsenal, New York

The marked difference in the formula for

$$\int_a^x t^y dt \qquad 0 < a < x$$

when $y = -1$ as compared with the case when $y \neq -1$ seems to imply that there is a "discontinuity." That this is not the case may be shown as follows:

Let

$$f(y) = \int_a^x t^y dt \qquad 0 < a < x$$

so that

$$f(y) = \frac{x^{y+1} - a^{y+1}}{y+1} \qquad \text{for } y \neq -1$$

and applying L'Hospital's Rule we find:

$$\lim_{y \to -1} f(y) = \lim_{y \to -1} \frac{(\ln x) x^{y+1} - (\ln a) a^{y+1}}{1}$$

$$= \ln x - \ln a$$

There are various ways given in the textbooks on calculus to arrive at

$$f(-1) = \int_a^x t^{-1} dt = \ln x - \ln a$$

Hence $f(y)$ is indeed continuous at $y = -1$.

---

* From MATHEMATICS MAGAZINE, vol. 35 (1962), p. 175.

$$\int \sec^3 x\, dx\,\dagger$$

JOSEPH D. E. KONHAUSER, HRB-Singer, Inc.

The standard method for evaluating $\int \sec^3 x\, dx$ is integration by parts. An alternative method is as follows:

$$\int \sec^3 x\, dx = \tfrac{1}{2}\int (\sec^3 x + \sec^3 x)\, dx = \tfrac{1}{2}\int (\sec^3 x + \sec x \tan^2 x + \sec x)\, dx$$

$$= \tfrac{1}{2}\int d(\sec x \tan x) + \tfrac{1}{2}\int \sec x\, dx$$

$$= \tfrac{1}{2}\sec x \tan x + \tfrac{1}{2}\ln(\sec x + \tan x) + C.$$

BIBLIOGRAPHIC ENTRIES: SPECIAL INTEGRALS

Except for the entry labeled MATHEMATICS MAGAZINE, the references below are to the AMERICAN MATHEMATICAL MONTHLY.

  1. Zerr, G. B. M., Integration by elliptic integrals, vol. 6, pp. 258, 302; vol. 7, pp. 67, 100, 218, 242, 285.
  2. Zerr, G. B. M., On the evaluation of certain definite integrals, vol. 11, p. 56.

$$\int_0^\infty e^{-(x^2 + c^2/x^2)}\, dx.$$

  3. Corey, S. A., The evaluation of $\int_0^\pi x^{-1} \sin mx\, dx$, vol. 13, p. 12.

  Uses Euler's summation formula but ignores questions of convergence and estimation of error.

  4. Moritz, R. E., On the quadrature of the parabola, vol. 26, p. 388.

  Compare with Dunkel, vol. 27, p. 116, reproduced on pp. 301–302. in this volume.

  5. Hathaway, A. S., A general type of reduction formula, vol. 29, p. 257.

  Transformations of

$$\int \cos^L \theta \sin^M \theta\, d\theta.$$

  6. Egan, M. F., On the evaluation of an elliptic integral, vol. 30, p. 258.

  Inequalities for

$$\int_0^{\pi/2} (1 - k^2 \sin^2 \phi)^{-1/2}\, d\phi.$$

  7. Bullard, J. A., On the evaluation of certain trigonometric integrals, vol. 40, p. 161.

$$\int_0^\alpha \sin^m x \cos^n x\, dx.$$

---

† From MATHEMATICS MAGAZINE, vol. 38 (1965), p. 45.

8. Rutledge, G., and Douglass, R. D., Evaluation of $\int_0^1 u^{-1} \log u \log^2(1+u)\, du$ and related definite integrals, vol. 41, p. 29.

9. Burton, W. W., and Miller, W. G., The integral of powers of the secant and cosecant, vol. 48, p. 544.

10. Wylie, C. R., Jr., New forms of certain integrals, vol. 49, p. 457.

$$\int x^n \cos x\, dx, \text{ etc.}$$

11. Sandham, H. F., A well-known integral, vol. 53, p. 587.

$$\int_0^\infty e^{-x^2}\, dx.$$

12. Taylor, A. E., On the need for care in using a certain integral formula, vol. 54, p. 156.

$$\int \frac{dx}{a + b \cos x}.$$

13. Hoyt, J. P., Evaluation of a trigonometric integral, vol. 54, p. 221.

$$\int \sin^m x \cos^n x\, dx, \ m,\ n \text{ even}.$$

14. Smith, H. W., Some integral formulas, vol. 56, p. 27.

$$\int \sec^k \theta\, e^{ip\theta}\, d\theta, \text{ etc.}$$

15. Lee, H. L., A different approach to a known integral, vol. 57, p. 413.

$$\int \frac{dx}{(ax+b)(cx+d)}.$$

16. Taylor, A. E., A note on the Poisson kernel, vol. 57, p. 478.

17. Spiegel, M. R., An elementary method for evaluating an infinite integral, vol. 58, p. 555.

$$\int_0^\infty \frac{\sin x}{x}\, dx.$$

18. Walker, A. W., An elementary integral, vol. 59, p. 103.

$$\int \frac{dx}{\sqrt{x^2+c}}.$$

19. Rao, S. K. Lakshmana, On the evaluation of Dirichlet's integral, vol. 61, p. 411.

20. Coleman, A. J., The probability integral, vol. 61, p. 710.

21. Campbell, J. W., A useful reduction integral, vol. 62, p. 117.

$$\int_{p\pi/2}^{q\pi/2} \sin^m \theta \cos^n \theta\, d\theta.$$

22. Olds, C. D., The evaluation of two infinite integrals, vol. 63, p. 575.

$$\int_0^\infty \frac{a \cos mx}{a^2 + x^2}\, dx \quad \text{and} \quad \int_0^\infty \frac{x \sin mx}{a^2 + x^2}\, dx.$$

23. Pennisi, L. L., A method for solving $\int \sin^{2n} ax\, dx$ and $\int \cos^{2n} ax\, dx$, MATHEMATICS MAGAZINE, vol. 29, p. 271.

(d)
APPLICATIONS

A NOTE ON AREAS

R. A. JOHNSON, Brooklyn College

It is well known (cf., for example, Goursat-Hedrick, *Mathematical Analysis*, v. 1, §94) that if x and y are differentiable functions of a parameter t, the integral

(1) $$I = \frac{1}{2} \int_{t_1}^{t_2} (xdy - ydx)$$

represents the vectorial area bounded by the curve $x=x(t)$, $y=y(t)$, and the lines OP_1, OP_2, where P_1, P_2 are the points of the curve corresponding to t_1 and t_2. It is a simple matter to transform this integral into the usual expression for area in polar coordinates.

In one of my classes we arrived at the following striking transformation of the integral, which (though I have made no extensive search) I think I have not seen in print.†

If the parameter t is the ratio y/x, the integral (1) takes the form

$$I = \frac{1}{2} \int_{t_1}^{t_2} x^2 dt.$$

For at once, if $y/x = t$, $xdy - ydx = x^2 dt$.

As an illustration, take the folium of Descartes, whose parametric equations

$$x = 3at(1 + t^3)^{-1}, \qquad y = 3at^2(1 + t^3)^{-1}$$

are known to every undergraduate. We have then for the area bounded by the loop of the curve

$$A = \frac{9}{2} a^2 \int_0^\infty (1 + t^3)^{-2} t^2 dt = \frac{3}{2} a^2.$$

The labor involved here compares favorably with that of the usual process of expressing $\int y dx$ in terms of t.

---

* From AMERICAN MATHEMATICAL MONTHLY, vol. 52 (1945), pp. 209–210.

† While this note is in press I learn that the result is given by de la Vallée Poussin, Cours d'analyse infinitesimale, 1938 edition, vol. 1, p. 320, ex. 7. I still think that it deserves to be better known.

The parameter $t=y/x$ is frequently used. It rationalizes any curve whose equation has the form

$$p_n(x, y) - p_{n-1}(x, y) = 0,$$

where p_r is a homogeneous polynomial of degree r; that is, any curve having a singularity of multiplicity $n-1$ at the origin.

Incidentally, will some reader contribute a historical note on the universal acceptance of the symbol

$$\int P dx + Q dy$$

without parentheses?

SOME REMARKS ON CENTROIDS*

R. B. DEAL AND W. N. HUFF, University of Oklahoma

Recently it was pointed out by two students in our integral calculus classes that the ordinate of the centroid of area of one arch of the cycloid and that of the centroid of the volume of revolution of the arch about the Y-axis were equal. The question was raised as to the reason for this and the present note concerns this problem.

Let us consider an area, A, in the first quadrant, bounded by the curves $y = f(x)$ above and $y = g(x)$ below, the curves intersecting at (a, b) and (c, d). Let $(\bar{x}, \bar{y})$ be the centroid of area A, $(\bar{x}_1, 0)$ the centroid of V_1, the volume of rotation about the x-axis, and $(0, \bar{y}_2)$ the centroid of V_2, the volume of rotation about the y-axis.

Then

$$A\bar{y} = \frac{1}{2}\int_a^c (f^2 - g^2)dx = \frac{V_1}{2\pi},$$

$$V_2\bar{y}_2 = \pi\int_a^c (f^2 - g^2)x\,dx.$$

Since by the theorem of Pappus $V_2 = 2\pi\bar{x}A$, it follows that

$$2\pi\bar{x}A\,\bar{y}_2 = \pi\int_a^c (f^2 - g^2)x\,dx = \bar{x}_1 V_1 = 2\pi A \bar{x}_1 \bar{y}.$$

Hence $\bar{x}\bar{y}_2 = \bar{x}_1\bar{y}$.

This can also be done by double integrals as follows:

$$2\pi A\,\bar{y}\bar{x}_1 = V_1\bar{x}_1 = \int_a^c \int_0^f 2\pi xy\,dy\,dx = V_2\bar{y}_2 = 2\pi A \bar{x}\bar{y}_2.$$

The above result then follows.

If the area in question possesses an axis of symmetry $x = k$, then $\bar{x}_1 = \bar{x} = k$ and $\bar{y}_2 = \bar{y}$. For the arch of the cycloid $k = \pi a$.

---

* From AMERICAN MATHEMATICAL MONTHLY, vol. 60 (1953), pp. 624–625.

SIMPLE PROBLEMS ON ARC LENGTHS

H. S. THURSTON, University of Alabama

The author of a popular text-book[†] in elementary calculus uses the determination of the circumference of a circle as an illustration of the use of the formula $s=\int\sqrt{1+y'^2}$. In a footnote he points out that a vicious circle is involved (although it develops that the girth of a vicious circle is the same as that of its gentle and saddle-broken cousin, viz. $c=2\pi r$) since it was necessary to know in advance that the circumference is $2\pi a$ in order to evaluate arcsin 1 which arises in the solution. He adds that the problem was chosen as an illustration "in order not to diminish the limited supply of simple problems."

A question arises as to the significance of the word "simple." One can easily infer that there is only a finite number of curves the determination of whose arc length is within the capabilities of the college student. On the other hand, the author may mean that there is a limited supply of problems which the average student can work in, let us say, fifteen minutes per problem. While the writer is inclined to agree with the latter interpretation, he wishes to point out that there is an unlimited supply of problems on this topic which involve only the usual techniques of integration.

It will be conceded, I am sure, that a simplification is achieved if $1+y'^2$ is a perfect square. Such functions as $y=\log \cos x$ and $y=\cosh x$ lend themselves to this type of simplification because of the identities $1+\tan^2 x=\sec^2 x$ and $1+\sinh^2 x=\cosh^2 x$. Problems involving the arc lengths of these curves are found in nearly every text-book in elementary calculus. The functions may be regarded as special cases of an infinite set of functions now to be considered.

Let $y=F(x)+G(x)$. Then $1+y'^2=1+F'^2+2F'G'+G'^2$. If we choose as F' and G' functions whose indefinite integrals can be found and such that $4F'G'+1=0$, then $1+y'^2=(F'-G')^2$ and $s=\int_a^b\sqrt{1+y'^2}=\int_a^b|F'(x)-G'(x)|dx$. The infinite variety of choices for F' and G' may well furnish both authors and instructors with a refreshingly new set of problems on the topic under consideration.

As noted above $y=\log \cos x$ and $y=\cosh x$ may be regarded as functions of the type just indicated. The definition of $\cosh x$ represents this function in the form $F(x)+G(x)$ with $4F'G'+1=0$, whereas $y=\log \cos x$ may be written as $y=\log (\cos \frac{1}{2}x-\sin \frac{1}{2}x)+\log (\cos \frac{1}{2}x+\sin \frac{1}{2}x)=F(x)+G(x)$, the condition involving F' and G' again being satisfied.

Of special interest, if simplicity is desired, are those functions for which $F'=ax^m$ and $G'=bx^n$. On imposing the condition $4F'G'+1=0$ we are led to the equations $m+n=0$, $4ab+1=0$, whence, if m is neither 1 nor -1,

(1) $$y = \frac{ax^{m+1}}{m+1} + \frac{x^{1-m}}{4a(m-1)}$$

---

*From AMERICAN MATHEMATICAL MONTHLY, vol. 60 (1953), pp. 705–706.

[†] Love, Differential and Integral Calculus, 4th edition, Macmillan, 1943.

and

(2) $$y = \frac{ax^2}{2} - \frac{1}{4a}\log x, \qquad y = a\log x - \frac{1}{8a}x^2$$

for $m = 1$ and $m = -1$, respectively.

In preparing a set of problems for classroom use the complexity of the functions may vary from one as simple as $y = 3x/4$ (by putting $a = 1$ and $m = 0$ in (1)) to something of the type

$$y = 1/4\left[\left(\log\frac{x-2}{\sqrt{x^2+4}} + \arctan \tfrac{1}{2}x\right) + (8\log x - 4x + x^2 - x^3/3)\right]$$

constructed from

$$F' = \frac{x}{(x-2)(x^2+4)}, \qquad G' = \frac{-(x-2)(x^2+4)}{4x}, \qquad x > 2.$$

It is readily seen that there is no limit to the complexity of the horrible monstrosities which may be constructed after this pattern, and no end to the supply.

BIBLIOGRAPHIC ENTRIES: APPLICATIONS

Except for the entries labeled MATHEMATICS MAGAZINE, the references below are to the AMERICAN MATHEMATICAL MONTHLY.

1. Zerr, G. B. M., Note on the centroid of plane areas, vol. 1, p. 120.

    For special curves.

2. Matz, F. P., The second hyperbolic integral, vol. 1, p. 341.

    Calculation of an integral for arc length of a hyperbola using a series expansion of the integrand. No discussion of the error.

3. Zerr, G. B. M., Note on areas and volumes, vol. 1, p. 380.

    Calculation of the area of the region bounded by the curve $(x/a)^{2/p} + (y/a)^{2/q} = 1$, where p and q are positive odd integers. Similar calculations for volumes of three-dimensional regions.

4. Nicholson, J. W., A proposition in reference to centre of gravity and its demonstration, vol. 2, p. 41.

5. Zerr, G. B. M., The centroid of areas and volumes, vol. 3, p. 46.

    Continuation of Zerr, vol. 1, p. 120 (see entry 1, above).

6. Zerr, G. B. M., Moments of inertia, vol. 4, p. 303.

    Continuation of Zerr, vol. 1, p. 120, and vol. 3, p. 46 (see entries 1 and 5 above).

7. Westlund, J., On the formula for the area of a curve in polar coordinates, vol. 13, p. 141.

Compare with Zeitlin, vol. 66, p. 135 (see entry 16, below).

8. Brown, B. H., Moment of inertia of a ring calculated by an elementary method, vol. 19, p. 70.

9. Dunkel, O., Integral inequalities with applications to the calculus of variations, vol. 31, p. 326.

10. Reynolds, J. B., A new formula for volume, vol. 35, p. 175.

A generalization of one of Pappus' theorems.

11. Mancill, J. D., On volumes bounded by cylindrical surfaces, vol. 45, p. 109.

12. Johnson, R. A., An ornithological note, vol. 54, p. 594.

An elementary problem in volumes of solids of revolution in which the parameters play an interesting role.

13. Pang, H-C., Areas of plane figures, vol. 55, p. 244.

Limits of areas of polygonal regions, using determinants.

14. Corliss, J. J., Volumes of revolution, vol. 59, p. 37.

15. Klamkin, M. S., On the uniqueness of the distribution function for the Buffon needle problem, vol. 60, p. 677.

16. Zeitlin, D., On plane area in polar coordinates, vol. 66, p. 135.

Compare with Westlund, vol. 13, p. 141 (see entry 7, above).

17. Porges, A., Solids of revolution, vol. 66, p. 302.

Axes not parallel to the coordinate axes.

18. Kefalas, C. N., A formula for the calculation of the inertia moment of some geometrical solids, MATHEMATICS MAGAZINE, vol. 26, p. 265.

Gives a general formula for calculating the moment of inertia of a number of solids whose horizontal cross sections are regular polygons of N sides. Limiting cases as $N \to \infty$ include cones, spheres, ellipsoids, paraboloids.

19. Hammer, P. C., Areas swept out by tangent line segments, MATHEMATICS MAGAZINE, vol. 28, p. 65.

20. Hoggatt, V. E., Multiple interpretations of some integrals, MATHEMATICS MAGAZINE, vol. 34, p. 207.

Centroids, moments of inertia, etc.

21. Sutcliffe, A., Waiting for a bus, MATHEMATICS MAGAZINE, vol. 38, p. 102.

Expected waiting time for a bus if the two independent bus lines do or do not coordinate their service.

(e)
MULTIPLE INTEGRALS AND LINE INTEGRALS

THE USE OF MODELS WHILE TEACHING TRIPLE INTEGRATION*

E. A. WHITMAN, Carnegie Institute of Technology

Let us suppose that the classroom topic is that of finding, by triple integration, the volumes bounded by certain surfaces. Let us further assume that some progress has been made in finding the volumes of such solids as circular cylinders and spheres, but that there is still some doubt as to what it all means. Here a model, such as one of those whose photographs are shown on adjacent pages, may help show what the integrand $dz\,dy\,dx$, or other integrand, represents. Also such a model may help show what is accomplished by each step of the integration. Thus the first integration, say with respect to x, sums the volume elements given by the integrand to give a column whose cross-sectional area is $dz\,dy$; the second integration sums these columns to give a slice of the solid, and the third integration sums the slices to give the solid.

Further along in the study of integration, especially as we are finding the volumes of solids bounded by more than one surface, the order in which we integrate with respect to the variables becomes important. Here again models may aid. Thus from the adjacent photographs of models, we may be aided in seeing why the same volume may be given by each of the following integrals:

$$2\int_0^6 \int_0^{\sqrt{36-y^2}} \int_0^{(36-y^2)/4} dz\,dx\,dy; \quad 2\int_0^6 \int_0^{(36-y^2)/4} \int_0^{\sqrt{36-y^2}} dx\,dz\,dy.$$

Also such models aid in showing why, in this case, integration first with respect to y would require two integrals but that one integral suffices for the integrands $dz\,dy\,dx$ and $dx\,dy\,dz$, as well as for the two shown above.

The two models which show the volume bounded by the surfaces $3r=2z$, $r=6\cos\theta$, and the plane $z=0$, are designed to aid in finding integrals for the bounded volume using cylindrical coördinates. They should also aid in seeing what is accomplished at each step in evaluating two such integrals as are given below:

$$2\int_0^{\pi/2} \int_0^{6\cos\theta} \int_0^{3r/2} r\,dz\,dr\,d\theta; \quad 2\int_0^{\pi/2} \int_0^{9\cos\theta} \int_{2z/3}^{6\cos\theta} r\,dr\,dz\,d\theta.$$

---

* From AMERICAN MATHEMATICAL MONTHLY, vol. 48 (1941), pp. 45–48.

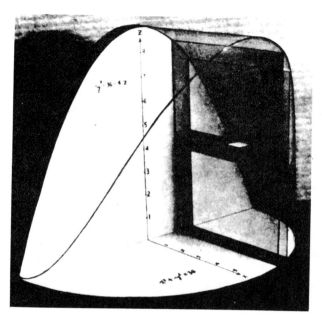

FIG. 1. This model illustrates the process of finding the volume of a solid by integrating, in turn, with respect to x, z, and y. The two cylindrical surfaces are defined by the equations $x^2+y^2=36$ and $y^2=36-4z$.

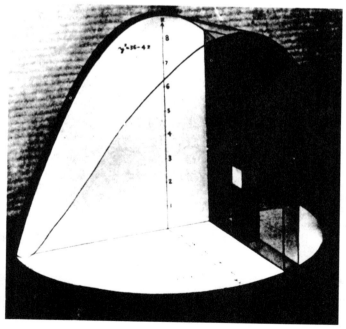

FIG. 2. This model illustrates the process of finding the same volume as in Figure 1, but by using variables x, y, z in a different order. Here one integrates, in turn, with respect to z, x, and y.

FIG. 3. This model illustrates the process of finding the volume of the solid within the cylinder $r = 6 \cos \theta$, above the plane $z = 0$, and below the cone $3r = 2z$. This volume is found by integrating, in turn, with respect to z, r, and θ.

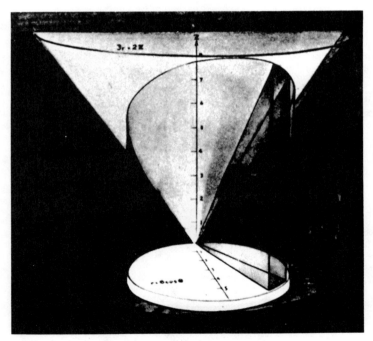

FIG. 4. This model illustrates the process of finding the same volume as in Figure 3, but by integrating, in turn, with respect to r, z, and θ.

In the use of models one danger appears. The models themselves, the material used, and the details of how to build them may become so interesting that we lose sight of the fact that such models are only a supplementary aid. In so far as the models help students in visualizing figures and drawing them, their use should be encouraged. But in the calculus, a working sketch of the particular solid under consideration, or a mental picture, is the important thing and any model however interesting or useful should not be allowed to obscure that objective. The author has found these models useful in attaining that objective, and also in teaching solid analytic geometry, particularly curves of intersections of surfaces.

These models represent the work of students of college freshman and sophomore level, and they were made as a part of an NYA project. The surfaces are made of celluloid, and four kinds were used: clear, opaque, and two colors. The supporting framework is of plywood. With these materials one is limited to planes, cones, and cylinders. By combining the use of these materials with that of string to represent lines on the surface, a model showing a hyperboloid of one sheet is easily constructed, while a hyperbolic paraboloid furnishes a more difficult task.

A SIMPLE PROBLEM IN CYLINDRICAL COORDINATES*

F. B. HILDEBRAND, Massachusetts Institute of Technology

An elementary calculus class is asked to employ double integration, using circular cylindrical coordinates, for the determination of the volume bounded above by a nappe of the cone $x^2+y^2=z^2$, below by the plane $z=0$, and laterally by the cylinder $x^2+y^2=x$. Many of the students set up the integral as

$$V = \int_{-\pi/2}^{\pi/2} \int_0^{\cos\theta} r^2 dr d\theta \quad \text{or} \quad 2\int_0^{\pi/2} \int_0^{\cos\theta} r^2 dr d\theta$$

and obtain a numerical answer which checks that given in the back of the book. Others, however, are led to the formulation

$$V = \int_0^{\pi} \int_0^{\cos\theta} r^2 dr d\theta,$$

and are less fortunate in the outcome of their calculation.

When the members of the class discuss this formulation, one student points out that here the variable r is not always positive, and hence that the equation of the nappe of the cone must be taken as $z=|r|$, so that, if the range $(0, \pi)$ in θ is insisted upon, the integrand should be written as $|r|r$. He proposes the formulation

$$V = \int_0^{\pi/2} \int_0^{\cos\theta} r^2 dr d\theta - \int_{\pi/2}^{\pi} \int_0^{\cos\theta} r^2 dr d\theta.$$

Calculation yields the correct result and all is well until another student insists that the element of plane area should be taken as $|r|dr d\theta$, so that the integrand truly should be written as $|r|^2 = r^2$ and the two errors appear to annul each other, leading again to the formulation under criticism.

While the remaining source of error eventually is discovered, to the satisfaction of the instructor, the varied reactions of the students afterward cause the instructor to wonder whether (1) the same discussion should be instigated in other classes or (2) such discussions should be averted in the future by a categorical statement that the range in θ should be so chosen that r is non-negative (presumably with a remark that the treatment of cases in which this is not possible is "beyond the scope of the course") or (3) there is a preferable third alternative.

---

* From AMERICAN MATHEMATICAL MONTHLY, vol. 64 (1957), pp. 194–195.

FUBINI IMPLIES LEIBNIZ IMPLIES $F_{yx} = F_{xy}$*

R. T. SEELEY, Harvey Mudd College

Three of the most common and important theorems on the interchange of limit operations are the rules for interchanging the order of integration in a repeated integral (Fubini's theorem), differentiating under the integral sign (Leibniz's rule), and taking mixed partial derivatives in either order. This note points out that the first of these leads easily to the other two. This has advantages for the course which aims to prove everything, by replacing two rather difficult arguments with what seem to be simpler ones; and for the course which does not want to handle all complications, by reducing the number of unproved statements. Moreover, the connections between these results seem interesting in themselves, and provide good results for Lebesgue integrals.

The note first assumes a simple form of Fubini's theorem for Riemann integrals (A), uses this to prove Leibniz's rule (B), and uses this in turn to prove one of the stronger forms of the theorem on mixed partials (C). It then states the corresponding results for Lebesgue integrals; the proofs remain the same. R denotes the rectangle $a \leq x \leq b$, $c \leq y \leq d$.

(A) *If $f(x, y)$ is continuous on R, then $\int_c^d f(x, y)dy$ is a continuous function of x, $\int_a^b f(x, y)dx$ is a continuous function of y, and*

$$\int_a^b \left[\int_c^d f(x, y)dy \right] dx = \int_c^d \left[\int_a^b f(x, y)dx \right] dy.$$

(B) *If $g(x, y)$ and its partial derivative $g_x(x, y)$ are continuous on R, then $\int_c^d g(x, y)dy$ has a derivative on $a < x < b$, which equals $\int_c^d g_x(x, y)dy$.*

Proof. By the fundamental theorem of calculus (FTC) and Theorem (A),

(1)
$$\int_c^d g(x, y)dy = \int_c^d \left[\int_a^x g_x(s, y)ds + g(a, y) \right] dy$$
$$= \int_a^x \left[\int_c^d g_x(s, y)dy \right] ds + \int_c^d g(a, y)dy.$$

By (A), $\int_c^d g_x(s, y)dy$ is continuous, so that the FTC asserts that the expression (1) has the derivative $\int_c^d g_x(x, y)dy$.

(C) *If f_y and f_{yx} are continuous on R, and $f_x(x, c)$ exists for $a < x < b$, then f_x and f_{xy} exist in the interior of R and f_{xy} equals f_{yx}.* (Here f_{yx} is $\partial f_y/\partial x$, and f_{xy} is $\partial f_x/\partial y$.)

Proof. By the FTC, $f(x, y) = \int_c^y f_y(x, t)dt + f(x, c)$. By Theorem (B)

(2)
$$f_x(x, y) = \int_c^y f_{yx}(x, t)dt + f_x(x, c).$$

---
* From AMERICAN MATHEMATICAL MONTHLY, vol. 68 (1961), pp. 56–57.

Again by the FTC, the expression on the right in (2) has a derivative with respect to y which equals $f_{yx}(x, y)$.

For the Lebesgue integral, we have the following formulation.

(A') *If f is integrable on R, then $\int_c^d [\int_a^b f(x, y)dx]dy = \int_a^b [\int_c^d f(x, y)dy]dx$, both repeated integrals existing.*

(FTC') *If f is integrable on $a<x<b$, then $F(x) = \text{constant} + \int_a^x f(t)dt$ has a derivative $F'(x) = f(x)$ almost everywhere. Any function F of this form is called absolutely continuous.*

(B') *If g is integrable on R and $G(x, y) = \int_a^x g(t, y)dt + G(a, y)$ for almost every y in $c \leq y \leq d$, and $\int_c^d G(a, y)dy$ exists, then $\int_c^d G(x, y)dy$ has a derivative equal to $\int_c^d g(x, y)dy$ for almost every x in $a \leq x \leq b$.*

(C') *If $f(x, y)$, defined on R, is absolutely continuous in y for almost every x, and $f(x, c)$ is absolutely continuous, and f_y is absolutely continuous in x for almost every y, and f_{yx} is integrable on R, then f_x is absolutely continuous in y for almost every x, and $f_{xy} = f_{yx}$ almost everywhere.*

In conclusion, we point out that (B) can be proved by (FTC) and (C), through the device of considering $f(x, y) = \int_c^y g(x, t)dt$; and (A) can be gotten in part from (B) by showing that the derivatives of $\int_a^x [\int_c^y f(s, t)dt]ds$ equal those of $\int_c^y [\int_a^x f(s, t)ds]dt$. But this sequence appears to be less useful than the one presented.

The proof of (B) given here can also be found in Kaplan, *Advanced Calculus*.

BIBLIOGRAPHIC ENTRIES: MULTIPLE INTEGRALS AND LINE INTEGRALS

Except for the entry labeled MATHEMATICS MAGAZINE, the references below are to the AMERICAN MATHEMATICAL MONTHLY.

  1. Lovitt, W. V., A geometrical interpretation of Green's formula vol. 22, p. 152.
  2. Franklin, P., The elementary character of certain multiple integrals connected with figures bounded by planes and spheres, vol. 33, p. 252.
  3. Morris, W. L., A new method for the evaluation of $\iint_A f(x, y)\,dy\,dx$, vol. 43, p. 358.
  4. Roever, W. H., A geometric representation of a line integral, vol. 44, p. 22.

  Geometric interpretation of $\int_C P\,dx + Q\,dy$ as change in height along a space curve.

  5. Brown, A. B., On transformation of multiple integrals, vol. 48, p. 29.
  6. Schwartz, J., The formula for change in variables in a multiple integral, vol. 61, p. 81.
  7. Christiano, J. G., Differentiation of the repeated integral, vol. 66, p. 127.
  8. Zimmerberg, H. J., Independence of path for line integrals, vol. 66, p. 303.
  9. Firey, W. J., Line integrals of exact differentials, vol. 68, p. 57.
  10. Zatzkis, H., Volume and surface of a sphere in an N-dimensional euclidean space, MATHEMATICS MAGAZINE, vol. 30, p. 155.

  Uses Gauss' theorem and induction.

10

NUMERICAL, GRAPHICAL, AND MECHANICAL METHODS AND APPROXIMATIONS

THE REMAINDER IN COMPUTING BY SERIES*

R. K. MORLEY, Worcester Polytechnic Institute

In computing by series it is important to have a bound for the remainder after a finite number of terms have been computed. The following method is simple enough to be presented to any class in which the integral test for convergence has been taken up, and it is applicable to any series on which that test may be used; for instance, the common series $\sum_{n=1}^{\infty} (1/n^p)$, commonly called the p-series.

Denote a series of positive terms by $\sum_{n=1}^{\infty} u_n$, and let $u_n = f(n)$ be such that $f(n)$ can be changed to $f(x)$ with x a continuous variable. Suppose that the sum of m terms of the series has been computed. We assume that $\int_m^{\infty} f(x)\,dx$ is finite and can be found; and that, for $m \leq x$, $f'(x)$ is negative and $f''(x)$ is positive.

Draw the curve $y = f(x)$ and, as in the usual proof of the integral test for convergence, represent the terms u_n by rectangles under the curve of base 1 whose right-hand sides are $y = f(n)$. Also, as in the usual proof of the integral test for divergence, draw rectangles of base 1 that extend above the curve whose left-hand sides are $y = f(n)$. Denote the points $[m, f(m)]$, $[m+1, f(m+1)]$, $[m+2, f(m+2)]$, $[m, f(m+1)]$, $[m+1, f(m+2)]$ by A, B, C, D, E, respectively. Draw the chord AB, and the chord CB, and produce the latter until it cuts $x = m$ at F.

Let $R = \sum_{n=m+1}^{\infty} u_n$. We shall find lower and upper bounds for R.
To find the lower bound we have

$$u_{m+1} > \int_m^{m+1} f(x)\,dx - \triangle ABD, \qquad \triangle ABD = \tfrac{1}{2}(u_m - u_{m+1})$$

and we have similar expressions for the other terms of R.
Hence

$$R > \int_m^{\infty} f(x)\,dx - \tfrac{1}{2}[(u_m - u_{m+1}) + (u_{m+1} - u_{m+2}) \cdots]$$

---

* From AMERICAN MATHEMATICAL MONTHLY, vol. 57 (1950), pp. 550–551.

or

$$R > \int_m^\infty f(x)dx - \tfrac{1}{2}u_m \quad \text{since} \quad \lim_{n\to\infty} u_n = 0.$$

To find the upper bound consider the triangle DBF = triangle BCE. The hypotheses assumed for $f(x)$ ensure that BF is entirely below the curve. Therefore

$$u_{m+1} < \int_m^{m+1} f(x)dx - \Delta DBF$$

or

$$u_{m+1} < \int_m^{m+1} f(x)dx - \Delta BCE, \qquad \Delta BCE = \tfrac{1}{2}(u_{m+1} - u_{m+2}).$$

Hence

$$R < \int_m^\infty f(x)dx - \tfrac{1}{2}[(u_{m+1} - u_{m+2}) + (u_{m+2} - u_{m+3}) \cdots]$$

or

$$R < \int_m^\infty f(x)dx - \tfrac{1}{2}u_{m+1}.$$

The above applies directly only to series with all positive terms, but an alternating series can often be changed by pairing the terms to one with all positive or all negative terms, and frequently this method can be used. It usually gives a much closer upper bound for R than $|u_{m+1}|$ which is commonly employed, and in addition it provides a lower bound.

A SIMPLE PROOF OF STIRLING'S FORMULA*

A. J. COLEMAN, University of Toronto

Stirling's formula for the asymptotic value of factorial N is widely used in physics and many different proofs of it are known. The following, which is a slight variation on an old approach, is so simple as to be useful as an exercise in integration and the convergence of montone, bounded sequences. It could therefore be used in a first course in Calculus.

Consider the graph of $y = \ln x$, where by $\ln x$ we mean the natural logarithm of x. Since $y' = 1/x$, the slope between x equal 1 and N is positive and decreas-

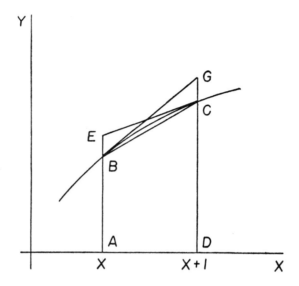

ing. Thus chords, such as BC in the accompanying figure, are below the curve and therefore the area of the trapezoid $ABCD$ is less than the area under the curve. The area of the trapezoid is

$$\tfrac{1}{2}[\ln x + \ln(x+1)],$$

and therefore the area of the $N-1$ trapezoids from x equal 1 to $N-1$ is

$$\tfrac{1}{2}\{[\ln 1 + \ln 2] + [\ln 2 + \ln 3] + \cdots + [\ln(N-1) + \ln N]\} = \ln N! - \tfrac{1}{2} \ln N.$$

Thus

(1) $\quad c_N = \displaystyle\int_1^N \ln x\, dx - \ln N! + \tfrac{1}{2} \ln N = (N + \tfrac{1}{2}) \ln N - N + 1 - \ln N!$

is a positive number equal to the area under the curve $y = \ln x$ from $x = 1$ to $x = N$

---

* From AMERICAN MATHEMATICAL MONTHLY, vol. 58 (1951), pp. 334–336.

minus the trapezoidal approximation to this area. The sequence c_N therefore increases monotonely with N.

On the other hand, the area under the curve is clearly less than the average of the areas of the trapezoids $AECD$ and $ABGD$, where BG is the tangent to the curve at B and CE is the tangent at C. From the known value of the slope, we have DG equal to $\ln x + 1/x$ and AE equal to $\ln(x+1) - 1/(x+1)$. Thus the average area of the two trapezoids is

$$\frac{1}{2}\left[\ln x + \ln(x+1) + \frac{1}{2}\left(\frac{1}{x} - \frac{1}{x+1}\right)\right].$$

Summing as before, we have

$$\int_1^N \ln x\, dx < \ln N! - \frac{1}{2}\ln N + \frac{1}{4} - \frac{1}{4N},$$

and, therefore,

$$c_N < \frac{1}{4} - \frac{1}{4N} < \frac{1}{4}.$$

Thus c_N, which is a monotone increasing sequence, is bounded above and therefore has a limit c such that $0 \leq c \leq \frac{1}{4}$. If we set $e^{1-c} = a$, so that $2.11 < a < 2.72$, it follows from (1) that

(2) $$\frac{N!}{\sqrt{N}\, N^N e^{-N}} \to a$$

as N tends to infinity. This is the essential part of Stirling's formula. However, we must still find the value of a. By geometrical intuition our method of approximation from below seems about twice as accurate as the approximation from above, so we shall expect a to be closer to 2.72 than to 2.11, perhaps around 2.5.

There is a well-known trick for finding the exact value of a, by making use of Wallis' formula. To prove the latter, we set $I_r = \int_0^{\pi/2} \sin^r \phi\, d\phi$, and since in the range $0 < \phi < \pi/2$,

$$\sin^{2k-1}\phi > \sin^{2k}\phi > \sin^{2k+1}\phi$$

we have

(3) $$I_{2k-1} > I_{2k} > I_{2k+1}.$$

Integrating I_r by parts, one easily shows that $rI_r = (r-1)I_{r-2}$, and therefore I_r may be evaluated by recursion. The inequalities (3) then immediately give

Wallis' Formula:

$$\frac{[2^k(k!)]^4}{[(2k)!]^2(2k+1)} \to \frac{\pi}{2}$$

as k approaches infinity.* But by (2) the left-hand side of this approaches $a^2/4$. Thus $a = \sqrt{2\pi} \doteq 2.50$, and we finally have:

$N!$ is asymptotically equal to $\sqrt{2\pi N}\, e^{-N} N^N$.

---

* See Courant and Robbins, What Is Mathematics?, pp. 509–510, for details.

A GRAPHICAL INTEGRATION

M. S. KLAMKIN, Polytechnic Institute of Brooklyn

The following is believed to be a new simple graphical method of integrating. Let the differential equation be given by

$$\text{(1)} \qquad \frac{dy}{dx} = F(x) \qquad \text{(where } x = x_0,\ y = y_0\text{)}.$$

One first plots the curve $y = xF(x)$ (assumed to be continuous). From the point $P_0(x_0, y_0)$, a vertical line is drawn intersecting the curve $y = xF(x)$ in the point A. Then the tangent at P_0 to the integral curve of equation (1) must be parallel to OA. A small segment $P_0 P_1$ of this tangent is drawn with a parallel ruler. From point P_1, the procedure is repeated. The proximity of the successive points P_0, P_1, etc. will determine the accuracy of the construction.

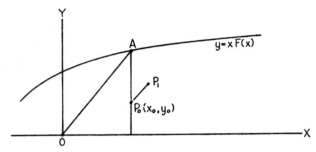

This method leads to a particularly simple construction for $y = \log x$.

Remarks: The above method can also be adapted to graphically solving the D.E., $dy/dx = F(x)/G(y)$, $(x = x_0, y = y_0)$. We let $dz/dx = F(x)$, $(x = x_0, z = z_0)$, and $dz/dy = G(y)$, $(y = y_0, z = z_0)$. Both of these latter equations are solved by the above method yielding the graphs of $z - z_0 = \int_{x_0}^{x} F(x)dx$, and $z - z_0 = \int_{y_0}^{y} G(y)dy$. It is now a simple matter to obtain the plot of y vs. x. Another method would be to specialize the functions in the graphical solution of

$$\frac{dy}{dx} = \frac{\phi(x) - G(y)}{\psi(x) - F(y)},$$

by letting $G(y) = \psi(x) = 0$. (See author's note, this MONTHLY, vol. 61, 1954, pp. 565–567).

THE MIDPOINT METHOD OF NUMERICAL INTEGRATION*

PRESTON C. HAMMER, University of Wisconsin

It is a noteworthy fact that while virtually all calculus texts and numerical analysis books discuss the trapezoidal method of numerical integration they ignore one which is simpler to use and generally superior to the trapezoidal—namely the midpoint rectangular method. The midpoint method also may be used with the trapezoidal method to obtain bounds for an integral in some cases. The midpoint method is an application of the Newton-Cotes "open" formulas with one point.

The trapezoidal and midpoint methods over one interval of length h are given respectively in the following equations:

$$\text{(1)} \quad \int_{x_0}^{x_1} f(x)dx = \frac{h}{2}(f(x_0) + f(x_1)) - \frac{f''(\xi)h^3}{12} \quad x_0 < \xi < x_1$$

$$\text{(2)} \quad \int_{x_0}^{x_1} f(x)dx = hf(x_{1/2}) + \frac{f''(\eta)h^3}{24} \quad x_0 < \eta < x_1.$$

In these equations, the second derivative of the integrand is assumed to exist on $[x_0, x_1]$. It is clear that the error in the midpoint formula is about $\frac{1}{2}$ the value of the error of the trapezoidal as $x_1 \to x_0$ if f'' is continuous on $[x_0, x_1]$. However, we now prove that (2) is superior to (1) always if f'' does not change signs on $[x_0, x_1]$.

Theorem: Let f be a function either concave† or convex, and continuous on $[x_0, x_1]$. Then

$$\left| \int_{x_0}^{x_1} f(x)\,dx - \frac{h}{2}(f(x_0) + f(x_1)) \right| \geq \left| \int_{x_0}^{x_1} f(x)\,dx - hf(x_{1/2}) \right|$$

and equality holds if and only if the graph of $f(x)$ is a line or a broken line with break point at $x = x_{1/2}$.

PROOF: If f'' exists, which we do not assume, and f is concave, then $f'' \leq 0$ and if f is convex $f'' \geq 0$. The argument for either case being the same, we will assume f is concave and since adding a constant function to f does not change the error for either formula we may assume $f > 0$ on the interval $[x_0, x_1]$.

Consider the graph in Figure 1. Since f is concave, the area bounded by $y = 0$, $x = x_0$, $x = x_1$, and $y = f(x)$ is a convex set. Now the midpoint formula gives the area under *any* line through $P_{1/2}$ over $[x_0, x_1]$ (more exactly, the integral of any

---

* From MATHEMATICS MAGAZINE, vol. 31 (1958), pp. 193–195.

† A function is *concave* if any line segment joining two points on its graph lies below or on the graph.

such linear function). Choosing m_0 as a supporting or tangent line to the curve at $P_{1/2}$ we have

$$\Delta_1 + \Delta_2 \geq hf(x_{1/2}) - \int_{x_0}^{x_1} f(x)dx \geq 0$$

where $\Delta_1 =$ area $P_0 P_{1/2} Q_0$ and $\Delta_2 =$ area $P_{1/2} P_1 Q_1$. Now, on the other hand, the trapezoidal formula gives an approximation which is less than the integral and

$$\int_{x_0}^{x_1} f(x)dx - \frac{h}{2}(f(x_0) + f(x_1)) \geq \Delta_3 = \text{area } P_0 P_1 P_{1/2}.$$

But if we now take line m_1 through $P_0 P_{1/2}$ we readily see $\Delta_4 = \Delta_1 + \Delta_2 =$ area $P_1 Q_2 P_{1/2} = \Delta_3$. Hence

$$\int_{x_0}^{x_1} f(x)dx - \frac{h}{2}(f(x_0) + f(x_1)) \geq hf(x_{1/2}) - \int_{x_0}^{x_1} f(x)dx.$$

Equality can happen only if f is linear or if the graph of f consists of two line segments joined at $P_{1/2}$. Hence, the midpoint formula is superior to the trapezoidal for either concave or convex functions and has an error opposite in sign.

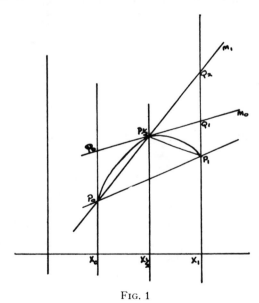

Fig. 1

Corollary: If f is concave and continuous on $[x_0, x_n]$ then

(3) $$h \sum_{i=0}^{n-1} f(x_{i+1/2}) \geq \int_{x_0}^{x_n} f(x)dx \geq h \sum_{i=0}^{n} f(x_i) - \frac{h}{2}(f(x_0) + f(x_1))$$

and

(4) $$\left| h \sum_{i=1}^{n-1} f(x_{i+1/2}) - \int_{x_0}^{x_n} f(x)dx \right| \leq \left| h \sum_{0}^{n} f(x_i) - \frac{h}{2}(f(x_0)+f(x_1)) - \int_{x_0}^{x_n} f(x)dx \right|$$

where the equality signs in the latter case hold only if the graph of f is a line or a broken line with vertices at $P_{i+1/2}$. If f is convex the inequality (4) holds and the inequality signs are reversed in (3).

Simpson's rule results if we take a weighted average of the midpoint and trapezoidal rules giving weight $\frac{2}{3}$ to the former and $\frac{1}{3}$ to the latter. That the midpoint rule is easier to use than the trapezoidal in general, either for hand calculation or by machine, is obvious unless there are reasons why $f(x_{i+1/2})$ is more difficult to compute. Calculating both, in case f is either convex or concave, would provide an error bound for the integral.

BIBLIOGRAPHIC ENTRIES: NUMERICAL, GRAPHICAL, AND MECHANICAL METHODS AND APPROXIMATIONS

Except for the entry labeled MATHEMATICS MAGAZINE, the references below are to the AMERICAN MATHEMATICAL MONTHLY.

1. Corey, S. A., A method of approximation, vol. 13, p. 137.

Formal use of Euler's summation formula, ignoring questions of convergence.

2. Corey, S. A., Certain integration formulae useful in numerical computation, vol. 19, p. 118.

A long list of integration formulas derived from Euler's summation formula, taking into account the error term in each case. No proofs are given.

3. Carmichael, R. D., On the remainder term in a certain development of $f(a+x)$, vol. 20, p. 20.

Proofs of the estimates for the error term in the paper by Corey, vol. 19, p. 118 (see entry 2, above).

4. Elmendorf, A., A differentiating machine, vol. 23, p. 292.

Describes a machine which plots the graph of $y=f'(x)$, given the graph of $y=f(x)$.

5. Daniell, P. J., New rules of quadrature, vol. 24, p. 109.

Three quadrature rules are deduced from Euler's summation formula. One is like Simpson's rule, but with only one fourth the error.

6. Corey S. A., Relating to new remainder terms for certain integration formulae vol. 25, p. 87.

A sequel to vol. 19, p. 118 (see entry 2, above).

7. Escott, E. B., Approximation to log $(1 \cdot 2 \cdot 3 \cdots x)$, vol. 30, p. 439.

An approximate formula for log $x!$ with no analysis of the error.

8. Bateman, H., On the occasional need of very accurate logarithms, vol. 32, p. 249.

9. Scarborough, J. B., Formulas for the error in Simpson's rule, vol. 33, p. 76.

10. Scarborough, J. B., On the relative accuracy of Simpson's rules and Weddle's rule, vol. 34, p. 135.

11. Garver, R., On the relative accuracy of Simpson's rules and Weddle's rule: a question, vol. 34, p. 369.

12. Scarborough, J. B., On the relative accuracy of Simpson's rules and Weddle's rule: a reply, vol. 34, p. 370.

13. Groat, B. F., Mean value of the ordinate of the locus of the rational integral algebraic function of degree n expressed as a weighted mean of $n+1$ ordinates and the resulting rules of quadrature, vol. 38, p. 212.

14. Sadowsky, M., A formula for approximate computation of a triple integral, vol. 47, p. 539.

   No discussion of error.

15. Ewing, G. M., On approximate cubature, vol. 48, p. 134.

   Prismoid formula and Simpson's rule for n-fold integrals.

16. Frame, J. S., Numerical integration, vol. 50, p 244

   Trapezoidal rule, Simpson's rule, and a quartic rule. Error is discussed.

17. Antosiewicz, H. A., and Hammersley, J. M., The convergence of numerical iteration, vol. 60, p. 604.

18. Goheen, Harry, On the remainder in formulas of numerical interpolation, differentiation and quadrature, vol. 61, p. 44.

19. Youse, B. K, Abridged series for numerical evaluation, vol. 61, p. 187.

20. Lowell, L. I., Comments on the polar planimeter, vol. 61, p. 467.

21. Kennedy, E. C., Approximation formulas for elliptic integrals, vol. 61, p. 613.

22. Frame, J. S., Some trigonometric hyperbolic and elliptic approximations, vol. 61, p. 623.

23. Aissen, M. I., Some remarks on Stirling's formula, vol. 61, p. 687.

24. Robbins, H., A remark on Stirling's formula, vol. 62, p. 26.

25. Morduchow, M., A note on Newton-Cotes quadrature formulas, vol 62, p. 33.

26. Seebeck, C. L., Jr., Note on linear interpolation error, vol. 62, p. 35.

27. Greenstein, D. S., Newton-Cotes quadrature formulas, vol. 62, p. 487.

28. Snyder, R. W., One more correction formula, vol. 62, p. 722.

29. Mott, T. E., Newton's method and multiple roots, vol. 64, p. 635.

30. Mosteller, F., and Richmond, D. E., Factorial $\frac{1}{2}$: a simple graphical treatment, vol. 65, p. 735.

31. Wolfe, J. M., An adjusted trapezoidal rule using function values within the range of integration, vol. 66, p. 125.

32. Shklov, N., Simpson's rule for unequally spaced ordinates, vol. 67, p. 1022.

33. Feller, W., A direct proof of Stirling's formula, vol. 74, p. 1223.

34. Weiner, L. M., Note on Newton's method, MATHEMATICS MAGAZINE, vol. 39, p. 143.

11

INFINITE SERIES AND SEQUENCES

(a)

THEORY

SUMMABILITY OF SERIES*†

C. N. MOORE, University of Cincinnati

1. *Introduction.* In the study of the various functions that occur in mathematical theory and its applications to science and engineering one of the most potent methods of investigation is expansion in series. The chief value of this method lies in the fact that it enables us to express new and more complicated functions in terms of simpler and better known functions. Examples of such expansions are met early in our study of mathematics. For instance the binomial expansion of $(1+x)^n$ for the case of negative or fractional exponents leads to an infinite series in powers of x. This expansion is a special case of the expansions due to Taylor and Maclaurin, which are treated in the usual first course in calculus. The functions that admit of such expansions form a wide class and they possess the fundamental property that when they are generalized to complex values they admit of similar expansions in the complex plane. This beautiful generalization of the theorems of Taylor and Maclaurin was one of the important discoveries due to Cauchy.

In order for a function to have a power series development in the neighborhood of a point it must possess derivatives of all orders at that point. While the class of functions that has this property is a very important one, it is nevertheless a highly restricted group when compared to functions in general. More general expansions are obviously necessary for the study of wider classes of functions. Moreover, even in the case of functions that possess a power series development, expansions in terms of other simple functions may be more useful for certain types of investigation.

Of coordinate importance with power series and next to them in simplicity

---

* From AMERICAN MATHEMATICAL MONTHLY, vol. 39 (1932), pp. 62–71.

† A lecture delivered by invitation at the S.P.E.E. Summer Session for Teachers of Mathematics to Engineering Students at Minneapolis, Sept. 4, 1931.

we have Fourier series or developments in sines and cosines of integral multiples of x. These developments were first studied in connection with certain problems of mathematical physics and they still rank as the expansions of greatest utility in the application of mathematics to other sciences. This arises from the frequent occurrence of periodic phenomena and the importance of wave motion as a hypothesis in explaining other phenomena.

Among the important problems of mathematical physics whose solution frequently involves Fourier series may be mentioned those that arise in the study of the flow of heat and electricity. Such problems also lead to the consideration of other related expansions, such as developments in Bessel's functions, Laplace's functions, Sturm-Liouville functions, and so on. These various functions have in common with the sine and cosine the properties of being oscillatory in character and satisfying differential equations of the second order. The expansions in terms of such functions have many properties analogous to those of Fourier series, but their study naturally presents greater difficulty.

Outside of power series, Fourier series and related developments, the type of series that has been most extensively studied is that known as Dirichlet's series. Such series have the form $\Sigma a_n e^{-\lambda_n s}$, where the λ's are a set of real constants such that λ_n becomes infinite with n, the a's are a set of constants real or complex, and s is a complex variable. So far the applications of such series have been primarily in the field of pure mathematics, and more particularly in the analytic study of the Theory of Numbers. However, like other conceptions[1] that first arose in the study of pure mathematics, they may prove later to have important applications in other sciences.

In order to apply mathematical theory to concrete situations we need always a method of obtaining numerical results. Thus in the case of an infinite series of constant terms we must have some convention as to a numerical value to be attached to the series. The simplest and most natural one is based on the conception known as convergence. Given a series

(1) $$u_0 + u_1 + u_2 + \cdots + u_n + \cdots,$$

we form

(2) $$s_n = u_0 + u_1 + \cdots + u_n$$

and consider the behavior of s_n with increasing n. If s_n approaches a definite limit s as n becomes infinite, we say that the series converges to the value s. If no limit exists we say that the series diverges. A series may diverge because the related s_n becomes positively or negatively infinite, or because it oscillates between finite or infinite limits. The latter type of divergent series is sometimes referred to as oscillatory.

2. *Use of divergent series in the eighteenth century.* The mathematicians of the eighteenth century found that the formal expansions of many simple functions

---

[1] Such as conic sections, non-Euclidean geometry, complex numbers, continued fractions.

failed to converge for values of the argument for which the function corresponding to the series was perfectly well defined. They found on the other hand that in many of these cases the divergent expansions could be used in various ways to obtain valid results just as well as if they were convergent. They therefore felt the need of attaching a numerical value to a divergent series that would be in agreement with the value of the function whose expansion gave rise to the series.

The procedure of two of the leading mathematicians of the period can be adequately illustrated in terms of the simple divergent series

$$1 - 1 + 1 - 1 + \cdots.$$

We note that the successive values of s_n are alternately 1 and 0. Since these two values occur with equal frequency, Leibnitz contended that on the basis of the Theory of Probability the value $\frac{1}{2}$ should be ascribed to the series. Suppose, on the other hand, that we introduce successive powers of x in the terms of the series, thus obtaining the power series

$$1 - x + x^2 - x^3 + \cdots.$$

For values of x numerically less than unity the series converges and has the value $1/(1+x)$. Now allow x to approach 1 from below. The function defined by the series approaches $\frac{1}{2}$; the series itself, if we take the limit term by term, approaches the series with which we started. For this reason Euler contended that the value $\frac{1}{2}$ should be ascribed to the original series. The surprising feature here is that these two entirely different methods of approach lead to the same value for the series. This coincidence is not an accident, however, but a special case of a general theorem proved by Frobenius in 1880, to which we will refer later.

It is apparent that the procedure of Leibnitz and Euler in the case of the simple series $1-1+1-1+\cdots$ is entirely out of harmony with present day notions of rigor in mathematical analysis. These notions are the result of a gradual development from ideas that first came into prominence in the mathematical world early in the nineteenth century as a consequence of the pronouncements of such outstanding mathematicians as Cauchy and Abel. The steadily rising standards of rigor, coupled with the fact that the use of divergent series occasionally led to serious contradictions, resulted in a gradually diminishing interest in such series during the first three fourths of the nineteenth century. At the same time the rigorous theory of convergent series was being actively developed by students of this field of mathematics, and thus the way was being paved for a rigorous treatment of divergent series.

3. *Beginnings of the modern theory.* The first definite advance in the direction of a rigorous theory to appear in a mathematical journal is found in a paper by Frobenius which was published in Crelle's Journal in 1880. The essential contribution of Frobenius was the proof of the following theorem.

If for the series (1) *the s_n defined by* (2) *are such that the limit*, $\lim_{n\to\infty}(s_0$

$+s_1+ \cdots +s_n)/(n+1)$, exists and is equal to s, then the series $u_0+u_1x+ \cdots +u_nx^n+ \cdots$ will converge for $|x|<1$ to a function $f(x)$ such that $\lim_{x \to 1-0} f(x) = s$. It is apparent that the agreement between the procedures of Leibnitz and Euler in the case of the series $1-1+1-1+ \cdots$ is a special case of Frobenius's theorem.

Two years after the publication of Frobenius's paper, his result was considerably generalized by Hölder. Let us represent the successive arithmetic means of the s's by $s_0^{(1)}, s_1^{(1)}, \ldots s_n^{(1)}, \ldots$. Then, if $s_n^{(1)}$ fails to approach a limit as n becomes infinite, it seems natural to repeat the process originally used by forming successive arithmetic means of the $s^{(1)}$'s, labeling them $s_0^{(2)}, s_1^{(2)}, \ldots s_n^{(2)}, \ldots$. If $s_n^{(2)}$ approaches a limit s we define that limit to be the generalized sum of the series. If $s_n^{(2)}$ does not approach a limit, we repeat the process, forming $s^{(3)}$'s and studying the behavior of $s_n^{(3)}$ as n becomes infinite. If for any integer r, $s_n^{(r)}$ approaches a limit s as n becomes infinite, we define that limit to be the generalized sum of the original series. In modern terminology we say that the series is summable (Hr) to the value s. We are now in a position to state Hölder's generalization of Frobenius's theorem. *If the series (1) is summable (Hr) to the value s for any integer r, then the series $\Sigma u_n x^n$ will converge for values of x numerically less than 1 to a function $f(x)$ such that $\lim_{x \to 1-0} f(x) = s$.*

In 1890, ten years after the publication of Frobenius's paper, Cesàro attacked a problem in series that is naturally suggested by the behavior of the Cauchy product of two convergent series. Given two infinite series, Σu_n and Σv_n, their Cauchy product is defined as the series $\Sigma(u_n v_0 + u_{n-1} v_1 + \cdots + u_0 v_n)$. If both the original series are convergent and at least one of them is absolutely convergent, it can be shown that the product series will always converge to a value that is the product of the values of the two given series. If, however, neither of the original series is absolutely convergent, it may happen that the product series diverges. If then there can be found methods of summing divergent series that will give to them values consistent with their mode of formation, it seems reasonable to expect that some one of these methods will serve to sum the product series of two convergent series to the proper value. Cesàro found that the simple method used by Frobenius had this interesting and important property. However, the problem suggested by the multiplication of convergent series was not completely elucidated by this discovery, nor did Cesàro's investigations stop at that point. Suppose that we are considering the product of three series. The product of two of them may diverge, and we are thus led to consider the behavior of the product of a convergent series and a series that can be summed by the method of Frobenius. If we start with four series, their products, taking two at a time, may each diverge, and we thus have the problem of multiplying two divergent series of a certain type, and so on. It was Cesàro's aim to answer all questions of this order, and thus to furnish the means for discussing the behavior of the product of any finite number of series. He succeeded brilliantly in his purpose, obtaining a complete solution of the problem in a six page paper that involved only analysis of a simple type.

Cesàro arrived at his goal by generalizing the process of forming an arithmetic mean of the first n partial sums of the series in a different manner from that adopted by Hölder. Instead of forming successive means from the means previously obtained, he conceived the idea of forming successive weighted means, the weights changing in a regular manner. Let us set

$$(4) \quad S_n^{(r)} = s_n + r s_{n-1} + \frac{r(r+1)}{2!} s_{n-2} + \cdots + \frac{r(r+1)\cdots(r+n-1)}{n!} s_0,$$

$$(5) \quad A_n^{(r)} = \frac{(r+1)(r+2)\cdots(r+n)}{n!}$$

$$(6) \quad \sigma_n^{(r)} = \frac{S_n^{(r)}}{A_n^{(r)}}.$$

The $A_n^{(r)}$ can be readily shown to be the sum of the coefficients of the s's in the expression for $S_n^{(r)}$, and therefore $\sigma_n^{(r)}$ is a weighted mean of $s_0, s_1, \cdots, s_n$. For $r=0$, $\sigma_n^0 = s_n$, and consequently convergence is included in our general scheme of summation for this value of the parameter. For $r=1$, $\sigma_n^{(1)} = \sum_{i=0}^{i=n} s_i/(n+1)$, and therefore summation by the ordinary arithmetic mean is included for the value unity of the parameter. For values of $r>1$, it is apparent from the above formula that the early s's are weighted more than the later ones, and therefore it seems probable that series which oscillate more strongly can be summed by using larger values of r. Since the general effect of forming the Cauchy product of two or more oscillating series is to increase the degree of oscillation, one would expect the general multiplication theorem to depend on the value of r in formula (6), and this expectation is realized. In modern terminology a series for which $\sigma_n^{(r)}$ approaches a limit σ as n becomes infinite is said to be summable (Cr). Using this definition, Cesàro's general theorem may be stated as follows. *The Cauchy product of a series summable (Cr) to value A by a series summable (Cs) to value B is always summable $(C, r+s+1)$ to value AB.* This includes the result regarding convergent series as the special case where $r=0=s$.

It is apparent at first sight that there is a rather close relationship between Hölder's generalization of Frobenius's method and that of Cesàro. Furthermore we readily find that a necessary condition for either summability (Hr) or (Cr) is $\lim_{n\to\infty} u_n/n^r = 0$. This suggests that the two methods are of the same general scope, and mathematicians working with them felt fairly certain that they were either equivalent or approximately so. Nevertheless, the discussion of the question of equivalence seemed at first to involve algebraic difficulties of a serious character. Finally, in 1907, Knopp proved that summability (Hr) implies summability (Cr) to the same value, and somewhat later Schur and W. B. Ford proved independently that summability (Cr) implies summability (Hr) to the same value. These early proofs involved rather complicated analysis, but it has since been found that the equivalence theorem can be proved in a quite elementary manner. The first simple proof was given by Schur in 1913.

4. Work of Borel and others on analytic extension.

We spoke at the beginning of the importance of developments in power series in the case of functions possessing such developments. For some functions, such as the sine and cosine, the corresponding power series converges for all values of the independent variable. In the case of other quite simple functions, however, the range of convergence of the power series may be quite limited. For example the power series development of the function $1/(1-z)$ is

(7) $$1 + z + z^2 + \cdots + z^n + \cdots,$$

and this series converges only within the unit circle. The Theory of Functions of a Complex Variable furnishes the reason for this restricted region of convergence. The function $1/(1-z)$, simple as it is, possesses an infinite discontinuity at the point $z=1$, and such a discontinuity on the circumference of a circle whose center is the point about which we develop the function always prevents convergence outside the circle. In spite of the failure of the convergence of the series for values of $z \geq 1$, there is obviously such a close relationship between the series and the function that we should expect to find the value of the function for such values of z by using a suitable method of summing the divergent series. Cesàro's success in summing the product of any finite number of convergent series suggested to Borel the possibility of summing power series outside their circle of convergence by using some similar scheme. It was apparent at once that neither the method of Cesàro nor that of Hölder would suffice, since the necessary condition for their applicability is not satisfied by a power series beyond the circle of convergence.

Borel found it useful, however, to take Cesàro's scheme of weighted means as a point of departure, but he introduced an important new element in this scheme. Instead of considering a weighted mean of a finite number of sums and then allowing the number of sums to become infinite, he considered a weighted mean of the infinite set of sums and allowed the weights to vary in a definite manner. Thus he formed

(8) $$F(x) = \frac{\sum_{n=0}^{\infty} s_n \frac{x^n}{n!}}{e^x}$$

and considered the limit of $F(x)$ as x becomes infinite. This definition is of interest on account of being a logical development from Cesàro's scheme, but a closely related definition which is obtained by a simple transformation of this one, is more useful in the analytic applications. We set

(9) $$u(x) = u_0 + u_1 x + u_2 \frac{x^2}{2!} + \cdots + u_n \frac{x^n}{n!} + \cdots$$

and form

(10) $$\int_0^\lambda e^{-x} u(x) dx.$$

If the limit of (10) as λ becomes infinite exists, we call this limit the generalized sum of the series. This definition has substantially the same scope as the earlier one.

The application of Borel's scheme to the power series development of $1/(1-z)$ is extremely simple and serves to show the scope of the method. We set

$$u(x, z) = 1 + zx + \frac{z^2 x^2}{2!} + \cdots = e^{zx}. \tag{11}$$

Then we form

$$\int_0^\lambda e^{-x} u(x, z) dx = \int_0^\lambda e^{(z-1)x} dx = \frac{e^{(z-1)\lambda}}{(z-1)} - \frac{1}{z-1}. \tag{12}$$

It is apparent that for all values of z in the complex plane such that the real part of z is less than 1, the first term on the right hand side of (12) approaches zero as λ becomes infinite. Therefore for all such values of z Borel's method serves to sum the series (7) to the appropriate value $1/(1-z)$. We have thus vastly extended the region in which the power series development can be utilized.

Shortly after Borel's investigations Leroy devised a more powerful method of summing a divergent series, which may be defined as follows. Given any series, $\Sigma u_n(z)$, we set

$$F(z, t) = \sum_{n=0}^\infty \frac{\Gamma(nt + 1)}{\Gamma(n + 1)} u_n(z) \qquad (0 < t < 1).$$

and define the value of the series as $\lim_{t \to 1-0} F(z, t)$, provided that this limit exists. It can readily be shown that Leroy's method will sum the series (7) to the value $1/(1-z)$ for all values of z in the complex plane except real values greater than one. In 1920, some twenty years after Leroy's work, Mittag-Leffler devised a method which will sum the series (7) to the value $1/(1-z)$ for all values of z for which the latter expression is defined, that is for all values of z except $z=1$.

We have seen how the various methods of summation serve to extend the region in which the series (7) can be summed to the value $1/(1-z)$. These methods are equally effective in the case of power series in general, provided that the function to which the power series corresponds is analytic in certain regions beyond the circle of convergence. They thus furnish an analytic extension of the function. To show the scope of the methods we shall state the facts in the case where the function is analytic throughout the complex plane, except for a finite number of singular points, none of which is at the origin. Suppose that we use Cauchy's generalization of Maclaurin's theorem to obtain the development $a_0 + a_1 z + a_2 z^2 + \cdots$. This series will converge within a circle whose center is at the origin and whose radius is the distance from the origin to the nearest singular point. Draw lines through the singular points perpendicular to the lines joining these points with the origin. These lines form a polygon, known as the Borel polygon, within which the series is summable to the value of the function by Borel's method. Draw lines from the singular points to infinity that are the prolongations of the lines from the origin to these singular points. The entire plane, except the lines extending out from the singular points, forms a region known as the Mittag-Leffler star. Leroy's method will sum the series to the value of the function everywhere in the Mittag-Leffler star. Finally, Mittag-Leffler's method will sum the series to the function everywhere, except at the singular points themselves.

5. *Summation of Fourier series and related developments.* We have seen what could be done with power series by the use of general methods of summation. Let us turn now to some other types of series and consider first of all the important class known as Fourier series. The Fourier series associated with a function $f(x)$, defined and integrable in the interval $(-\pi, \pi)$, may be written in the form

(13) $\quad \frac{1}{2}a_0 + (a_1 \cos x + b_1 \sin x) + \cdots + (a_n \cos nx + b_n \sin nx) + \cdots,$

where

(14) $\qquad a_n = \frac{1}{\pi} \int_{-\pi}^{\pi} f(x) \cos nx \, dx, \quad b_n = \frac{1}{\pi} \int_{-\pi}^{\pi} f(x) \sin nx \, dx.$

If the function satisfies certain fairly general conditions, the series will converge to the value of the function at all points of continuity of the function. However, there exist continuous functions for which the series fails to converge at points that are densely distributed in every sub-interval of the interval $(-\pi, \pi)$. Shortly after Borel's work on power series, it occurred to Fejér that some of the methods for summing divergent series might be applied to advantage in the case of Fourier series. An investigation of the matter showed that this was indeed the case and led to the following noteworthy result. *If the function of which the Fourier series is a development is finite except at a finite number of points and is integrable in the sense of Riemann, throughout the interval $(-\pi, \pi)$, then the Fourier series is summable (C1) to the value of the function at all points of continuity and to the value $\frac{1}{2}[f(x+0)+f(x-0)]$ at all points of discontinuity where this expression has a meaning.*

We have spoken earlier of developments analogous to Fourier series, such as developments in Laplace's functions or Bessel's functions, which likewise have important applications in mathematical physics. Shortly after his investigation of Fourier series Fejér made an analogous study of developments in Laplace's functions. A little later C. N. Moore investigated the developments in Bessel's functions, and Haar studied the developments in Sturm-Liouville functions. In all these cases it was the method of summation due to Cesàro that proved fruitful. In general the results were analogous to those already obtained for Fourier series, but in the case of developments in Laplace's functions and Bessel's functions certain important differences were found in the neighborhood of the point where the differential equation defining these functions has a singular point.

The investigations just described were merely the first applications of general methods of summation to the developments in question. Many subsequent studies have been made and the field of research opened up by the initial investigations is still far from exhausted. It is not possible to give any detailed account of the various results that have been obtained in a general lecture on the whole subject of summability of series. It may be stated, however, that the papers in which these results are found form an important part of the mathematical literature of the present century.

6. *Summation of Dirichlet's series.* There is one further type of series to which we have referred in the introduction, namely series of the form $\Sigma a_n e^{-\lambda_n s}$, known as Dirichlet's series. A brief account of some of the investigations of these series seems desirable, since it will show the importance of modifying various methods of summability in order to adapt them to the specific needs of certain types of series. The first investigations of the summability of Dirichlet's series were applications of Cesàro's method to certain special types of Dirichlet's series by Bohr and M. Riesz. In passing to the case of the general Dirichlet's series, Riesz found it desirable to replace Cesàro's method by a more general one, which he termed summation by typical means. For the type of series initially studied, summation by Riesz's typical mean is equivalent to summation by a corresponding Cesàro mean, but even in this case the method due to Riesz is a more effective weapon in studying the behavior of Dirichlet's series.

7. *Conclusion.* The preceding account of what has been accomplished by the use of various methods of summation in the investigation of divergent series is necessarily somewhat sketchy. I trust that it will serve to give some idea of the very notable success that has attended the application of these methods. It is easy to see that among series in general a convergent series is a very exceptional case. For convergence it is necessary first of all that the general term of the series approach zero as a limit, and many important series where the general term does approach zero fail to converge. It is apparent that if we are to achieve any comprehensive knowledge of series we must make use of methods of evaluating divergent series. The methods which we have described have amply proved their worth in the studies that have been made. We have every reason to expect that further application of these methods and others yet to be devised will vastly extend our knowledge of the important field of mathematics known as the theory of infinite series.

ON A CERTAIN TRANSFORMATION OF INFINITE SERIES[*][†]

J. A. SHOHAT, University of Pennsylvania

Introduction. The actual evaluation of the sum of a given convergent series is very often greatly hampered, if not rendered impossible, by the slowness of the convergence (example: $\pi/4 = 1 - 1/3 + 1/5 - 1/7 + \cdots$). Euler, Markoff and others[‡] have given practical methods for improving the convergence, which, however, are omitted even in many advanced courses on infinite series. Thus, it seems worthy of interest to give an extremely simple transformation of infinite series which, in case of convergence, often renders it more rapid, and thus the series in question becomes more useful for practical applications. In case of divergence, the same transformation often yields the asymptotic expression (for $n \to \infty$) of S_n—sum of its first n terms. We give concrete illustrations for both cases, for the latter by means of Stirling's formula.

I. *Transformation of S_n.* The transformation referred to above is:

$$(1) \qquad S_n \equiv \sum_{i=1}^{n} u_i = \sum_{i=1}^{n-1} i(u_i - u_{i+1}) + n u_n,$$

an identity holding true for any $u_1, u_2, \cdots, u_n$ (special case of Abel's transformation). Thus we are led to consider two series:

$$(U) \sum_{n=1}^{\infty} u_n, \quad (V) \sum_{n=1}^{\infty} v_n \qquad (v_n = n(u_n - u_{n+1})).$$

We know that, if in the series $\sum_{n=1}^{\infty} a_n$, $\lim_{n=\infty} n a_n$ exists and $\neq 0$, then the series diverges (the converse is not true). This, in conjunction with (1), leads at once to several interesting conclusions.

THEOREM I. *If $\lim_{n=\infty} n u_n = l$ exists, then the two series (U) and (V) are either both convergent or both divergent, according as $l = 0$ or $l \neq 0$. In case of convergence, the two series have the same sum.*

THEOREM II. *If (U) and (V) both converge, then $\lim_{n=\infty} n u_n = 0$; hence (U) and (V) have the same sum.*

It is the latter property which makes the transformation (1) very valuable as the illustrations below indicate. The following theorem, not directly connected with the main line of the present discussion, is also of frequent use in the theory of infinite series.

THEOREM III. *If $\{u_n\}$ is a monotonic decreasing sequence and (U) is convergent, then necessarily $\lim_{n=\infty} n u_n = 0$.*

In fact, here (V) consists of positive terms; hence, it either converges or diverges to $+\infty$. In the first case our statement is proved by Theorem I; in the second

---

[*] From AMERICAN MATHEMATICAL MONTHLY, vol. 40 (1933), pp. 226–229.

[†] Read before the Pennsylvania Chapter of Pi Mu Epsilon, in October, 1932.

[‡] Cf. K. Knopp, *Theorie und Anwendungen der Unendlichen Reihen*, 2nd ed. (1924), pp. 242–249.

case, (1) shows, since (U) converges, that $\lim_{n=\infty} nu_n = -\infty$, which is impossible.

2. *Illustrations in case of convergence.* (i) Consider the series

(2) $$\frac{\pi^2}{6} = \sum_{n=1}^{\infty} \frac{1}{n^2}.$$

Here

$$u_n = \frac{1}{n^2}, \quad v_n = n\left(\frac{1}{n^2} - \frac{1}{(n+1)^2}\right) = \frac{2}{(n+1)^2} + \frac{1}{n(n+1)^2}$$

$$\frac{\pi^2}{6} = 2 \sum_{n=1}^{\infty} \frac{1}{(n+1)^2} + \sum_{n=1}^{\infty} \frac{1}{n(n+1)^2} = 2 \sum_{n=1}^{\infty} \frac{1}{n^2} + \sum_{n=1}^{\infty} \frac{1}{n(n+1)^2} - 2$$

$$\frac{\pi^2}{6} = 2 \cdot \frac{\pi^2}{6} + \sum_{n=1}^{\infty} \frac{1}{n(n+1)^2} - 2$$

(3) $$\frac{\pi^2}{6} = 2 - \sum_{n=1}^{\infty} \frac{1}{n(n+1)^2}.$$

If we apply once more the transformation (1), we get:

$$\sum_{n=1}^{\infty} \frac{1}{n(n+1)^2} = 3 \sum_{n=1}^{\infty} \frac{1}{(n+1)(n+2)^2} + \sum_{n=1}^{\infty} \frac{1}{(n+1)^2(n+2)^2}$$

$$= 3 \sum_{n=1}^{\infty} \frac{1}{n(n+1)^2} + \sum_{n=1}^{\infty} \frac{1}{(n+1)^2(n+2)^2} - \frac{3}{4}$$

(4) $$\frac{\pi^2}{6} = \frac{13}{8} + \frac{1}{2} \sum_{n=1}^{\infty} \frac{1}{(n+1)^2(n+2)^2}.$$

We thus have transformed (2) *into more rapidly convergent series* (3) *and* (4), the remainders, i.e. the sums of all terms starting with the $(n+1)$st being respectively of order of magnitude n^{-1}, n^{-2}, n^{-3}. With little computation, we could repeat the same transformation again. We also notice that, using the series (3) and (4), we get numerical values *limiting π both from above and below*. Thus,

$$\frac{\pi^2}{6} > \frac{13}{8} + \frac{1}{2} \cdot \frac{1}{2^2 \cdot 3^2}, \quad \frac{\pi^2}{6} < 2 - \frac{1}{4} - \frac{1}{2 \cdot 9}$$

(5) $$\tfrac{1}{6}\sqrt{354} < \pi < \tfrac{1}{6}\sqrt{366}; \quad 3.13 \cdots < \pi < 3.18 \cdots.$$

Hence, taking one term in (4) and two terms in (3), we obtain the value of π with an error not exceeding 0.05. It can be shown (making use of $\int_1^n x^{-2} dx$) that

$$\lim_{n=\infty} n \left[\frac{1}{(n+1)^2} + \frac{1}{(n+2)^2} + \cdots \right] = 1.$$

It follows that about *twenty* terms are needed in (2), if we wish to obtain π with the above accuracy.

(ii) Apply the transformation (1) to

$$\text{(6)} \qquad \frac{\pi}{4} = 1 - \frac{1}{3} + \frac{1}{5} - \frac{1}{7} \pm \cdots = \sum_{n=1}^{\infty} \frac{(-1)^{n-1}}{2n-1}.$$

We get, if n is even: $n = 2m$,

$$S_{2m} = \sum_{i=1}^{2m-1} (-1)^{i-1} \left[1 + \frac{1}{4i^2 - 1} \right] - \frac{2m}{4m - 1},$$

and a similar formula for S_{2m+1}. Letting $m \to \infty$, we thus obtain:

$$\text{(7)} \qquad \frac{\pi}{4} = \frac{1}{2} + \frac{1}{1 \cdot 3} - \frac{1}{3 \cdot 5} + \frac{1}{5 \cdot 7} \mp \cdots = \frac{1}{2} + \sum_{n=1}^{\infty} \frac{(-1)^{n-1}}{4n^2 - 1}.$$

The series (7), being an alternating one, yields both upper and lower bound for for π. It converges far more rapidly than (6). If, for example, an accuracy of 0.01 for $\pi/4$ is desired, we need 50 terms in (6), and only 5 terms in (7).

3. *Illustration: some elementary summations.* It is, we believe, of some interest to show that the same transformation (1) yields, in many cases, with very little or no computation, the *precise expression* for certain $S_n = \sum_{i=1}^{n} u_i$. As an example, the arithmetic series $\sum_{i=1}^{n} i$ may serve. Here, by (1),

$$\sum_{i=1}^{n} i = - \sum_{i=1}^{n-1} i + n^2 = - S_{n-1}^{(1)} + n^2 = - S_n^{(1)} + n^2 + n; \quad S_n^{(1)} = \frac{n(n+1)}{2}.$$

Similarly,

$$S_n^{(2)} \equiv \sum_{i=1}^{n} i^2 = - \sum_{i=1}^{n-1} i(2i+1) + n^3 = - 2S_n^{(2)} - S_{n-1}^{(1)} + n^3 + 2n^2$$

$$S_n^{(2)} = \frac{1}{3} \left[n^3 + 2n^2 - \frac{(n-1)n}{2} \right] = \frac{n(n+1)(2n+1)}{6}.$$

In the same way we can evaluate $S_n^{(3)} \equiv \sum_{i=1}^{n} i^3$, $S_n^{(4)}$, $\cdots$. In general, if the precise expression of $\sum_{i=1}^{n} u_i$ is known, (1) gives at once the expression for $\sum_{i=1}^{n} i(u_i - u_{i+1})$. For example, applying (1) to $\sum_{i=1}^{n} q^i [=(1-q^{n+1})/(1-q)]$, we get immediately:

$$\sum_{i=1}^{n} i q^{i-1} = \frac{1 - q^{n+1}}{(1-q)^2} - \frac{(n+1)q^n}{1-q}.$$

Of course, we cannot expect such an elementary transformation as (1) to work successfully in all cases, as is shown by the series $\log 2 = 1 - 1/2 + 1/3 - 1/4 + \cdots$.

4. *Illustration in case of divergence.* We close our discussion by showing that

(8) $$\lim_{n=\infty} \frac{n!}{n^{n+1/2}e^{-n}} \text{ exists, } = l, \text{ with } 0 < l < \infty.$$

[Stirling's formula gives for l the value $(2\pi)^{1/2}$]. Applying the transformation (1) to $\sum_{i=1}^{n} \log i$, we get:

(9) $$\sum_{i=1}^{n} \log i = \sum_{i=1}^{n-1} i \log \frac{i}{i+1} + n \log n = -\sum_{i=1}^{n-1} i \log\left(1 + \frac{1}{i}\right) + n \log n.$$

Taking 3 terms in the expansion for $\log(1+x)$, we have:

$$\log\left(1 + \frac{1}{i}\right) = \frac{1}{i} - \frac{1}{2i^2} + \frac{\theta_i}{3i^3} \quad (i \geq 1; \ 0 < \theta_i < 1),$$

which, substituted in (8), leads to

(10) $$\sum_{i=1}^{n} \log i = n \log n - (n-1) + \frac{1}{2} \sum_{i=1}^{n-1} \frac{1}{i} - \frac{1}{3} \sum_{i=1}^{n-1} \frac{\theta_i}{i^2}.$$

Making use of the convergence of the series $\sum_{i=1}^{\infty} \theta_i/i^2$ and of the existence of $\lim_{n=\infty} (\sum_{i=1}^{n} 1/i - \log n)$ (which is readily established, by means of $\int_1^n dx/x$), we obtain from (10) immediately:

$$\lim_{n=\infty}\left(\sum_{i=1}^{n} \log i - n \log n - \tfrac{1}{2} \log n + n\right) \text{ exists, } = B, \text{ with } 0 < B < \infty,$$

which is but another way of writing (8).

AN EXAMPLE ON DOUBLE SERIES*

J. E. BROCK, Washington University

It is known that the convergence of a double series $\sum_{i=1, j=1}^{\infty, \infty} a_{ij}$ i.e., the existence of the double limit, $\lim_{m\to\infty, n\to\infty} S_{mn}$ (where $S_{mn} = \sum_{i=1, j=1}^{m, n} a_{ij}$), does not imply the convergence of the series $\sum_{i=1}^{\infty} a_{ij}$ or $\sum_{j=1}^{\infty} a_{ij}$. Bromwich[†] cites an example of a convergent double series, a finite number of whose rows and columns diverge. It might be of interest, particularly for instructional purposes, to have a simple example in which the series $\sum_{i=1}^{\infty} a_{ij}$ and $\sum_{j=1}^{\infty} a_{ij}$ diverge for *all* j and i.

Consider the double series $\sum_{i=1, j=1}^{\infty, \infty} a_{ij}$ where $a_{ij} = (-1)^j b_i + (-1)^i b_j$ and $\sum_{n=1}^{\infty} b_n$ converges and has the sum 0. It may easily be seen that the double limit exists, for the partial sum $\sum_{i=1, j=1}^{m, n} a_{ij} = \sum_{i=1, j=1}^{m, n} (-1)^j b_i + \sum_{i=1, j=1}^{m, n} (-1)^i b_j$ and the sums on the right are either equal to zero or to $-\sum_{i=1}^{m} b_i$ or to $-\sum_{j=1}^{n} b_j$. Since $\sum_{n=1}^{\infty} b_n = 0$, these latter sums approach zero as a limit as $m \to \infty$, $n \to \infty$.

However, $\sum_{i=1}^{\infty} a_{ij}$ and $\sum_{j=1}^{\infty} a_{ij}$ both diverge, since the general term a_{ij} does not approach zero as a limit, when for fixed i, $j \to \infty$, or when for fixed j, $i \to \infty$.

---

* From AMERICAN MATHEMATICAL MONTHLY, vol. 50 (1943), p. 619.
† Bromwich, T. J. I'A., Theory of Infinite Series, Macmillan (1908), p. 90.

REVERSION OF SERIES WITH APPLICATIONS*

J. B. REYNOLDS, Lehigh University

It has been the experience of the author that men doing research work in applied mathematics are frequently baffled by a difficulty that could be resolved by the method of reversion of series. It is his conviction that every student who is trained for engineering or other scientific work should be taught this subject at some time in his courses in mathematics. Texts on algebra usually present this subject in connection with the theory of undetermined coefficients but it will be presented here as a topic in differential calculus. This approach has certain advantages over the algebraic method. Applications will be given to show how the method may be used.

Let $y=f(x)$ be such that $f(0)=0$ and $f(x)$ is expansible in a series in powers of x; then by Maclaurin's theorem.

$$(1) \quad y = px + qx^2/2! + rx^3/3! + sx^4/4! + tx^5/5! + \cdots, \quad (p \neq 0),$$

in which p, q, r, s, and t stand respectively for the values of the first, second, third, fourth, and fifth derivatives of y with respect to x, each evaluated for $x=0$.

If this series be reverted we may write

$$(2) \quad x = Py + Qy^2/2! + Ry^3/3! + Sy^4/4! + Ty^5/5! + \cdots$$

in which P, Q, R, S, and T stand respectively for the first, second, third, fourth, and fifth derivatives of x with respect to y evaluated for $x=0$ for which value y is, also, zero.

From the relation $dy/dx=1/(dx/dy)$, we find by successive differentiations the following relations

$$(3) \quad \begin{array}{l} P = 1/p, \quad Q = -q/p^3, \quad R = (3q^2 - pr)/p^5. \\ S = (10pqr - 15q^3 - p^2s)/p^7, \\ T = (15p^2qs + 10p^2r^2 - 105pq^2r + 105q^4 - p^3t)/p^9. \end{array}$$

These relations determine the values of the coefficients of the powers of y in the reverted series (2).

One advantage of using the calculus approach is that $y=f(x)$ does not have to be expanded into a series in order to obtain the reverted series giving the value of x in a power series in y as it does in the method of undetermined coefficients.

APPLICATIONS

1. Revert the series

$$y = x + x^2 + x^3/2 + x^4/6 + x^5/24 + \cdots$$

and find the value of x that satisfies $10xe^x=1$. From the successive derivatives of

---

* From AMERICAN MATHEMATICAL MONTHLY, vol. 51 (1944), pp. 578–580.

y with respect to x, evaluated for $x=0$, we find the values $p=1$, $q=2$, $r=3$, $s=4$, $t=5$; and by means of equations (3) $P=1$, $Q=-2$, $R=9$, $S=-64$, $T=625$; so that by equation (2)

$$x = y - y^2 + 3y^3/2 - 8y^4/3 + 125y^5/24 \cdots$$

which is the required reverted series. Since the given series is the expansion of xe^x, the value of y that makes x satisfy $10xe^x=1$ is 0.1. This value put into the reverted series gives $x=0.09128$.

2. Compute the root of $x^3+10x^2+8x-120=0$ that lies between 2 and 3. In order to have the reverted series converge as rapidly as possible we reduce the roots by 3 rather than by 2 and get the equation $x^3+19x^2+95x+21=0$. If now we let

$$y = x^3 + 19x^2 + 95x$$

so that $y=0$ for $x=0$, the equation will be satisfied when $y=-21$. From the assumed value of y we find

$$p = 95, \qquad q = 38, \qquad r = 6, \qquad s = 0;$$

whence

$$P = 1/95, \qquad Q = -38/(95)^3, \qquad R = 3762/(95)^5, \qquad S = -606480/(95)^7.$$

These values give by equation (2)

$$x = y/95 - (1/5)(y/95)^2 + (33/475)(y/95)^3 - (14/475)(y/95)^4 \cdots$$

and for $y=-21$, $x=-0.23165$; so that the corresponding root of the original equation is $3-0.23165=2.76835$.

3. Solve the equation

$$x + e^x + \sin x + \tan x = 2.$$

Let

$$y = x + e^x + \sin x + \tan x - 1,$$

so that $y=0$ for $x=0$; then the given equation will be satisfied when $y=1$. From the assumed value of y we find

$$p = 4, \qquad q = 1, \qquad r = 2, \qquad s = 1, \qquad t = 18;$$

whence

$$P = 1/4, \qquad Q = -1/64, \qquad R = -5/1024, \qquad S = 49/(4)^7, \qquad T = -1007/(4)^9.$$

When these values are substituted into equation (2) we get

$$x = y/4 - 2(y/16)^2 - (10/3)(y/16)^3 + (49/6)(y/16)^4 - (1007/30)(y/16)^5 \cdots.$$

When y is set equal to 1 in this, we find $x=0.24147$ which is the required root.

A SLOWLY DIVERGENT SERIES*

R. P. AGNEW, Cornell University

It is well known, and is easily proved by the integral test for convergence and divergence of series, that the series

(1) $$\sum \frac{1}{n}, \quad \sum \frac{1}{n \log n}, \quad \sum \frac{1}{n \log n \log \log n}, \cdots$$

are all divergent. These are classic examples of slowly divergent series, each one after the first diverging more slowly than its predecessor. The following theorem, which is a simple corollary of a theorem of B. Pettineo,† gives a series which diverges more slowly than any of those in (1).

THEOREM. *For each $x > 0$, let $P(x)$ denote the product of x and all of the numbers*

(2) $$\log x, \quad \log \log x, \quad \log \log \log x, \cdots$$

which are greater than 1. *Then the series*

(3) $$\sum_{n=1}^{\infty} \frac{1}{P(n)}$$

is divergent.

Following a method different from that of Pettineo we shall use the integral test, proving that (3) is divergent by proving that

$$\int_1^\infty \frac{1}{P(x)} dx = \infty.$$

The function $1/P(x)$ is continuous and decreasing over the infinite interval $x \geq 1$. Over the interval $1 \leq x \leq e$, $1/P(x)$ is $1/x$; hence

$$\int_1^e \frac{1}{P(x)} dx = \log x \Big]_1^e = 1.$$

Over the interval $e \leq x \leq e_2 = e^e$, $1/P(x)$ is $1/(x \log x)$; hence

$$\int_e^{e_2} \frac{1}{P(x)} dx = \log \log x \Big]_e^{e_2} = 1.$$

Over the interval $e_2 \leq x \leq e_3 = \exp e_2$, $1/P(x)$ is $1/[x(\log x) \log \log x]$; hence

---

* From AMERICAN MATHEMATICAL MONTHLY, vol. 54 (1947), pp. 273–274.

† B. Pettineo. Estensione di una classe di serie divergenti. Atti della Reale Accademia Nationale dei Lincei, Rendiconti, Classe di Scienze Fisiche, Matematiche e Naturali. Series 8, vol 1 (1946), pp. 680–685.

$$\int_{e_2}^{e_3} \frac{1}{P(x)}\, dx = \log \log \log x \Big]_{e_2}^{e_3} = 1.$$

Continuation of this process gives an infinite set of intervals (of rapidly increasing lengths) over each of which the integral of $1/P(x)$ is 1. The results follow from this.

TERM-WISE DIFFERENTIATION OF POWER SERIES*

T. M. APOSTOL, California Institute of Technology

Let the function $f(x)$ be defined by a power series

$$f(x) = \sum_{n=0}^{\infty} a_n x^n \tag{1}$$

in the convergence-interval $|x| < r$. The purpose of this note is to present an elementary proof of the fact that the derivative $f'(x)$ exists at every point in $|x| < r$ and that it may be obtained by differentiating the series in (1) term-by-term. The usual proofs of this fact make use of the concept of uniform convergence or else involve rearrangement of double series, and consequently are not given in most beginning calculus texts. Hardy's *Pure Mathematics* does not treat differentiation or integration of power series since he avoids "the inclusion of anything that involves really difficult ideas." The proof given below makes use only of the mean-value theorem of differential calculus and it is felt that the proof is simple enough to be presented to students in sophomore calculus.

It is assumed that the existence of an interval of convergence has been established for power series and that it is known that power series converge absolutely inside the interval. We first prove:

THEOREM 1. *Let the series in equation* (1) *have the interval of convergence* $|x| < r$. *Then the series*

$$\sum_{n=0}^{\infty} n a_n x^{n-1} \tag{2}$$

has the same interval of convergence.

Proof. Let us first take a positive value of x, $0 < x < r$, and let h be a small positive number such as $0 < x < x+h < r$. Then the series for $f(x)$ and for $f(x+h)$ are each absolutely convergent and hence so is the series

$$\frac{f(x+h) - f(x)}{h} = \sum_{n=0}^{\infty} a_n \frac{(x+h)^n - x^n}{h}. \tag{3}$$

Applying the mean-value theorem to each term of this series we have

$$\frac{(x+h)^n - x^n}{h} = n\xi_n^{n-1}$$

---

* From AMERICAN MATHEMATICAL MONTHLY, vol. 59 (1952), pp. 323–326.

where $x < \xi_n < x+h$. Hence the series in (3) is identical to the series

(4) $$\sum_{n=0}^{\infty} na_n \cdot \xi_n^{n-1},$$

which must be absolutely convergent, since the series in (3) is. The series in (4) is no longer a power series but it dominates the power series (2) so that the series (2) must be absolutely convergent for this x. Hence the interval of convergence of (2) includes the interval $|x| < r$, and it cannot extend beyond this interval since (2) dominates (1). This proves the theorem.

If we apply the theorem to series (2) we obtain the following:

COROLLARY. *The series*

$$\sum_{n=0}^{\infty} n(n-1)a_n x^{n-2}$$

has the same interval of convergence as the series in (1).

The next theorem can be obtained as a consequence of Theorem 3 but a direct proof is simple and it also serves as an introduction to the proof of Theorem 3.

THEOREM 2. *The function defined by* (1) *is continuous for each x inside the interval of convergence.*

Proof. Let $0 \leq |x| < r$ and choose x_1 so that $0 \leq |x| < |x_1| < r$. Next choose h so that $0 \leq |x+h| < |x_1| < r$. We will prove that

$$\lim_{h \to 0} |f(x+h) - f(x)| = 0.$$

Proceeding as in the proof of Theorem 1, we have

(5) $$f(x+h) - f(x) = h \sum_{n=0}^{\infty} na_n \xi_n^{n-1},$$

where each number ξ_n lies between x and $x+h$. By Theorem 1, the series

$$\sum_{n=0}^{\infty} na_n x_1^{n-1}$$

is absolutely convergent and since $|\xi_n| < |x_1|$, equation (5) leads to the inequality

(6) $$0 \leq |f(x+h) - f(x)| \leq |h| \sum_{n=0}^{\infty} n |a_n| |x_1|^{n-1}.$$

The series multiplying $|h|$ in (6) represents a number independent of h so that if we now let $h \to 0$ the theorem follows.

THEOREM 3. *If the function $f(x)$ is given by the power series*

$$f(x) = \sum_{n=0}^{\infty} a_n x^n$$

in the interval $|x| < r$, then for any x inside this interval the derivative $f'(x)$ exists and is given by the power series

$$f'(x) = \sum_{n=0}^{\infty} n a_n x^{n-1}.$$

Proof. Take x and then x_1 so that $0 \leq |x| < |x_1| < r$. Let

$$F(x) = \sum_{n=0}^{\infty} n a_n x^{n-1}.$$

This series converges absolutely in $|x| < r$ by Theorem 1, and our aim is to prove that $f'(x)$ exists and equals $F(x)$ in this interval. Proceeding as in the previous theorem we choose h so that $|x+h| < |x_1| < r$ and obtain

(7) $$F(x) - \frac{f(x+h) - f(x)}{h} = \sum_{n=0}^{\infty} n a_n (x^{n-1} - \xi_n^{n-1}),$$

where each number ξ_n lies between x and $x+h$. Now we apply the **mean-value theorem** again to obtain

$$x^{n-1} - \xi_n^{n-1} = (n-1) \eta_n^{n-2} (x - \xi_n)$$

where each number η_n lies between x and ξ_n. Hence we have

(8) $$F(x) - \frac{f(x+h) - f(x)}{h} = \sum_{n=0}^{\infty} n(n-1) a_n \eta_n^{n-2} (x - \xi_n),$$

the series on the right of (8) being absolutely convergent since it is identical to the one in (7). But we have $|x - \xi_n| < |h|$ and $|\eta_n| < |x_1|$ so that we have

(9) $$0 \leq \left| F(x) - \frac{f(x+h) - f(x)}{h} \right| \leq |h| \sum_{n=0}^{\infty} n(n-1) |a_n| |x_1|^{n-2},$$

the series in (9) being convergent by the corollary to Theorem 1. Letting $h \to 0$ in (9) we see that $f'(x)$ exists and equals $F(x)$.

For the sake of completeness we include

THEOREM 4. *Let*

$$f(x) = \sum_{n=0}^{\infty} a_n x^n$$

have the interval of convergence $|x|<r$. *Let*

$$G(x) = \sum_{n=0}^{\infty} \frac{a_n}{n+1} x^{n+1}.$$

Then the series for $G(x)$ has the same interval of convergence and

$$G'(x) = f(x).$$

Proof. Since the power series for $G(x)$ has *some* interval of convergence, Theorem 3 tells us that it must be the same as the interval for $f(x)$ and that $G'(x) = f(x)$.

A NOTE ON ALTERNATING SERIES*

PHILIP CALABRESE, Illinois Institute of Technology

It is not often that a classroom assignment develops into something worth writing about. Most of the time it just develops into something worth putting off until after dinner. So when Professor Curme, my calculus teacher, asked the class to determine how many terms of the alternating harmonic series one would have to add in order that the sum be within .00005 of the limit, my reaction was, "Okay, I'll do it after my accounting homework."

Well, true to the mathematics majors' code, I looked for an easy way to do it, and at the next session gave my answer of 10,000 terms. Unfortunately, Professor Curme, substantiated by Love and Rainville's calculus book,† said my answer was one half the number of terms needed to insure the necessary accuracy. Well, I went home, thought about it some more, and wrote up this proof. I hope there aren't too many holes in it!

Given a convergent alternating series, $\sum_{n=1}^{\infty} (-1)^{n+1} U_n$, that satisfies the added condition $(U_n - U_{n+1}) > (U_{n+1} - U_{n+2})$, one can estimate the limit of the series within any accuracy ϵ by adding up the first n terms, where $U_n \leq 2\epsilon$. That is, the sum of the first n terms of the series is within an ϵ of the limit if $U_n \leq 2\epsilon$. This is equivalent to saying that $|S_n - L| \leq \epsilon$ if $U_n \leq 2\epsilon$, where S_n is the sum of n terms, and L the limit of the series.

Proof. For simplicity, we introduce the following notation. Let $S_n = \sum_{i=1}^{n} (-1)^{i+1} U_i$, $a = U_{n+1} - U_{n+2}$, $b = U_{n+3} - U_{n+4}$, $\cdots$, $a' = U_n - U_{n+1}$, $b' = U_{n+2} - U_{n+3}$, $\cdots$. We can write the alternating series as follows:

$$S_n + (-1)^n (a + b + \cdots),$$

or

$$S_{n-1} + (-1)^{n+1} (a' + b' + \cdots).$$

By our assumption, $a < a'$, $b < b'$, $\cdots$. Therefore

$$(a + b + \cdots) < (a' + b' + \cdots).$$

But both groupings yield the same limit L. That is,

$$S_n + (-1)^n (a + b + \cdots) = L,$$
$$S_{n-1} + (-1)^{n+1} (a' + b' + \cdots) = L,$$

and

$$|S_n - L| = (a + b + \cdots),$$
$$|S_{n-1} - L| = (a' + b' + \cdots).$$

---

* From AMERICAN MATHEMATICAL MONTHLY, vol. 69 (1962), pp. 215–217.
† Love and Rainville, Differential and Integral Calculus, 5th ed., Macmillan, p. 369, No. 23.

Therefore
$$|S_n - L| < |S_{n-1} - L|.$$

Now, since in every alternating series the limit L must lie between any two successive sums S_{n-1} and S_n, and since $U_n = |S_n - S_{n-1}|$, it follows that $U_n = |S_n - L| + |S_{n-1} - L|$.

Now if we find a term for which $2\epsilon \geq U_n$, $U_n = |S_n - L| + |S_{n-1} - L| \leq 2\epsilon$. But $|S_n - L| < |S_{n-1} - L|$. Therefore $|S_n - L| + |S_n - L| < 2\epsilon$, and finally, $|S_n - L| < \epsilon$. Also, since $|S_{n+1} - L| < |S_n - L|$, we have $|S_{n+1} - L| < \epsilon$. So, all the sums following S_n are within an ϵ of the limit.

Furthermore, if one can find a term U_n which is exactly equal to 2ϵ, S_n is the first sum that meets the necessary conditions. For $U_n = |S_n - L| + |S_{n-1} - L| = 2\epsilon$. But $|S_n - L| < |S_{n-1} - L|$. Therefore $|S_n - L| < \epsilon < |S_{n-1} - L|$.

For an example, let us consider the well-known alternating harmonic series $\sum_{n=1}^{\infty} (-1)^{n+1}/n$. The condition $a < a'$ is easily verified. Let us now take $\epsilon = .00005 = 1/20{,}000$. Then using the relation $U_n \leq 2\epsilon = 2(1/20{,}000)$ we get $1/n \leq 1/10{,}000$ or $n \geq 10{,}000$. Therefore, $S_{10{,}000}$ is within $.00005$ of the limit of the series. It is also the first sum within that accuracy because we found a U_n exactly equal to 2ϵ.

In conclusion, I would like to thank Professor G. L. Curme for encouraging me to write this paper, and for helping to direct me into mathematics.

ON A CONVERGENCE TEST FOR ALTERNATING SERIES*

R. LARIVIERE

Many calculus texts give the following test for the convergence of an alternating series without giving any examples of series which do not satisfy both its requirements.

An alternating series $u_1 - u_2 + u_3 - u_4 + \cdots$, in which the u's are all positive, converges if (a) $\lim_{n\to\infty} u = 0$, and if (b) for terms beyond a certain kth term, $u_n \geq u_{n+1}$, $n = k, k+1, k+2, \cdots$.

The beginner often concludes that the alternating series which do not satisfy both (a) and (b) are necessarily divergent. This error may be obviated by showing him examples of both convergent and divergent series that do not satisfy (b), preferably examples that seem to have uniform general terms such as the following:

$$(1) \quad S_1 = \sum_{1}^{\infty} (-1)^{n+1} \frac{|\sin n|}{n^2},$$

$$(2) \quad S_2 = \sum_{1}^{\infty} (-1)^{n+1} \frac{(\sqrt{2})^{(-1)^{n+1}}}{n}.$$

In (1) S_1 is obviously convergent, but

$$\frac{u_{n+1}}{u_n} = \frac{n^2}{(n+1)^2} \frac{|\sin(n+1)|}{|\sin n|},$$

$$\frac{u_{n+1}}{u_n} = |\cos 1 + \sin 1 \cot n| \cdot \frac{n^2}{(n+1)^2}.$$

Hence (b) is not satisfied.

$$\frac{u_{n+1}}{u_n} = \left(\frac{n}{n+1}\right)(\sqrt{2})^{[(-1)^{n+2}-(-1)^{n+1}]} = \left(\frac{n}{n+1}\right) 2^{(-1)^n}.$$

This ratio is of greater inequality when n is even, and of lesser inequality when n is odd. Hence (b) is not satisfied. S_2 may be shown to be divergent as follows:

$$S_2 = \frac{\sqrt{2}}{1} - \frac{\sqrt{2}^{-1}}{2} + \frac{\sqrt{2}}{3} - \frac{\sqrt{2}^{-1}}{4} + \frac{\sqrt{2}}{5} - \frac{\sqrt{2}^{-1}}{6} + \cdots$$

Since

$$\lim_{n \to \infty} u_n = 0,$$

we may group these terms in pairs, and number them $2n + 1$, $2n + 2$ in pairs,

---

* From MATHEMATICS MAGAZINE, vol. 29 (1956), p. 88.

with $n = 0, 1, 2 \cdots$, as follows:

$$S_2 = \left[\frac{\sqrt{2}}{1} - \frac{\sqrt{2}}{4}\right] + \left[\frac{\sqrt{2}}{3} - \frac{\sqrt{2}}{8}\right] + \left[\frac{\sqrt{2}}{5} - \frac{\sqrt{2}}{12}\right] + \cdots$$

$$+ \left[\frac{\sqrt{2}}{2n+1} - \frac{2}{4n+4}\right] + \cdots,$$

$$S_2 = \sum_0^\infty \left[\frac{\sqrt{2}}{2n+1} - \frac{\sqrt{2}}{4n+4}\right] = \sum_0^\infty \frac{\sqrt{2}}{4} \cdot \frac{1}{n+1} \cdot \frac{2n+3}{2n+1}.$$

Therefore S_2 is divergent.

BIBLIOGRAPHIC ENTRIES: THEORY

Except for the entries labeled MATHEMATICS MAGAZINE, the references below are to the AMERICAN MATHEMATICAL MONTHLY.

1. Callecot, O. L., The approximate summation of n terms of any harmonic series, vol. 13, p. 97.

2. Porter, M. B., Note on Cauchy's integral test, vol. 18, p. 37.

3. Kempner, A. J., A curious convergent series, vol. 21, p. 48.

Delete from the harmonic series those terms $1/n$ for which n contains no 9 in its decimal expansion. The resulting series converges to a sum less than 90.

4. Irwin, F., A curious convergent series, vol. 23, p. 149.

A generalization of Kempner, vol. 21, p. 48 (see entry 3, above).

5. James, G., A note on the sum of the remainders of a series, vol. 23, p. 333.

6. Sensenig, W., A proof of the definite integral formula, vol. 27, p. 209.

Proof that power series can be integrated term by term.

7. Miller, N., Infinite series formed by associating the terms of a series of functions with the points of a sequence, vol. 37, p. 224.

A scheme for constructing slowly convergent or slowly divergent series. Many examples.

8. Brink, R. W., A simplified integral test for the convergence of infinite series, vol. 38, p. 205.

9. Gurney, M., A non-uniformly convergent series, vol. 39, p. 108.

10. Libby, W. F., A convergence test and a remainder theorem, vol. 40, p. 216.

11. Herschfeld, A., On infinite radicals, vol. 42, p. 419.

A discussion of "right" infinite radicals

$$\sqrt{a_1 + \sqrt{a_2 + \sqrt{a_3 + \cdots}}}$$

and "left" infinite radicals

$$\cdots \sqrt{a_3 + \sqrt{a_2 + \sqrt{a_1}}}.$$

12. Barrow, D. F., Infinite exponentials, vol. 43, p. 150.

Given $\{a_n\}$, study $\{e_n\}$, where $e_0 = a_0$, $e_n = e_{n-1}^{a_n}$. A possible source of exercises.

13. MacDonald, J. K. L., and Sharpe, F. R., On some series arising from a definition of the exponential function, vol. 44, p. 312.

14. Chand, H., On some generalizations of Cauchy's condensation and integral tests, vol. 46, p. 338.

15. Garabedian, H. L., Hausdorff matrices, vol. 46, p. 390.

An exposition of material in two papers of Hausdorff on matrix methods of summation of series. See also Moore, vol. 39, p. 62, reproduced on pp. 333–341, this book.

16. Lennes, N. J., The ratio test for convergence of series, vol. 46, p. 434.

17. Bradshaw, J. W., Modifies series, vol. 46, p. 486.

Various simple ways of estimating the remainder of a series. Errors are discussed.

18. Perlin, I. E., Series with deleted terms, vol. 48, p. 93.

Another generalization of Kempner, vol. 21, p. 48 (See entry 3, above).

19. Rajagopal, C. T., Remarks on some generalizations of Cauchy's condensation and integral tests, vol. 48, p. 180.

An extension of Chand, vol. 46, p. 338 (see entry 14, above).

20. Bellman, R., A note on the divergence of a series, vol. 50, p. 318.

Divergence of the series of reciprocals of the primes.

21. Bradshaw, J. W., More modified series, vol. 51, p. 389.

Sequel to Bradshaw, vol. 46, p. 486 (see entry 17, above).

22. Hamming, R. W., Convergent monotone series, vol. 52, p. 70.
23. Sheffer, I. M., Convergence of multiply-infinite series, vol. 52, p. 365.
24. Hamming, R. W., Subseries of a monotone divergent series, vol. 54, p. 462.
25. Green, L. C., Uniform convergence and continuity, vol. 54, p. 541.

A source for examples.

26. Duffin, R. J., A generalization of the ratio test for series, vol. 55, p. 153.
27. Ward, M., A generalized integral test for convergence of series, vol. 56, p. 170.

Useful mostly in the complex domain.

28. Stalley, R., A generalization of the geometric series, vol. 56, p. 325.
29. Parker, S. T., Summable series and integrals, vol. 56, p. 678.

Essentially on Hölder summability of divergent series.

30. Morley, R. K., Further note on the remainder in computing by series, vol. 58, p. 410.

Sequel to Morley, vol. 57, p. 550, reproduced on pp. 324–325, this book.

31. Norris, M. J., Some necessary conditions for convergence of infinite series and improper integrals, vol. 60, p. 96.

32. Andress, W. R., The expansion of a function in terms of its values and derivatives at several points, vol. 60, p. 394.

33. Meyer, Burnett, On the convergence of alternating double series, vol. 60, p. 402.

34. Foster, R. M., and Klamkin, M. S., On the convergence of the p-series. vol. 60, p. 625.

35. Yates, R. C., Developments in power series, vol. 61, p. 256.

Connections with differential equations.

36. Newton, T. A., A note on a generalization of the Cauchy-Maclaurin integral test, vol. 61, p. 331.

37. Greenstein, D. S., On the iteration of continuous nonincreasing functions, vol. 61, p. 637.

The recurrence relation $x_{n+1} = f(x_n)$.

38. Meyer, Burnett, On the Cauchy convergence criterion, vol. 62, p. 488.
39. Epstein, L. F., and French, N. E., Improving the convergence of series: application to some elliptic integrals, vol. 63, p. 698.
40. Shanks, E. B., Convergence of series with positive terms, vol. 64, p. 338.
41. Wilansky, A., On the Cauchy criterion for the convergence of an infinite series, vol. 64, p. 469.
42. Moser, Leo, On the series $\Sigma\ 1/p$, vol. 65, p. 104.
43. Munro, W. D., Note on the Euler-Maclaurin formula, vol. 65, p. 201.
44. Michelow, J., A note on two convergence tests, vol. 67, p. 581.
45. Schwerdtfeger, H., On the convergence of the series $\Sigma\ n^{-\alpha}$, vol. 68, p. 361.
46. Stanaitis, O. E., On approximation of slowly convergent series, vol. 71, p. 186.
47. Tull, J. P., and Rearick, David, A convergence criterion for positive series, vol. 71, p. 294.
48. Cargeo, G. T., Some extensions of the integral test, vol. 73, p. 521.
49. Feingold, A., New geometric representation of an old infinite series, vol. 73, p. 528.
50. Howard, A., On the convergence of the binomial series, vol. 73, p. 760.
51. Taylor, M. E., Some divergence properties of series, vol. 73, p. 872.
52. Yong-Jeng, Lee, Change of variables test for convergence of series, vol. 73, p. 996.
53. Reyner, S. W., Convergence tests based on $\sqrt[n]{a_n}$, vol. 73, p. 998.
54. James, R. C., Infinite series and Taylor and Fourier expansions, MATHEMATICS MAGAZINE, vol. 25, p. 269.
55. Boas, R. P., Jr., Tannery's theorem, MATHEMATICS MAGAZINE, vol. 38, p. 66.

(b)
TRIGONOMETRIC SERIES

A SIMPLE DISCUSSION OF THE REPRESENTATION OF FUNCTIONS BY FOURIER SERIES*

PHILIP FRANKLIN, Massachusetts Institute of Technology

1. Introduction. Some time ago Professor Birkhoff (*1921*, 200) gave a brief treatment of the Fourier expansion of functions which were periodic and had three continuous derivatives. As most of the elementary examples of Fourier expansions correspond to functions with discontinuities, at least at the end points of the fundamental period interval, it is thought that the following discussion, which covers such cases, may be of interest. While our proof reduces everything to first principles, it is much briefer than those usually given.

2. Theorem. Having defined the Fourier coefficients of a function $f(x)$ by the equations

$$a_n = \frac{1}{\pi}\int_{-\pi}^{\pi} f(t) \cos ntdt; \qquad b_n = \frac{1}{\pi}\int_{-\pi}^{\pi} f(t) \sin ntdt,$$

we shall now establish the following theorem:

THEOREM. *Let $f(x)$ be a periodic function, period 2π, continuous with a continuous first derivative, except for a finite number of points in each period interval, and let it have forward and backward derivatives at the points of discontinuity. Moreover, let it be defined as equal to the average value of its right and left hand limit values at those points. Then $f(x)$ is represented by the series*

$$\tfrac{1}{2}a_0 + \sum_{n=1}^{\infty}(a_n \cos nx + b_n \sin nx).$$

PROOF. We wish to prove that

$$f(x) = \lim_{n=\infty} S_n,$$

where

$$S_n = \frac{1}{2}a_0 + \sum_{m=1}^{n}(a_m \cos mx + b_m \sin mx)$$

$$= \frac{1}{2\pi}\int_{-\pi}^{\pi} f(t)dt + \sum_{m=1}^{n}\left(\frac{1}{\pi}\int_{-\pi}^{\pi} f(t) \cos mtdt \cos mx + \frac{1}{\pi}\int_{-\pi}^{\pi} f(t) \sin mtdt \sin mx\right)$$

$$= \frac{1}{\pi}\int_{-\pi}^{\pi} f(t)\left[\frac{1}{2} + \sum_{m=1}^{n} \cos m(t-x)\right]dt.$$

---

* From AMERICAN MATHEMATICAL MONTHLY, vol. 31 (1924), pp. 475–478.

We first change the variable of integration from t to u where
$$t = u + x.$$
Since the integrand is periodic, we may integrate over any complete period, and hence keep the limits unchanged, obtaining
$$S_n = \frac{1}{\pi}\int_{-\pi}^{\pi} f(u+x)\left[\frac{1}{2} + \sum_{m=1}^{n} \cos mu\right] du = \frac{1}{\pi}\int_{-\pi}^{\pi} f(u+x)s_n du.$$

We may evaluate the trigonometric sum s_n by the following device:
$$(2\sin u/2)s_n = \sin u/2 + \sum_{m=1}^{n} 2\cos mu \sin u/2$$
$$= \sin u/2 + \sum_{m=1}^{n}[\sin(m+\tfrac{1}{2})u - \sin(m-\tfrac{1}{2})u]$$
$$= \sin(n+\tfrac{1}{2})u,$$
since the remaining terms appear twice with opposite signs. This gives:
$$s_n = \frac{1}{2} + \sum_{m=1}^{n} \cos mu = \frac{\sin(n+\tfrac{1}{2})u}{2\sin u/2}.$$

Our problem now is to show that
$$f(x) = \lim_{n=\infty} \frac{1}{\pi}\int_{-\pi}^{\pi} f(u+x)s_n du.$$
From the definition of $f(x)$ at the points of discontinuity, at all points,
$$f(x) = \tfrac{1}{2}f(x+) + \tfrac{1}{2}f(x-).$$
Since
$$\frac{1}{\pi}\int_{-\pi}^{\pi} f(u+x)s_n du = \frac{1}{\pi}\int_{0}^{\pi} f(u+x)s_n du + \frac{1}{\pi}\int_{-\pi}^{0} f(u+x)s_n du,$$
the conclusion will follow if
$$\tfrac{1}{2}f(x+) = \lim_{n=\infty}\frac{1}{\pi}\int_{0}^{\pi} f(u+x)s_n du; \text{ and } \tfrac{1}{2}f(x-) = \lim_{n=\infty}\frac{1}{\pi}\int_{-\pi}^{0} f(u+x)s_n du.$$

We shall merely prove the first in full, as the second can be proved either by similar reasoning, or by applying the first result to the function $F(y) = f(2x - y)$.
Since
$$\frac{1}{\pi}\int_{0}^{\pi} s_n du = \frac{1}{\pi}\int_{0}^{\pi}\left[\frac{1}{2} + \sum_{m=1}^{n}\cos mu\right] du = \frac{1}{2},$$
$$\tfrac{1}{2}f(x+) = \frac{1}{\pi}\int_{0}^{\pi} f(x+)s_n du = \lim_{n=\infty}\frac{1}{\pi}\int_{0}^{\pi} f(x+)s_n du.$$

Thus the equation in question will follow from

$$\lim_{n=\infty} \frac{1}{\pi} \int_0^\pi f(x+)s_n du = \lim_{n=\infty} \frac{1}{\pi} \int_0^\pi f(u+x)s_n du,$$

or

$$\lim_{n=\infty} \int_0^\pi [f(u+x) - f(x+)]s_n du = 0.$$

To discuss this equation, we break up the interval of integration $(0, \pi)$ into smaller intervals: $(0, p_0), (p_0, p_1), \cdots, (p_{D-1}, \pi)$; where $p_1, \cdots, p_{D-1}$ are the values of u for which $f(u + x)$ is discontinuous (exclusive of 0, which may be such a value), and where p_0 is a value between 0 and p_1 to be specified presently.
The integral in the first interval is:

$$\int_0^{p_0} [f(u+x) - f(x+)]s_n du = \int_0^{p_0} \frac{f(u+x) - f(x+)}{u} \cdot \frac{u/2}{\sin u/2} \sin\left(n + \frac{1}{2}\right)u\, du,$$

as appears upon replacing s_n by its value, and inserting the factor u in both numerator and denominator. We may now obtain upper bounds for each factor in succession. Thus,

$$\left|\frac{f(x+u) - f(x+)}{u}\right| = f'(x + \theta u) \leq D, \qquad (|\theta| < 1),$$

as appears upon using the law of the mean, letting D represent an upper bound for the derivative of the given function $f(x)$. Moreover,

$$\frac{u/2}{\sin u/2} \leq \frac{\pi}{2},$$

since its derivative $\dfrac{(\cos u/2)(\tan u/2 - u/2)}{2 \sin^2 u/2}$ is positive in the interval $(0, \pi)$,

showing that its maximum value in that interval is that for $u = \pi$, or $\pi/2$.
Finally,

$$|\sin (n + \tfrac{1}{2})u| \leq 1.$$

Consequently,

$$\left|\int_0^{p_0} [f(u+x) - f(x+)]s_n du\right| \leq \left|\int_0^{p_0} D \cdot \frac{\pi}{2} \cdot 1 du\right| \leq \frac{D\pi p_0}{2}.$$

This can be made small, say less than $\epsilon/2$, by taking p_0 small.

The remaining integrals may now be treated by integrating by parts. We have:

$$I_k = \int_{p_{k-1}}^{p_k} [f(u+x) - f(x+)]s_n du$$

$$= \int_{p_{k-1}}^{p_k} \frac{f(u+x) - f(x+)}{2 \sin u/2} \sin\left(n + \frac{1}{2}\right)u\, du$$

$$= -\frac{f(u+x) - f(x+)}{2 \sin u/2} \frac{\cos(n+\tfrac{1}{2})u}{n+\tfrac{1}{2}}\bigg|_{p_{k-1}}^{p_k}$$

$$+ \frac{1}{n+\tfrac{1}{2}} \int_{p_{k-1}}^{p_k} \left[\frac{f'(u+x)\cos(n+\tfrac{1}{2})u}{2 \sin u/2}\right.$$

$$\left. - \frac{[f(u+x) - f(x+)]\cos(n+\tfrac{1}{2})u \cos u/2}{4 \sin^2 u/2}\right] du.$$

But, if F is an upper bound for $f(x)$,

$$|f(u+x) - f(x+)| \le 2F,$$

while, since all the p_i are greater than p_0,

$$\sin u/2 > \sin p_0/2.$$

It follows from these inequalities that

$$|I_k| \le \frac{1}{n+\tfrac{1}{2}} \left[\frac{2F}{\sin p_0/2} + (p_{k+1} - p_k)\left(\frac{D}{2 \sin p_0/2} + \frac{F}{2 \sin^2 p_0/2}\right)\right].$$

As the quantity in the braces is independent of n, the sum of all the integrals I_k will be numerically less than some expression independent of n (though involving ϵ through p_0), divided by $n + \tfrac{1}{2}$. Consequently, by taking n sufficiently large ($\ge N$), it can be made small, say less than $\epsilon/2$.

We have shown that, for any ϵ, an n can be found for which

$$\int_0^\pi [f(u+x) - f(x+)]s_n du < \epsilon \quad \text{if} \quad n \ge N.$$

Consequently, when $n = \infty$, the limit of the left member must be zero, thus completing the proof.

NOTE ON THE CONVERGENCE OF FOURIER SERIES*†

DUNHAM JACKSON, University of Minnesota

In a recent number of the MONTHLY, Franklin has given a particularly clear and simple proof of the convergence of Fourier series under hypotheses general enough to cover the most important applications.‡ The conditions on the function to be represented may be essentially paraphrased by saying that it has the period 2π, and that a period interval can be divided into a finite number of subintervals, in each of which, considered by itself, the function has a continuous derivative, if suitably defined at the ends of the subinterval, though there may be a discontinuity in passing from one subinterval to the next.

The purpose of this note is to arrive at the same result by a proof which is believed to be still simpler in the character of its reasoning, though perhaps not very much shorter when spread on paper.

It is convenient to keep the notation and the identical language of Franklin's paper to a point a little above the middle of page 477, through the lines **
"Thus the equation in question will follow from ....

$$\lim_{n=\infty} \int_0^\pi [f(u+x)-f(x+)]s_n du = 0 \text{ ."}$$

Then the proof proceeds as follows:

Let $\varphi(x)$ be an arbitrary function which is bounded and integrable from $-\pi$ to π (but not necessarily periodic nor defined in any way outside this interval), let α_n, β_n be its Fourier coefficients:

$$\alpha_n = \frac{1}{\pi}\int_{-\pi}^\pi \varphi(t) \cos nt\, dt, \qquad \beta_n = \frac{1}{\pi}\int_{-\pi}^\pi \varphi(t) \sin nt\, dt,$$

and let

$$\sigma_n = \tfrac{1}{2}\alpha_0 + \sum_{m=1}^n (\alpha_m \cos mx + \beta_m \sin mx).$$

Then

$$\frac{1}{\pi}\int_{-\pi}^\pi [\varphi(x)-\sigma_n(x)]^2 dx = \frac{1}{\pi}\int_{-\pi}^\pi [\varphi(x)]^2 dx - \left[\tfrac{1}{2}\alpha_0^2 + \sum_{m=1}^n (\alpha_m^2 + \beta_m^2)\right].$$

---

* From AMERICAN MATHEMATICAL MONTHLY, vol. 33 (1926), pp. 39–40.
† Presented to the American Mathematical Society, October 31, 1925.
‡ P. Franklin, A simple discussion of the representation of functions by Fourier series, this MONTHLY (1924), pp. 475–478. [*Ed. note:* Reprinted on pp. 405–408, this book.]
***Ed. note:* See p. 407 in this book.

As the left-hand member can not be negative, the sum on the right must remain bounded as n increases indefinitely, the series

$$\sum_{m=1}^{\infty} (\alpha_m^2 + \beta_m^2)$$

must converge, and it must be that*$\lim_{n=\infty} \alpha_n = 0$, $\lim_{n=\infty} \beta_n = 0$. Since $\varphi(x)\cos\tfrac{1}{2}x$ and $\varphi(x)\sin\tfrac{1}{2}x$ satisfy the conditions imposed on $\varphi(x)$, the integrals

$$\int_{-\pi}^{\pi} \varphi(t) \sin\tfrac{1}{2}t \cos nt\, dt\ , \qquad \int_{-\pi}^{\pi} \varphi(t) \cos\tfrac{1}{2}t \sin nt\, dt$$

approach zero as n becomes infinite, and their sum

$$\int_{-\pi}^{\pi} \varphi(t) \sin(n+\tfrac{1}{2})t\, dt$$

does the same.

If $\psi(x)$ is bounded and integrable from 0 to π, the function $\varphi(x)$ which is equal to $\psi(x)$ for $0 < x \leq \pi$, equal to $-\psi(-x)$ for $-\pi \leq x < 0$, and equal to 0, say, for $x = 0$, is bounded and integrable from $-\pi$ to π, and the integral

$$\int_{-\pi}^{\pi} \varphi(t) \sin(n+\tfrac{1}{2})t\, dt = 2\int_{0}^{\pi} \psi(t) \sin(n+\tfrac{1}{2})t\, dt$$

still has the limit zero.

But, under the hypotheses of the theorem, the function

$$\psi(u) = \frac{f(x+u)-f(x+)}{2\sin u/2} = \frac{f(x+u)-f(x+)}{u} \cdot \frac{u/2}{\sin u/2}$$

is bounded and integrable for $0 < u \leq \pi$, since it has essentially the same discontinuities as $f(x+u)$ in the open interval, and approaches a limit as u approaches zero from the right. Consequently†

$$\lim_{n=\infty} \int_{0}^{\pi} [f(u+x)-f(x+)]s_n du = \lim_{n=\infty} \int_{0}^{\pi} \psi(u) \sin(u+\tfrac{1}{2})u\, du = 0\ .$$

It may be pointed out that this proof does not require that the derivative of $f(x)$ be continuous.

---

* Cf., e. g., M. Bôcher, Introduction to the theory of Fourier's series, *Annals of Mathematics*, vol. 7 (1906), pp. 81-152; p. 86.

† Cf. de la Vallée Poussin, *Leçons sur l'approximation des fonctions d'une variable réelle*, Paris, 1919; pp. 19-20.

BIBLIOGRAPHIC ENTRIES: TRIGONOMETRIC SERIES

The references below are to the AMERICAN MATHEMATICAL MONTHLY.

1. Birkhoff, G. D., An elementary treatment of Fourier's series, vol. 28, p. 200.
2. McShane, E. J., An application of Fourier's series and a theorem on definite integrals, vol. 33, p. 421.

   Applies Fourier series and the Cauchy-Schwarz inequality for integrals to a physical problem on alternating currents.

3. Jackson, D., The convergence of Fourier series, vol. 41, p. 67.
4. Szasz, O., On Fourier series of continuous functions, vol. 42, p. 37.
5. Taylor, A. E., Differentiation of Fourier series and integrals, vol. 51, p. 19.

(c)
SUMS OF SPECIAL SERIES

A FORMULA FOR THE SUM OF A CERTAIN TYPE OF INFINITE POWER SERIES*†

ELBERT H. CLARKE, Purdue University

Introduction. The problem to be considered is that of finding a definite, finite formula which will give the sum of any convergent infinite series whose terms are such that their numerators form an arithmetical progression of any order‡ and whose denominators form a geometric progression.

Since the nth term of an arithmetic progression of the kth order may be reduced to the form

$$a_n = b_0 n^k + b_1 n^{k-1} + \cdots + b_k,$$

our problem is to evaluate the expression

$$T = \sum_{n=1}^{\infty} \frac{b_0 n^k + b_1 n^{k-1} + \cdots + b_k}{ar^n}$$

in which the b's are independent of n. But this may be written

$$T = \frac{b_0}{a} \sum_{n=1}^{\infty} \frac{n^k}{r^n} + \frac{b_1}{a} \sum_{n=1}^{\infty} \frac{n^{k-1}}{r^n} + \cdots + \frac{b_k}{a} \sum_{n=1}^{\infty} \frac{1}{r^n}$$

and so the problem is at once reduced to that of finding a formula for

$$\sum_{n=1}^{\infty} \frac{n^k}{r^n} \quad (|r| > 1, \ k \text{ a positive integer or zero})$$

which will be denoted in what follows by $S_{k,r}$.

The Form of the Sum. The series $S_{k,r}$, being a power series in $1/r$, may be differentiated term by term with respect to r in its region of convergence.

$$\frac{dS_{k,r}}{dr} = \sum_{n=1}^{\infty} - \frac{n^{k+1}}{r^{n+1}} = -\frac{1}{r} \sum_{n=1}^{\infty} \frac{n^{k+1}}{r^n}.$$

---

* From AMERICAN MATHEMATICAL MONTHLY, vol. 21 (1914), pp. 292–296.

† The author wishes to acknowledge criticisms and suggestions from Professors R. D. Carmichael and A. C. Lunn.

‡ See *Text Book of Algebra*, CHRYSTALL, Vol. 1, page 484.

This gives at once a fundamental relation,

(1) $$S_{k+1,r} = -r\frac{dS_{k,r}}{dr}.$$

Furthermore, $S_{0,r}$ is none other than the ordinary geometric series, so we have at once

$$S_{0,r} = \frac{1}{r-1}$$

and we can derive any particular $S_{k,r}$ by k applications of the fundamental relation (1).

By an inspection of the forms of $S_{1,r}$; $S_{2,r}$ $\cdots$; we are led to expect

$$S_{k,r} = \frac{r}{(r-1)^{k+1}} F_{k-1}(r),$$

where $F_{k-1}(r)$ is a polynomial of degree $(k-1)$ in r. Applying (1) we obtain, after a few simple reductions,

(2) $$S_{k+1,r} = \frac{r}{(r-1)^{k+2}} [(kr+1)F_{k-1}(r) + (r-r^2)F'_{k-1}(r)].$$

If $F_{k-1}(r)$ is a polynomial of degree $(k-1)$ in r then the bracketed quantity is a polynomial of degree k in r and may be called $F_k(r)$. In this way the form of $S_{k,r}$ is determined except for the particular coefficients occurring in the polynomial $F_{k-1}(r)$.

Determination of the Coefficients. Let us write

$$F_{k-1}(r) = \alpha_{k-1,1}r^{k-1} + \alpha_{k-1,2}r^{k-2} + \alpha_{k-1,3}r^{k-3} + \cdots + \alpha_{k-1,k-1}r + \alpha_{k-1,k}.$$

Then by carrying out the operations indicated inside the brackets in (2) we obtain

(3) $$\begin{aligned}F_k(r) = &\alpha_{k-1,1}r^k + [2\alpha_{k-1,2} + k\alpha_{k-1,1}]r^{k-1} + \cdots \\ &+ [t\alpha_{k-1,t} + (k-t+2)\alpha_{k-1,t-1}]r^{k-t+1} + \cdots \\ &+ [k\alpha_{k-1,2} + 2\alpha_{k-1,k-1}]r + \alpha_{k-1,k}.\end{aligned}$$

The following relation is seen to exist between the coefficients of $F_k(r)$ and $F_{k-1}(r)$:

$$\alpha_{k,t} = t\alpha_{k-1,t} + (k-t+2)\alpha_{k-1,t-1}.$$

By inspection of $F_1(r)$, $F_2(r)$ $\cdots$ we notice that the coefficients are symmetrically arranged and that the first and last coefficients are always 1. By our work we see that the first and last coefficients in $F_{k-1}(r)$ and $F_k(r)$ are the same, therefore they must be 1. Let us assume further a symmetrical arrangement of the coefficients in $F_{k-1}(r)$. That is to say $\alpha_{k-1,t} = \alpha_{k-1,k-t+1}$. This assumption, in connection with the above recurrence relation between the coefficients of $F_k(r)$ and

$F_{k-1}(r)$, gives us immediately

(4)
$$\alpha_{k,t} = t\alpha_{k-1,t} + (k - t + 2)\alpha_{k-1,t-1}$$
$$= (k - t + 2)\alpha_{k-1,k-t+2} + t\alpha_{k-1,k-t+1} = \alpha_{k,k-t+2}.$$

So that symmetry of arrangement of the coefficients in $F_{k-1}(r)$ implies symmetry of arrangement of the coefficients in $F_k(r)$. That the coefficients in $F_k(r)$ are symmetrically arranged follows at once from an inspection of a particular $F_k(r)$, e.g., $F_4(r) = r^4 + 26r^3 + 66r^2 + 26r + 1$.

It is an interesting fact that the sum of the coefficients in $F_k(r)$ is $(k+1)!$ The reader may easily derive the proof from (3).

As already stated, the *first* coefficient satisfies the relation

$$\alpha_{k,1} = \alpha_{k-1,1} = \cdots = 1.$$

By carrying out the recurrence relations for the *second* coefficient we obtain

$$\alpha_{2,2} = 2\alpha_{2,1} + 2, \qquad (\alpha_{2,1} = 1)$$
$$\alpha_{3,2} = 2^2 \cdot 1 + 2 \cdot 2 + 3,$$
$$\alpha_{4,2} = 2^3 \cdot 1 + 2^2 \cdot 2 + 2 \cdot 3 + 4,$$
$$\cdots \cdots \cdots \cdots \cdots \cdots \cdots \cdots$$
$$\alpha_{k,2} = 2^{k-1} \cdot 1 + 2^{k-2} \cdot 2 + 2^{k-3} \cdot 3 + \cdots$$
$$+ 2^2(k-2) + 2(k-1) + k.$$

If we write $\alpha_{k,2}$ in rows and add by columns we shall obtain without difficulty

$$\alpha_{k,2} = 2^{k+1} - (k+2).$$

To determine the *third* coefficient we may proceed as above making use of the following formulas:

$$\sum_{k=0}^{e} 3^{e-k} \cdot 2^k = 3^{e+1} - 2^{e+1},$$

$$\sum_{k=0}^{e} 3^{e-k} \cdot k^2 = \sum_{k=0}^{e} 3^{e-k} \cdot k + \frac{1}{2} \sum_{k=0}^{e} 3^{e-k} - \frac{1}{2}(e+1)^2,$$

$$\sum_{k=0}^{e} 3^{e-k} \cdot k = \frac{1}{2} \sum_{k=0}^{e} 3^{e-k} - \frac{1}{2}(e+1).$$

When we have gone through with the necessary work, which is not difficult but is tedious, we obtain the following result

$$\alpha_{k,3} = 2^2[3^{k-2} \cdot 1 + 3^{k-3} \cdot 2 \cdot 2 + \cdots + 3(k-2)2^{k-3} + (k-1)2^{k-2}]$$
$$- [3^{k-2} \cdot 1 \cdot 3 + 3^{k-3} \cdot 2 \cdot 4 + \cdots + 3(k-2)k + (k-1)(k+1)]$$
$$= 3^{k+1} - 2^{k+1}(k+2) + \frac{(k+2)(k+1)}{2}.$$

SUM OF A CERTAIN TYPE OF INFINITE POWER SERIES

It is quite possible, of course, to proceed in the same way to obtain the fourth coefficient and so on. But the work will be extremely tedious and will not be likely to render much easier the generalization by inspection which must be made in any case. By an inspection of the results for $\alpha_{k,2}$ and for $\alpha_{k,3}$, together with (3) we are led to assume for the tth coefficient in $F_k(r)$,

$$\alpha_{k,t} = \left[t^{k+1} - (t-1)^{k+1} \binom{k+2}{1} + \cdots \right.$$
$$\left. + (-1)^{s-1}(t-s+1)^{k+1} \binom{k+2}{s-1} + \cdots + (-1)^{t-1} \binom{k+2}{t-1} \right],$$

which gives the proper forms for $\alpha_{k,2}$ and $\alpha_{k,3}$ above and which further yields proper numerical values for $\alpha_{1,4}$, $\alpha_{2,4}$, $\cdots$. All that is needed to complete the induction is a proof that $\alpha_{k,t}$ will satisfy the recurrence relation (4). That is to say,

$$t^{k+1} - (t-1)^{k+1}\binom{k+2}{1} + \cdots + (-1)^{s-1}(t-s+1)^{k+1}\binom{k+2}{s-1} + \cdots$$
$$+ (-1)^{t-2} 2^{k+1}\binom{k+2}{t-2} + (-1)^{t-1}\binom{k+2}{t-1}$$

must be equal to

$$t\left[t^k - (t-1)^k\binom{k+1}{1} + \cdots + (-1)^{s-1}(t-s+1)^k\binom{k+1}{s-1} + \cdots \right.$$
$$\left. + (-1)^{t-1}\binom{k+1}{t-1} \right] + (k-t+2)\left[(t-1)^k - (t-2)^k\binom{k+1}{1} + \cdots \right.$$
$$\left. + (-1)^{s-1}(t-s)^k\binom{k+1}{s-1} + \cdots + (-1)^{t-2}\binom{k+1}{t-2} \right].$$

Carrying out the work, the latter expression becomes

$$t^{k+1} - (t-1)^k t\binom{k+1}{1} + \cdots + (1)^{s-1}(t-s+1)^k t\binom{k+1}{s-1} + \cdots$$
$$+ (-1)^{t-1} t\binom{k+1}{t-1} + (t-1)^k(k-t+2) + \cdots$$
$$+ (-1)^{s-2}(t-s+1)^k(k-t+2)\binom{k+1}{s-2} + \cdots$$
$$+ (-1)^{t-2}(k-t+2)\binom{k+1}{t-2}.$$

Combining like terms, we have

$$t^{k+1} - (t-1)^k[t(k+1) - k + t - 2] + \cdots + (-1)^{s-1}(t-s+1)^k \binom{k+1}{s-2}$$
$$\times \left[t\frac{k-s+3}{s-1} - k + t - 2\right] + \cdots$$
$$+ (-1)^{t-1}\binom{k+1}{t-2}\left[t\frac{k-t+3}{t-1} - k + t - 2\right]$$

which reduces to the left member above:

$$t^{k+1} - (t-1)^{k+1}\binom{k+2}{1} + \cdots + (-1)^{s-1}(t-s+1)^{k+1}\binom{k+2}{s-1} + \cdots$$
$$+ (-1)^{t-1}\binom{k+2}{t-1}.$$

If we replace k by $k-1$ we shall have the tth term in $F_{k-1}(r)$. Hence we have the desired formula for $S_{k,r}$:

$$S_{k,r} = \sum_{n=1}^{\infty} \frac{n^k}{r^n} = \frac{r}{(r-1)^{k+1}}\Big\{r^{k-1} + [2^k - (k+1)]r^{k-2} + \cdots$$
$$+ \left[t^k + \cdots + (-1)^{s-1}(t-s+1)^k\binom{k+1}{s-1} + \cdots\right.$$
$$\left. + (-1)^{t-1}\binom{k+1}{t-1}\right]r^{k-t} + \cdots + 1\Big\}.$$

The general series of the introductory paragraph may now be written

$$T = \frac{b_0}{a}S_{k,r} + \frac{b_1}{a}S_{k-1,r} + \cdots + \frac{b_r}{a}S_{0,r}.$$

A SIMPLE DERIVATION OF THE LEIBNITZ-GREGORY SERIES FOR $\pi/4$*

D. K. KAZARINOFF, University of Michigan

The Leibnitz-Gregory formula,

$$\frac{\pi}{4} = \sum_{0}^{\infty} \frac{(-1)^k}{2k+1},$$

is usually obtained from the series for $\tan^{-1} x$ by applying Abel's theorem of limits. In this note we give a simpler and more elegant derivation of this result and the series

$$\ln 2 = \sum_{1}^{\infty} \frac{(-1)^{k-1}}{k}.$$

We consider the integral

$$I_n = \int_0^{\pi/4} \tan^n x \, dx, \qquad n \geq 2,$$

and note that it is monotone decreasing with n. It is obvious that

$$I_n + I_{n-2} = \frac{1}{n-1},$$

and

$$I_n + I_{n+2} = \frac{1}{n+1}.$$

Since I_n is monotone decreasing,

$$\frac{1}{2(n-1)} > I_n > \frac{1}{2(n+1)}.$$

On the other hand, for n a positive integer, by repeated application of the reduction formula

$$\int_0^{\pi/4} \tan^n x \, dx = \frac{1}{n-1} - \int_0^{\pi/4} \tan^{n-2} x \, dx,$$

we know that

$$I_{2n} = \frac{1}{2n-1} - \frac{1}{2n-3} + \cdots \pm 1 \mp \frac{\pi}{4},$$

---

* From AMERICAN MATHEMATICAL MONTHLY, vol. 62 (1955), pp. 726–727.

and
$$I_{2n+1} = \frac{1}{2n} - \frac{1}{2n-2} + \cdots \pm \frac{1}{2} \mp \frac{1}{2} \ln 2.$$

Thus
$$\frac{1}{2(2n+1)} < \left| \sum_0^{n-1} \frac{(-1)^k}{2k+1} - \frac{\pi}{4} \right| < \frac{1}{2(2n-1)}, \qquad n = 1, 2, \cdots,$$

and
$$\frac{1}{4(n+1)} < \left| \sum_1^n \frac{(-1)^{k-1}}{2k} - \frac{1}{2} \ln 2 \right| < \frac{1}{4(n-1)}, \qquad n = 1, 2, \cdots.$$

Letting n become infinite, we obtain the series for $\pi/4$ and $\ln 2$.

We remark that the inequalities above provide a sharper estimate of the error made by using n terms of these series than the usual estimate of "less than the first neglected term."

PROBABILITY AND SUMS OF SERIES*

G. B. THOMAS, JR., Massachusetts Institute of Technology and Stanford University

Many interesting problems in probability stem from the simple model of repeated Bernoulli trials, where the probability of success is the same at each trial. If the trials are independent, but the probability p_n of success at trial n is allowed to vary with n, the trials are called Poisson trials [1, p. 233]). When we consider an infinite sequence of Poisson trials we are led quite naturally to certain associated series (for example, generating functions, cf. Feller, Chapter 11). Here we consider the probability of eventually obtaining a success. This is given by

$$(1) \qquad Pr \text{ (success on some trial)} = \sum_{n=1}^{\infty} f_n,$$

where f_n is the probability of *first* success at trial n,

$$(2) \qquad f_1 = p_1, \qquad f_n = (1 - p_1)(1 - p_2) \cdots (1 - p_{n-1})p_n, \qquad n > 1.$$

In particular, we are interested in circumstances under which the probability of getting at least one success is one. The following theorem gives the answer. By specializing the p_n we also obtain the sums of certain infinite series in explicit form.

THEOREM. *Let $\{p_n\}$ be a sequence of real numbers, $0 \leq p_n < 1$. Let f_n be given by (2). Then*

$$(3) \qquad \sum_{n=1}^{\infty} f_n \leq 1,$$

and equality holds in (3), if and only if

$$(4) \qquad \prod_{n=1}^{\infty} (1 - p_n) \text{ diverges to zero,}$$

or (equivalently), if and only if

$$(5) \qquad \sum_{n=1}^{\infty} p_n \text{ diverges to plus infinity.}$$

Proof. Let N be a positive integer. Let

$$(6) \qquad P_N = \sum_{1}^{N} f_n, \qquad Q_N = \prod_{1}^{N} (1 - p_n).$$

---

* From AMERICAN MATHEMATICAL MONTHLY, vol. 64 (1957), pp. 586–589.

Then P_N is the probability of at least one success in the first N trials, while Q_N is the probability of no successes in the same N trials. Hence

(7) $$P_N + Q_N \equiv 1.$$

Equation (7) is an identity for all sequences $\{p_n\}$ and for all positive integers N. In case $0 \leq p_n \leq 1$, the sequence $\{Q_N\}$ is monotone nonincreasing and bounded below by zero. Hence Q_N tends to a limit as N increases indefinitely. Likewise P_N has a limit. We denote these limits by Q and P respectively and have from (7)

(8) $$P + Q = 1,$$

identically for all sequences $\{p_n\}$ such that $0 \leq p_n \leq 1$, with

(9) $$P = \sum_1^\infty f_n \leq 1, \qquad Q = \prod_1^\infty (1 - p_n) \geq 0.$$

From (8) it follows at once that $P = 1$ if and only if $Q = 0$. And $Q = 0$ if and only if either (i) $p_n = 1$ for some n, or (ii) $\sum p_n$ diverges to plus infinity [2]. The theorem is thus established.

Example 1. Let $p_n = 1/(n+1)$ and obtain

$$\sum_1^\infty \frac{1}{n(n+1)} = 1.$$

Example 2. Let $p_n = n/(n+1)$ and obtain

$$\sum_1^\infty \frac{n}{(n+1)!} = 1.$$

Example 3. Let $p_n = n/(n+k)$, $k > 0$, and obtain

$$\sum_{n=1}^\infty \frac{nk^{n-1}}{\Gamma(n+k+1)} = \frac{1}{\Gamma(k+1)}.$$

In particular, for $k = 1/2$,

$$\sum_{n=1}^\infty n \bigg/ \left(\prod_{k=1}^n (2k+1)\right) = 1/2.$$

Remarks. (a) The identity (7) may be of interest to algebra teachers looking for another easy exercise in mathematical induction. Feller's book provides many more. Their probabilistic interpretations may provide additional interest.

(b) The series in Example 2 above is clearly closely related to the series for e^x. In fact, if we look at the generating function for the sequence $\{f_n\}$ we have

$$F(x) \equiv \sum_1^\infty f_n x^n = \sum_1^\infty n x^n / (n+1)!,$$

and we soon recognize this as being x times the derivative of

$$G(x) = \sum_{1}^{\infty} x^n/(n+1)! = x^{-1}(e^x - 1 - x).$$

Hence

$$F(x) = x\frac{d}{dx} G(x) = e^x - x^{-1}(e^x - 1),$$

and the series in Example 2 is evaluated by taking $x=1$; $F(1)=1$. To obtain the expected number of trials until first success occurs, we would compute

$$F'(1) = \sum_{n=1}^{\infty} nf_n = e - 1.$$

(c) One of the Borel-Cantelli lemmas [1, p. 155] asserts: *If the events A_k are mutually independent, and if $\sum Pr(A_k)$ diverges, then with probability one infinitely many A_k occur.*

We may take A_k to be the event, success at trial k, and apply this lemma and our theorem to obtain the following interesting corollary.

COROLLARY. *Let p_n be the probability of success on the nth trial in a sequence of Poisson trials. If $0 \leq p_n < 1$, and the probability of obtaining at least one success is one, then the probability of obtaining infinitely many successes is also one.*

References

1. W. Feller, An Introduction to Probability and its Applications, New York, 1950.
2. J. M. Hyslop, Infinite Series, New York, 1942, p. 94.

AN ELEMENTARY PROOF OF THE FORMULA $\sum_{k=1}^{\infty} 1/k^2 = \pi^2/6$*

YOSHIO MATSUOKA, Kagoshima University, Japan

The formula in the title has been well known, but its various known proofs are less elementary, (see for example, [1], p. 219, 360, [3], p. 237, 267, 324, [4], Problem 99, p. 196, [5], p. 379). The following proof is quite elementary in character.

For any positive integer n, we consider

$$\int_0^{\pi/2} \cos^{2n} t\, dt.$$

Applying integration by parts twice, we obtain

$$\int_0^{\pi/2} \cos^{2n} t\, dt = [t \cos^{2n} t]_0^{\pi/2} + 2n \int_0^{\pi/2} t \cos^{2n-1} t \sin t\, dt$$

$$= n[t^2 \cos^{2n-1} t \sin t]_0^{\pi/2} - n \int_0^{\pi/2} t^2 [-(2n-1)\cos^{2n-2} t \sin^2 t + \cos^{2n} t]\, dt$$

$$= -2n^2 \int_0^{\pi/2} t^2 \cos^{2n} t\, dt + n(2n-1) \int_0^{\pi/2} t^2 \cos^{2n-2} t\, dt$$

$$= -2n^2 I_{2n} + n(2n-1) I_{2n-2},$$

where $I_{2n} = \int_0^{\pi/2} t^2 \cos^{2n} t\, dt$. Hence

$$-2n^2 I_{2n} + n(2n-1) I_{2n-2} = \int_0^{\pi/2} \cos^{2n} t\, dt.$$

As is known, (see, for example, [2], p. 226)

$$\int_0^{\pi/2} \cos^{2n} t\, dt = \frac{(2n-1)!!}{(2n)!!} \cdot \frac{\pi}{2},$$

where, as usual,

$$(2n)!! = 2 \cdot 4 \cdots (2n-2)(2n), \quad 0!! = 1;$$
$$(2n+1)!! = 1 \cdot 3 \cdots (2n-1)(2n+1), \quad (-1)!! = 1.$$

Thus we have

$$-2n^2 I_{2n} + n(2n-1) I_{2n-2} = \frac{(2n-1)!!}{(2n)!!} \cdot \frac{\pi}{2},$$

$$\frac{(2n)!!}{(2n-1)!!} I_{2n} - \frac{(2n-2)!!}{(2n-3)!!} I_{2n-2} = -\frac{\pi}{4} \cdot \frac{1}{n^2}.$$

---

* From AMERICAN MATHEMATICAL MONTHLY, vol. 68 (1961), pp. 485–487.

This implies that

$$\frac{(2n)!!}{(2n-1)!!}I_{2n} - \frac{0!!}{(-1)!!}I_0 = \sum_{k=1}^{n}\left[\frac{(2k)!!}{(2k-1)!!}I_{2k} - \frac{(2k-2)!!}{(2k-3)!!}I_{2k-2}\right]$$

$$= -\frac{\pi}{4}\sum_{k=1}^{n}\frac{1}{k^2},$$

and hence that

$$\frac{(2n)!!}{(2n-1)!!}I_{2n} = \frac{\pi^3}{24} - \frac{\pi}{4}\sum_{k=1}^{n}\frac{1}{k^2} = \frac{\pi}{4}\left[\frac{\pi^2}{6} - \sum_{k=1}^{n}\frac{1}{k^2}\right].$$

It is sufficient to prove that

(1) $$\lim_{n\to\infty}\frac{(2n)!!}{(2n-1)!!}I_{2n} = 0.$$

Now we have

$$I_{2n} = \int_0^{\pi/2} t^2 \cos^{2n} t\, dt \leq \left(\frac{\pi}{2}\right)^2 \int_0^{\pi/2} \sin^2 t \cos^{2n} t\, dt$$

$$= \frac{\pi^2}{4}\left[\int_0^{\pi/2} \cos^{2n} t\, dt - \int_0^{\pi/2} \cos^{2n+2} t\, dt\right]$$

$$= \frac{\pi^3}{8}\left[\frac{(2n-1)!!}{(2n)!!} - \frac{(2n+1)!!}{(2n+2)!!}\right] = \frac{\pi^3}{8}\frac{(2n-1)!!}{(2n+2)!!}.$$

Therefore

$$0 < \frac{(2n)!!}{(2n-1)!!}I_{2n} \leq \frac{\pi^3}{8}\frac{1}{2n+2}.$$

Thus we establish (1); hence the formula is proved.

Finally, the author wishes to express his thanks to a referee for valuable criticism.

References

1. T. J. Bromwich, An Introduction to the Theory of Infinite Series (2nd ed.), London, 1926.
2. J. Edwards, The Integral Calculus, vol. 1, New York, 1954.
3. K. Knopp, Theory and Application of Infinite Series, London and Glasgow, 1928.
4. A. Ostrowski, Vorlesungen über Differential- und Integralrechnung, vol. 2, Basel, 1951.
5. ———, Vorlesungen über Differential- und Integralrechnung, vol. 3, Basel, 1954.

BIBLIOGRAPHIC ENTRIES: SUMS OF SPECIAL SERIES

Except for the entry labeled MATHEMATICS MAGAZINE, the references below are to the AMERICAN MATHEMATICAL MONTHLY.

1. Zerr, G. B. M., Summation of series, vol. 5, p. 128.
    Results stated without proof.

12

SPECIAL NUMBERS

(a)
e

METHODS OF PRESENTING e AND π*†

KARL MENGER, University of Notre Dame

1. Introduction. Initiating a student into calculus is about what sailing through the straits of Messina used to be: On one side the Charybdis dragging the boat into her whirlpool, on the other side the Scylla waiting for the vessel to shatter on her rock. The whirlpool engulfing so many teachers consists of the false statements concerning infinitely small quantities, the rock on which the beginner goes to pieces is the solid foundation of analysis. The proper initiation into calculus must painstakingly avoid all senseless statements and at the same time avoid unduly rigorous reasoning. If we add that even a first introduction should aim at conveying to the student the understanding of calculus rather than a mere mechanical ability to handle formulae, then we have about described the difficulties confronting the teacher.

The solution is to present only statements and arguments which the student can easily visualize and which are capable of rigorous proofs, but to present them without any attempt at rigorously proving them beyond what may come up in answering the questions of intelligent students. As an example of a presentation in this spirit, in what follows I outline methods of introducing e and π which for years I have found useful in teaching.

2. Concerning e. We start by plotting the exponential curves E_a given by the equations $y = a^x$ for particular values of $a(>1)$, especially for the bases $a = 2$ and $a = 4$. Then we compare the curve E_a with an auxiliary line L given by the equation $y = x + 1$. Clearly, each of the curves E_a has the point $(0, 1)$ in common with L. The curve $E_2(y = 2^x)$ has a second point with positive abscissa in common with L, viz., the point $(1, 2)$, the curve E_4 a second point with negative abscissa, viz., the point $(-1/2, 1/2)$.

The student easily realizes that if we let a increase beyond 2, the abscissa of the second point of intersection of E_a and L comes closer to 0 from the right side, and that if we let a decrease below 4, the abscissa of this point comes closer to 0 from the left. In fact, if we preassign the abscissa $\xi(>-1)$ of the second

---

* From AMERICAN MATHEMATICAL MONTHLY, vol. 52 (1945), pp. 28–33.

† An address to the teachers of calculus in the V-12 program at the University of Notre Dame.

point of intersection, it is easy to find exactly one value of a for which the curve E_a intersects L at the point $(\xi, \xi+1)$. We have to satisfy the equation $a^\xi = \xi + 1$, which yields $a = (1+\xi)^{1/\xi}$. The student actually computes the values of a corresponding to some small positive and negative values of ξ (for very small values of ξ by means of decadic or natural logarithmic tables), e.g., for $\xi = 1/2, 1/10, \cdots ; -1/4, -1/10, \cdots$. On the basis of this experience he will readily believe that as ξ approaches 0, the value of a approaches a number whose first five decimals have been computed as 2.71828. He is told that this number has been given the name e while the curve E_e is called the natural exponential curve, and e^x the natural exponential function.

The student will further admit:

that for $a > e$ the curve E_a intersects L at a second point, with a negative abscissa, which is the closer to 0 the closer a is to e; that each secant joining the point $(0, 1)$ to a close neighboring point on E_a is appreciably steeper than L, and that consequently the tangent to E_a at $(0, 1)$ has a slope >1;

that for $a < e$ the curve E_a intersects L at a second point, with a positive abscissa, which is the closer to 0 the closer a is to e; that each secant joining the point $(0, 1)$ to a close neighboring point on E_a is appreciably flatter than L, and that consequently the tangent to E_a at $(0, 1)$ has a slope <1;

that for $a = e$ the curve E_e with the equation $y = e^x$ does not intersect L in any second point, that the secants joining the point $(0, 1)$ to neighboring points on E_e come the closer to L the closer the neighboring point is to $(0, 1)$, and that consequently L is the tangent to E_e at $(0, 1)$; in other words, that the tangent to E_e at $(0, 1)$ has the slope 1.

Thus, in addition to introducing e, our method has made plausible this last statement, which we shall reformulate after the introduction of the concept of the derivative, by saying that the derivative of the function e^x at $x = 0$ is $= 1$, a fact of fundamental importance for the entire calculus. If a rigorous proof of the fact that $(e^h - 1)/h$ approaches 1 as h approaches 0 is desired, it can easily be supplemented for the case that h assumes the values $1, 1/2, 1/3, \cdots$.

Clearly, from $a = b + (a-b)$ and $a > b > 1$ it follows that $a^m > b^m + mb^{m-1}(b-a) > b^m + m(b-a)$. Thus we obtain the following

LEMMA

$$|\alpha - \beta| > m |\alpha^{1/m} - \beta^{1/m}| \quad \text{if} \quad \alpha > \beta > 1.$$

In order to show that with increasing m the numbers $(e^{1/m} - 1)/(1/m)$ approach 1 we form the auxiliary numbers $e_m = (1 + 1/m)^m$ for which $e_m^{1/m} = 1 + 1/m$. By our lemma,

$$|e^{1/m} - e_m^{1/m}| < |e - e_m|/m,$$

that is,

$$|e^{1/m} - 1 - 1/m| < |e - e_m|/m.$$

Dividing this inequality by $1/m$ we obtain

$$\left|\frac{e^{1/m}-1}{1/m}-1\right|<|e-e_m|,$$

from which our contention follows since, by the definition of e, the numbers e_m approach e, and thus the numbers $|e-e_m|$ approach 0.

The reason for the importance of the fact that $(e^h-1)/h$ approaches 1, is that in conjunction with the functional equation $e^{x+y}=e^x\cdot e^y$ this fact enables us to differentiate e^x:

$$\frac{e^{x+h}-e^x}{h}=e^x\cdot\frac{e^h-1}{h}\to e^x \quad\text{as}\quad h\to 0;$$

and that from the formula $(e^x)'=e^x$ the entire differential calculus of elementary functions, excluding the trigonometric functions, can be derived by virtue of the two rules concerning the differentiation of a sum of two functions, and the differentiation of a function of a function. (For a detailed development of this "Algebra of Derivation" the reader is referred to the author's booklet *Algebra of Analysis*, Notre Dame Mathematical Lectures No. 3, 1944.)

3. Concerning π. In introducing the beginner into the differentiation of the trigonometric functions, the teacher will, of course, use the concept of π familiar to the student from elementary geometry, and the trigonometric functions as they are known from trigonometry, including the concept of radian measure. Again the essential fact which he has to learn, is that for $x=0$ the derivative of the sine, or still better of the tangent, is equal to 1. In view of the fact that $\tan 0=0$, this amounts to the statement that $\tan h/h$ approaches 1 as h approaches 0.

Again, the reason for the importance of the fact that $\tan h/h$ approaches 1, is that in conjunction with the functional equation,

$$\tan(x+y)=\frac{\tan x+\tan y}{1-\tan x\cdot\tan y},$$

it enables us to differentiate $\tan x$:

$$\frac{\tan(x+h)-\tan x}{h}=\frac{1}{h}\left[\frac{\tan x+\tan h}{1-\tan h\cdot\tan x}-\tan x\right]$$

$$=\frac{1}{1-\tan h\cdot\tan x}\cdot\frac{\tan h}{h}(1+\tan^2 x)\to(1+\tan^2 x)=\sec^2 x;$$

and that by means of the formula $(\tan x)'=\sec^2 x$ all the trigonometric and arc functions can be differentiated. Thus to complete the differentiation of all elementary functions we merely have to use the formulae

$$\sin x=\frac{\tan(x/2)}{1+\tan^2(x/2)} \quad\text{and}\quad \cos x=\frac{1-\tan^2(x/2)}{1+\tan^2(x/2)},$$

which obviously are important also for other reasons, and

$$\tan(\arctan x) = x, \quad \sin(\arcsin x) = x.$$

(Again for details, the reader is referred to the author's *Algebra of Analysis*.)

In order to prove that $\tan h/h \to 1$ as $h \to 0$, the following idea is useful. It applies to the case that h assumes the values $\pi/3, \pi/4, \cdots, \pi/n, \cdots$ but can easily be adapted to the general case, which, in the opinion of the author, should not be presented to the beginner. If we circumscribe to the circle of radius 1 a regular polygon of n sides, then from elementary trigonometry it is clear that each side has the length $2\tan(\pi/n)$ where π is the radian measure of two right angles. Hence the length of the polygon is $2n \tan(\pi/n)$. As n gets larger, the lengths of the polygons approach that of the circle (elementary geometry!), that is, 2π. Dividing the formula $2n \tan(\pi/n) \to 2\pi$ by 2π, we obtain

$$\frac{\tan(\pi/n)}{\pi/n} \to 1.$$

The same reasoning applied to inscribed polygons would, of course, yield

$$\frac{\sin(\pi/n)}{\pi/n} \to 1.$$

While this procedure is quite satisfactory as a first introduction, the teacher should realize that the use of radian measure in developing the differentiation of the trigonometric functions is objectionable. The radian measure is based on the concept of the length of a circle which is the limit of the lengths of inscribed and circumscribed polygons. Whenever we consider $\tan x$ where x is measured in radians, we really presuppose a process of the same logical order as the formation of e^x. Now we certainly would not start the exposition of exponential functions with the study of e^x. We start with 2^x, 4^x, etc., then proceed to the idea of a one-parameter family of functions a^x. In fact, we begin with rational values of x and then, on the basis of considerations of continuity, extend a^x to all values of x. Subsequently a procedure described in the first part of this paper singles out the natural exponential curve E_e from the curves E_a (and e from the numbers a) in such a way that for $x=0$ the derivative of the natural exponential function is $=1$.

4. Analogous methods of approach. It may thus be of interest that in a completely analogous way we may start with a one-parameter family of trigonometric functions T_p, among which those corresponding with rational values, like 2 and 4, can be as easily handled as 2^x and 4^x. Then a procedure can be developed which is completely analogous to that which we developed for exponential functions and which singles out a "natural" trigonometric function T_π from the T_p (and π from the numbers p) in such a way that for $x=0$ the derivative of the natural trigonometric function is $=1$.

We shall describe this development for the tangential functions. We call tangential curve of period p, and denote by T_p, the curve with the equation

$$y = \tan(2Rx/p).$$

The ordinate of T_p at the abscissa x is the trigonometric tangent of an angle which is $2x/p$ times a right angle. Clearly, this definition is independent of the concept of length. In order to construct T_p for a given number p, we declare that on the X-axis the point with the abscissa $p/2$ will symbolize the right angle, and the points with other abscissae will stand for proportional multiples of the right angle, e.g., the point $p/4$ for one-half of a right angle, $p/6$ for one-third of a right angle, the point x for $2x/p$ times a right angle. As ordinates at these abscissae we lay off the tangents of $R/2$, $R/3$, $2Rx/p$, respectively. For instance, we can construct T_4 exactly as we constructed E_4. We start with rational abscissae and then extend the curve T_4 by virtue of continuity considerations to all values of x.

Next we might compare the curves T_p with the auxiliary line L_0 given by the equation $y=x$. Clearly, each curve T_p has the origin in common with L_0. The curve T_4 intersects L_0 in another point with an abscissa between 0 and $p/2$ (and only this interval will be considered in what follows). The curve T_3 has only the origin in common with L_0. One readily sees that if we let p decrease below 4, the abscissa of the second point of intersection of T_p and L_0 approaches 0. In fact, if we preassign the abscissa ξ of the second point of intersection, it is easy to compute by means of trigonometric tables (in degrees or in radians) a value of p for which T_p and L_0 intersect at the abscissa ξ. We have to satisfy the equation $\tan(2R\xi/p) = \xi$, which yields

$$p = \frac{2R\xi}{\arctan \xi},$$

so that T_p has the equation

$$y = \tan \frac{\arctan \xi}{\xi} x.$$

Unfortunately, while in the theory of exponential functions the expression of the base a in terms of ξ leads to the ordinary definition of e when ξ approaches 0, for tangential functions the expression of the period p in terms of ξ does not give us a clear insight into the nature of the limit which p approaches as $\xi \to 0$. We thus shall define as "natural" tangential curve the curve T_p which has L_0 as tangent at the origin. This procedure has the double advantage of guaranteeing the existence and unicity of the natural tangent, and of implying that the natural tangential curve is T_π where π denotes the length of a semicircle of radius 1.

In fact, since every curve T_p can be obtained from any one of them by a proper change of the unit on the X-axis, thus every tangential function $t(x)$ can be obtained from any one of them, say, from $t_0(x)$, by a transformation

$t(x) = t_0(cx)$ for a proper constant c, it is clear that the tangents to the various lines T_p at the origin form the pencil of straight lines through this point with the exception of the two coördinate axes; in other words, that each finite slope $\neq 0$ is associated with exactly one number p. There is exactly one curve, let us call it T_q, with the slope 1 at 0. It is easily seen that for $p < q$ the line T_p has no point other than the origin in common with L_p while in this case the sine curve S_p with the equation

$$y = \sin \frac{2R}{p} x$$

does intersect L_0 in another point, and that for $p > q$ the line T_p has another point in common with L_0.

In order to prove that $q = \pi$, we repeat our former reasoning about the regular polygons with n sides in slightly changed form. We remark that with increasing n the lengths $2n \tan (2R/n)$ approach the length of the circle. If we call this latter length 2π, then

$$\frac{\tan (2R/n)}{\pi/n} \to 1 \quad \text{as} \quad n \to \infty.$$

It clearly follows that

$$\frac{\tan \dfrac{\pi}{\pi} \dfrac{2R}{n}}{\pi/n} \to 1 \quad \text{as} \quad n \to \infty.$$

Now only for $p = q$ do we have

$$\frac{(\tan 2Rx/p)}{x} \to 1 \quad \text{as} \quad x \to 0.$$

Setting $x = \pi/n$ we see: only for $p = q$ do we have

$$\frac{\tan \dfrac{\pi}{p} \dfrac{2R}{n}}{\pi/n} \to 1 \quad \text{as} \quad n \to \infty.$$

We have seen that the above formula does hold for $p = \pi$. Hence $q = \pi$.

AN APPLICATION OF A FAMOUS INEQUALITY*

N. S. MENDELSOHN, University of Manitoba

The proofs usually given in elementary textbooks of the fact that the sequence $S_n = (1+1/n)^n$ approaches a limit as $n \to \infty$ are either incorrect or quite messy. The correct proofs are usually based on the following facts:

(1) The sequence $S_n = (1+1/n)^n$ is an increasing function of n for positive integral n.

(2) The sequence $T_n = (1+1/n)^{n+1}$ is a decreasing function of n for positive integral n.

(3) $T_n > S_n$ for every positive integer n.

The proof of (3) is quite trivial. In this note we address ourselves to the problem of obtaining interesting proofs of (1) and (2). The proofs are based on the famous inequality which states that the arithmetic mean of a set of k positive numbers is greater than their geometric mean.

Proof of (1). Consider the set of $(n+1)$ numbers

$$1, \ 1+\frac{1}{n}, \ 1+\frac{1}{n}, \ 1+\frac{1}{n}, \ \cdots, \ 1+\frac{1}{n}.$$

These have an arithmetic mean of $1+1/(n+1)$ and a geometric mean of $(1+1/n)^{n/(n+1)}$. Hence

$$1 + \frac{1}{n+1} > \left(1+\frac{1}{n}\right)^{n/(n+1)} \quad \text{or} \quad \left(1+\frac{1}{n+1}\right)^{n+1} > \left(1+\frac{1}{n}\right)^n.$$

Hence S_n increases with n.

Proof of (2). Consider the set of $(n+2)$ numbers.

$$1, \ \frac{n}{n+1}, \ \frac{n}{n+1}, \ \frac{n}{n+1}, \ \cdots, \ \frac{n}{n+1}.$$

These have an arithmetic mean of $(n+1)/(n+2)$ and a geometric mean of $(n/n+1)^{(n+1)/(n+2)}$. Hence

$$\frac{n+1}{n+2} > \left(\frac{n}{n+1}\right)^{(n+1)/(n+2)}$$

On taking reciprocals this becomes

$$1 + \frac{1}{n+1} < \left(1+\frac{1}{n}\right)^{(n+1)/(n+2)} \quad \text{or} \quad \left(1+\frac{1}{n+1}\right)^{n+2} < \left(1+\frac{1}{n}\right)^{n+1}.$$

Hence T_n is a decreasing function of n.

The proof of (1) was first given by a first year student at the University of Toronto about 10 years ago. The proof of (2) rests with the author of this note.

---

* From AMERICAN MATHEMATICAL MONTHLY, vol. 58 (1951), p. 563.

THE LIMIT FOR e*

RICHARD LYON AND MORGAN WARD, California Institute of Technology

The following way of showing that $(1 + 1/n)^n$ tends to e as n tends to infinity seems simpler than the current proofs in elementary texts.

Define e as $\lim_{n\to\infty} \sum_0^n 1/r!$. Then the result is evident from the inequality

$$\sum_0^n \frac{1}{r!} > \left(1 + \frac{1}{n}\right)^n > \sum_0^n \frac{1}{r!} - \frac{3}{2n}, \qquad n \geq 3.$$

This inequality may be proved as follows. Assume $n \geq 3$. Then by the binomial theorem,

$$\left(1 + \frac{1}{n}\right)^n = 2 + \sum_2^n \left(1 - \frac{1}{n}\right)\left(1 - \frac{2}{n}\right) \cdots \left(1 - \frac{r-1}{n}\right) \frac{1}{r!}.$$

Clearly

$$1 > \left(1 - \frac{1}{n}\right)\left(1 - \frac{2}{n}\right) \cdots \left(1 - \frac{r-1}{n}\right).$$

Hence

$$\sum_0^n \frac{1}{r!} = 2 + \sum_2^n \frac{1}{r!} > \left(1 + \frac{1}{n}\right)^n.$$

On the other hand,

$$\left(1 - \frac{1}{n}\right)\left(1 - \frac{2}{n}\right) = 1 - \frac{(1+2)}{n} + \frac{2}{n^2} > 1 - \frac{(1+2)}{n}.$$

Hence

$$\left(1 - \frac{1}{n}\right)\left(1 - \frac{2}{n}\right)\left(1 - \frac{3}{n}\right) > \left(1 - \frac{(1+2)}{n}\right)\left(1 - \frac{3}{n}\right)$$

$$> 1 - \frac{(1+2+3)}{n},$$

and so on. Consequently if $r \geq 2$,

$$\left(1 - \frac{1}{n}\right)\left(1 - \frac{2}{n}\right) \cdots \left(1 - \frac{r-1}{n}\right) \geq 1 - \frac{(1+2+3+\cdots+r-1)}{n}$$

$$= 1 - \frac{r(r-1)}{2n}$$

---

* From AMERICAN MATHEMATICAL MONTHLY, vol. 59 (1952), pp. 102–103.

with equality only when $r=2$. Consequently since $n \geq 3$

$$\left(1+\frac{1}{n}\right)^n > 2 + \sum_2^n \left(1 - \frac{r(r-1)}{2n}\right)\frac{1}{r!} = \sum_0^n \frac{1}{r!} - \frac{1}{2n}\sum_0^{n-2}\frac{1}{r!}$$

$$> \sum_0^n \frac{1}{r!} - \frac{3}{2n}, \quad \text{since } \sum_0^{n-2}\frac{1}{r!} \text{ is less than 3.}$$

ELEMENTARY PROOF THAT e IS IRRATIONAL*

L. L. PENNISI, University of Illinois

The following variation on the usual proof of the irrationality of e is perhaps slightly simpler. Suppose that e is rational, say $e=a/b$. Then

$$b/a = e^{-1} = \sum_{n=0}^{\infty} (-1)^n/n!,$$

and multiplication by $(-1)^{a+1}a!$ and transposition of terms gives

$$(-1)^{a+1}\left\{b(a-1)! - \sum_{n=0}^{a}(-1)^n a!/n!\right\}$$

$$= \frac{1}{(a+1)} - \frac{1}{(a+1)(a+2)} + \frac{1}{(a+1)(a+2)(a+3)} - \cdots.$$

The right side has a value between 0 and 1 since the alternating series clearly converges to a value between its first term and the sum of its first two terms. But the left side is an integer, so we have a contradiction.

---

* From AMERICAN MATHEMATICAL MONTHLY, vol. 60 (1953), p. 474.

A NOTE ON THE BASE OF NATURAL LOGARITHMS*

ERNEST LEACH, Massachusetts Institute of Technology

A quick approach to the theory of natural logarithms starts from the defining formula

$$\ln x = \int_1^x t^{-1} dt. \tag{1}$$

The base e is then defined implicitly by:

$$\int_1^e t^{-1} dt = 1. \tag{2}$$

It is intuitively evident that if we vary the exponent in the integrand slightly, we shall have an approximation to e. That is, if we define E_α by:

$$\int_1^{E_\alpha} t^{-1+\alpha} dt = 1 \tag{3}$$

we should then have the relation

$$e = \lim_{\alpha \to 0} E_\alpha. \tag{4}$$

If $\alpha \neq 0$, we may solve for E_α quite easily, and find that

$$E_\alpha = (1 + \alpha)^{1/\alpha}.$$

Thus (4) is completely equivalent to the familiar relation

$$e = \lim_{\alpha \to 0} (1 + \alpha)^{1/\alpha}. \tag{4'}$$

We can use equation (3) to obtain a proof of the equivalent relations (4) and (4'). One method is to follow the classical proof of (4'), which is quite simple after it is known that E_α is a monotone function of α. But equation (3) gives this monotoneness as a trivial consequence of the monotoneness of the integrand in (3) as a function of α.

Another proof of (4) can be obtained by a simple continuity argument using (2) and (3). For, given $\epsilon > 0$, we have easily

$$\int_1^{e+\epsilon} t^{-1} dt > 1. \tag{5}$$

---

*From AMERICAN MATHEMATICAL MONTHLY, vol. 60 (1953), p. 622–623.

Then, by uniform continuity of $t^{-1+\alpha}$, we have for sufficiently small α,

$$(7) \qquad \int_1^{e+\epsilon} t^{-1+\alpha} dt > 1,$$

from which we conclude that $E_\alpha < e+\epsilon$, for small α. In the same manner we prove that $E_\alpha > e-\epsilon$, for sufficiently small α, from which the relation (4) follows, since ϵ is arbitrary.

Although the elementary theory of the natural logarithm and the exponential function can be developed without recourse to (4'), this relation is useful in giving sidelights of the theory. As well-known examples, we remark that the series expansion for e^x can be derived from (4') and the binomial theorem, and that the Stirling approximation to $\sqrt[n]{n!}$ can be derived from (4') with the aid of the multiplicative analogue of Césaro summation.

ELEMENTARY PROOF THAT e IS NOT QUADRATICALLY ALGEBRAIC*

S. BEATTY, University of Toronto

We propose to indicate how the elementary means employed by Pennisi† to show that e is not rational can be applied to prove that e is not quadratically algebraic, or, in other words, that

(1) $$ae^2 + be + c = 0$$

is impossible, for a, b, and c integral but not all zero.

From the series for e and $1/e$, it appears that

(2) $$e(n!) = I + \frac{1}{n+\theta}, \quad \text{for all positive integers } n,$$

(3) $$e^{-1}(n!) = J + \frac{1}{n+1+\phi}, \quad \text{for all odd, positive integers } n,$$

where I and J are integers and θ and ϕ proper fractions, all four depending on n for their actual values. It is only necessary to verify that the value of

$$\frac{1}{n+1} + \frac{1}{n+1}\frac{1}{n+2} + \cdots$$

is of the form $1/(n+\theta)$, while the value of

$$\frac{1}{n+1} - \frac{1}{n+1}\frac{1}{n+2} + \cdots$$

is of the form $1/(n+1+\phi)$. Taking n odd from here on, and substituting (2) and (3) in the equation obtained from (1) by multiplying its left side by $e^{-1}(n!)$, we see that

$$\frac{a}{n+\theta} + \frac{c}{n+1+\phi}$$

must be an integer, which integer must be zero, indeed, for n sufficiently large. This implies, in turn, that $a+c$ must be zero and that θ and $1+\phi$ must be equal, since, by hypothesis, a and c cannot be zero themselves. To say, however, that θ and $1+\phi$ are equal is absurd, seeing that unity lies between them.

---

*From AMERICAN MATHEMATICAL MONTHLY, vol. 62 (1955), pp. 32–33.

†Pennisi, L. L., Elementary proof that e is irrational, this MONTHLY, 1953, vol. 60, p. 474. [*Ed. note:* Reprinted on p. 434, this book.]

BIBLIOGRAPHIC ENTRIES: e

The references below are to the AMERICAN MATHEMATICAL MONTHLY.

1. Sandham, H. F., An approximate construction for e, vol. 54, p. 215.

   A ruler and compass construction giving a length differing from e by less than 16×10^{-11}.

2. Isbell, J. R., The Maclaurin series for e^x, vol. 71, p. 1033.
3. Bird, M. T., Approximation of e, vol. 75, p. 286.

(b)
π

A PROOF OF THE IRRATIONALITY OF π*

ROBERT BREUSCH, Amherst College

Assume $\pi = a/b$, a and b integers. Then, with $N = 2a$, $\sin N = 0$, $\cos N = 1$, and $\cos(N/2) = \pm 1$.

If m is zero or a positive integer, then

$$A_m(x) \equiv \sum_{k=0}^{\infty} (-1)^k (2k+1)^m \frac{x^{2k+1}}{(2k+1)!} = P_m(x) \cos x + Q_m(x) \sin x$$

where $P_m(x)$ and $Q_m(x)$ are polynomials in x with integral coefficients. (Proof by induction on m: $A_{m+1} = x \, dA_m/dx$, and $A_0 = \sin x$.)

Thus $A_m(N)$ is an integer for every positive integer m.

If t is any positive integer, then

$$B_t(N) \equiv \sum_{k=0}^{\infty} (-1)^k \frac{(2k+1-t-1)(2k+1-t-2)\cdots(2k+1-2t)}{(2k+1)!} N^{2k+1}$$

$$= \sum_{k=0}^{\infty} (-1)^k \frac{(2k+1)^t - b_1(2k+1)^{t-1} + \cdots \pm b_t}{(2k+1)!} N^{2k+1}$$

$$= A_t(N) - b_1 A_{t-1}(N) + \cdots \pm b_t A_0(N).$$

Since all the b_i are integers, $B_t(N)$ must be an integer too.

Now

$$B_t(N) = \sum_{k=0}^{[(t-1)/2]} + \sum_{k=[(t+1)/2]}^{t-1} + \sum_{k=t}^{\infty} ..$$

In the first sum, the numerator of each fraction is a product of t consecutive integers, therefore it is divisible by $t!$, and therefore by $(2k+1)!$ since $2k+1 \leq t$. Thus each term of the first sum is an integer. Each term of the second sum is zero. Thus the third sum must be an integer, for every positive integer t.

---

* From AMERICAN MATHEMATICAL MONTHLY, vol. 61 (1954), pp. 631–632.

This third sum is

$$\sum_{k=t}^{\infty} (-1)^k \frac{(2k-t)!}{(2k+1)!(2k-2t)!} N^{2k+1}$$

$$= (-1)^t \frac{t!}{(2t+1)!} N^{2t+1} \left(1 - \frac{(t+1)(t+2)}{(2t+2)(2t+3)} \frac{N^2}{2!} \right.$$

$$\left. + \frac{(t+1)(t+2)(t+3)(t+4)}{(2t+2)(2t+3)(2t+4)(2t+5)} \frac{N^4}{4!} - \cdots \right).$$

Let $S(t)$ stand for the sum in the parenthesis. Certainly

$$|S(t)| < 1 + N + \frac{N^2}{2!} + \cdots = e^N.$$

Thus the whole expression is absolutely less than

$$\frac{t!}{(2t+1)!} N^{2t+1} e^N < \frac{N^{2t+1}}{t^{t+1}} e^N < (N^2/t)^{t+1} e^N;$$

for $t > t_0$, this is certainly less than 1.

Therefore necessarily $S(t) = 0$ for every integer $t > t_0$. But this is impossible, because

$$\lim_{t \to \infty} S(t) = 1 - \frac{1}{2^2} \cdot \frac{N^2}{2!} + \frac{1}{2^4} \cdot \frac{N^4}{4!} - \cdots = \cos(N/2) = \pm 1.$$

It can be proved similarly that the natural logarithm of a rational number must be irrational: From $\log(a/b) = c/d$ would follow $e^c = a^d/b^d = A/B$. Then

$$B \cdot \sum_{k=0}^{\infty} \frac{(k-t-1)(k-t-2) \cdots (k-2t)}{k!} c^k$$

would have to be an integer for every positive integer t, and a contradiction results, as before.

BIBLIOGRAPHIC ENTRIES: π

The references below are to the AMERICAN MATHEMATICAL MONTHLY.

1. Heal, W. E., Quadrature of the circle, vol. 3, p. 41.
    Reproduces Gordan's proof that both π and e are transcendental.
2. Bennett, A. A., Two new arctangent relations for π, vol. 32, p. 253.
3. Camp, C. C., A new calculation of π, vol. 33, p. 472.
    Uses an arctangent relation for π.

4. Frame, J. S., A series useful in the computation of π, vol. 42, p. 499.

   A method for computing $\arctan(1/239)$ in powers of $1/(240)$.

5. Wrench, J. W., On the derivation of arctangent equalities, vol. 45, p. 108.

   Via the complex logarithm.

6. Gaba, M. G., A simple approximation for π, vol. 45, p. 373.

   A geometric derivation of a rational sequence with limit π.

7. Lehmer, D. H., On arccotangent relations for π, vol. 45, p. 657.

   Estimates of the labor involved in approximating π by using power series approximations for arctangents.

8. Niven, I., The transcendence of π, vol. 46, p. 469.
9. Ballantine, J. P., The best (?) formula for computing π to a thousand places, vol. 46, p. 499.

   A postscript to Lehmer, vol. 45, p. 657 (see entry 7, above).

10. Pennisi, L. L., Expansions for π and π^2, vol. 62, p. 653.
11. Gurland, John, On Wallis' formula, vol. 63, p. 643.

(c)
EULER'S CONSTANT

A NOTE ON THE CALCULATION OF EULER'S CONSTANT*

GOLDIE HORTON, University of Texas

Euler's constant, which plays an important rôle in the theory of gamma functions, is usually defined by the relation

$$\gamma = \lim_{n=\infty} \sum_{1}^{n}\left[\frac{1}{m} - \log\left(1 + \frac{1}{m}\right)\right].$$

Its direct calculation from this definition is impracticable, and it is actually computed by means of the asymptotic expansion

$$\gamma = \lim_{n=\infty} \sum_{1}^{n}\frac{1}{m} - \log n - \frac{1}{2n} + \frac{B_1}{2n^2} - \frac{B_2}{2n^4} + \cdots, \qquad (1)$$

where the B's are the Bernoullian numbers, the error being less than the first term omitted.† The proof of (1) is a delicate matter, however, and it seems therefore worth while to call attention to the fact that γ can be calculated as readily as π or $\log n$ to three or four places of decimals, if one makes use of an idea employed in Cauchy's integral test. This method, which is applicable to all convergent series of positive monotone decreasing terms, depends upon the fact that from Cauchy's test it follows immediately that both

$$\sum_{1}^{n} u_m + \int_{n}^{\infty} u_m dm \quad \text{and} \quad \sum_{1}^{n} u_m + \int_{n+1}^{\infty} u_m dm$$

approximate $\sum_{1}^{\infty} u_m$ with an error less than $\int_{n}^{n+1} u_m dm$, and hence less than u_n.‡

Applying this to

$$\gamma = \sum_{1}^{\infty}\left[\frac{1}{m} - \log\left(1 + \frac{1}{m}\right)\right],$$

---

* From AMERICAN MATHEMATICAL MONTHLY, vol. 23 (1916), p. 73.

† See Wm. Shank's paper, "On the Calculation of the Numerical Value of Euler's Constant." *Proceedings of the Royal Society of London*, Vol. XV, p. 429.

‡ In these integrals the definition of the function u_m is extended so that it relates to the continuous variable m in such wise that u_m is a monotone decreasing function of the continuous variable m.

we see that

$$\sum_1^n \left[\frac{1}{m} - \log\left(1 + \frac{1}{m}\right)\right] + \int_n^\infty \left[\frac{1}{m} - \log\left(1 + \frac{1}{m}\right)\right] dm$$

or

$$\sum_1^n \left[\frac{1}{m} - \log\left(1 + \frac{1}{m}\right)\right] + \left[-1 - \log\left(\frac{n}{n+1}\right)^{n+1}\right]$$

approximates to Euler's constant with an error less than

$$\log\left(\frac{n+1}{n+2}\right)^{n+2} \left(\frac{n+1}{n}\right)^{n+1}$$

For $n=5$, this gives $\gamma = .58249$ with an error less than $.01487$; for $n=10$, $\gamma = .57948$ with an error less than $.00429$; for $n=20$, $\gamma = .57779$ with an error less than $.00115$.

ON THE SEQUENCE FOR EULER'S CONSTANT*

S. K. LAKSHMANA RAO, Indian Institute of Science

The convergence of the sequence $a_n = 1 + \frac{1}{2} + \cdots + 1/n - \log n$ as $n \to \infty$ is proved below by expressing a_n as an infinite integral. This procedure has the advantage of leading at once to the integral representation for $\lim a_n = \gamma$ (Euler's Constant) and we may also examine the behaviour of $a_n - \gamma$ for large n.

Noting that $1/r = \int_0^\infty e^{-rx} dx$ $(r > 0)$ and $\log n = \int_0^\infty (e^{-x} - e^{-nx}) dx/x$ (Frullani's integral†), we write

$$a_n = \int_0^\infty (e^{-x} + e^{-2x} + \cdots + e^{-nx}) dx - \int_0^\infty \frac{e^{-x} - e^{-nx}}{x} dx$$

$$= \int_0^\infty \left(\frac{e^{-x} - e^{-(n+1)x}}{1 - e^{-x}} - \frac{e^{-x} - e^{-nx}}{x} \right) dx$$

which can now be split up into a sum of two convergent infinite integrals in the form

$$a_n = \int_0^\infty \left(\frac{e^{-x}}{1 - e^{-x}} - \frac{e^{-x}}{x} \right) dx + \int_0^\infty e^{-nx} \left(\frac{1}{x} - \frac{1}{e^x - 1} \right) dx.$$

The following inequality

(1) $$\frac{1}{2} - \frac{x}{8} < \frac{1}{x} - \frac{1}{e^x - 1} < \frac{1}{2} \qquad (x > 0)$$

can be easily proved by showing that it depends on the elementary inequality $\tanh x < x \, (x > 0)$. Therefore

$$\int_0^\infty e^{-nx} \left(\frac{1}{x} - \frac{1}{e^x - 1} \right) dx < \int_0^\infty \frac{e^{-nx}}{2} dx = \frac{1}{2n}.$$

This proves that a_n converges to a limit γ given by

$$\gamma = \int_0^\infty \left(\frac{e^{-x}}{1 - e^{-x}} - \frac{e^{-x}}{x} \right) dx.$$

Further,

$$a_n - \gamma = \int_0^\infty e^{-nx} \left(\frac{1}{x} - \frac{1}{e^x - 1} \right) dx$$

---

* From AMERICAN MATHEMATICAL MONTHLY, vol. 63 (1956), pp. 572–573.

† Referee's comment: "Frullani's integral" is a consequence of integration from 1 to n of the previous integral.

and by the inequality (1), we have

$$\int_0^\infty e^{-nx}\left(\frac{1}{2}-\frac{x}{8}\right)dx < a_n - \gamma < \int_0^\infty e^{-nx}\cdot\frac{1}{2}\,dx$$

or

$$\frac{1}{2n}-\frac{1}{8n^2} < a_n - \gamma < \frac{1}{2n}.$$

Therefore $a_n - \gamma$ behaves like $1/2n$ and $n(a_n-\gamma) = n(1+\frac{1}{2}+ \cdots +1/n - \log n -\gamma)$ converges to the limit $\frac{1}{2}$.

BIBLIOGRAPHIC ENTRIES: EULER'S CONSTANT

The references below are to the AMERICAN MATHEMATICAL MONTHLY.

1. Addison, A. W., A series representation for Euler's constant, vol. 74, p. 823.
2. Leighton, W., Remarks on certain Eulerian constants, vol. 75, p. 283.

AUTHOR INDEX

Numbers in italic type refer to bibliographic entries.

Adamson, Iain T., *170*
Addison, A. W., *445*
Agnew, R. P., 208, *316*, 393
Aissen, M. I., *376*
Allendoerfer, C. B., 136, 205
Anderson, A. G., *129*
Andress, W. R., *403*
Antosiewicz, H. S., *376*
Apostol, T. M., 395
Aull, C. E., *170*
Ayres, W. L., *159*

Baez, A. V., 283
Baker, F. E., *341*
Ballantine, J. P., *99*, 153, 237, *341*, *441*
Ballard, W. R., *253*
Barrett, L. C., 228
Barrett, Louis C., 219
Barrow, D. F., *403*
Bateman, H., *375*
Beardon, A. F., *160*
Beatley, Ralph, 52
Beatty, S., 437
Beesack, P. R., *170*, *253*
Begle, E. G., *159*
Bellman, R., *403*
Bennett, A. A., *144*, 228, *306*, *440*
Bernau, S. J., *160*
Bers, L., 225
Bird, M. T., *438*
Birkhoff, G. D., *411*
Bliss, G. A., *305*
Blumenthal, L. M., *252*
Boas, R. P. Jr., *404*
Borman, J. L., 319
Boyer, C. B., 32
Bradshaw, J. W., *403*
Brand, Louis, *211*, 307
Breusch, Robert, 439
Brink, R. W., *402*
Britton, J. R., *306*
Brock, J. E. 390
Brookes, C. J., *170*
Brown, A. B., *99*, *316*, *366*
Brown, B. H., *359*
Brown, James, W., 335
Brown, O. E., *306*
Bullard, J. A., *352*
Burton, W. W., *353*
Bush, K. A., *59*, 228
Bussey, W. H., 259

Cairns, W. D., *44*
Cajori, F., *44*
Calabrese, Philip, 399
Callecot, O. L., *402*
Camp, C. C., *440*
Campbell, J. W., *207*, *353*
Cargeo, C. T., *404*
Carmichael, R. D., *375*
Chand, H., *403*
Christiano, J. G., *366*
Church, Alonzo, 188
Clarke, Elbert H., 412
Coe, C. J., *306*
Cohen, L. W., 228
Coleman, A. J., *353*, 369
Coolidge, J. L., 8, 36, *44*
Corey, S. A., *352*, *375*
Corliss, J. J., *359*
Courant, R., *306*
Cunningham, F., Jr., 315

Daniell, P. J., *375*
Davis, Allan, 158
Davis, Philip J., *44*
Deal, R. B., 356
Dean, G. R., *316*
DeCou, E. E., *305*
Diananda, P. H., *253*
Diaz, J. B., *160*
Doole, H. P., *341*
Douglas, R. D., *353*
Downing, H. H., 228
Dresden, A., *169*
Dresden, Arnold, 261
Duffin, R. J., *403*
Duncan, D. G., *341*
Dunkel, O., *341*, *359*
Dunkel, Otto, 134, 342
Dushnik, B., *169*
Dyer-Bennet, J., *184*

Eaves, J. E., *170*
Eberlein, W. F., 138
Egan, M. F., *352*
Elder, F. S., *144*
Elmendorf, A., *375*
Epstein, L. F., *404*
Erdélyi, A., 67
Escott, E. B., *375*
Evans, Jacqueline P., 217

447

Eves, Howard, 174
Ewing, G. M., 171, *376*

Fain, Bill W., *424*
Feingold, A., *404*
Feller, W., *376*
Ficken, F. A., *341*
Firey, W. J., *366*
Firey, William J., 249
Fish, E., *305*
Fleshler, A. D. 65
Flett, T. M., *147*
Folley, K. W., 322
Ford, W. B., *99*
Fort, M. K., Jr., *159, 170*, 193
Foster, R. M., *403*
Frame, J. S., *376, 441*
Franklin, P., *146, 366*
Franklin, Philip, 405
French, N. E., *404*
Frink, O., Jr., *169*
Frumveller, A. F., *129*
Furstenberg, H., 236

Gaba, M. G., *441*
Gans, David, 165
Garabedian, H. L., *403*
Garrett, J. A., *44*
Garver, R., *306, 376*
Garver, Raymond, 148
Ghurye, S. G., *144*
Goheen, Harry, *376*
Goldberg, R. R., *147*
Goodner, D. B., *228, 253*
Gould, S. H., 41
Graham, B., *99*
Graves, L. M., 318
Green, J. W., *179*
Green, L. C., *403*
Greenstein, D. S., 132, *376, 404*
Groat, B. F., *376*
Gunderson, N. G., *306*
Gurland, John, *441*
Gurney, M., *402*

Haddock, A. G., *170*
Halperin, Israel 163
Halsted, G. B., *44*
Hamilton, H. J., *184*, 199, *252*
Hammer, P. C., *359*
Hammer, Preston C., 373
Hammersley, J. M., *376*
Hamming, R. W., *403*
Hathaway, A. S., *352*
Hawthorne, Frank, 290
Heal, W. E., *440*

Hedrick, E. R., *169, 258*
Heineman, E. R., 251
Helsel, R. G., 48
Hendler, A. S., 291
Herschfeld, A., *402*
Hight, D. W., *170*
Hildebrand, F. B., 364
Hildebrandt, T. H., 111, *159, 305, 316*
Hille, E., *159*
Hoffman, Stephen, 124
Hoggatt, V. E., *359*
Hoggatt, Vern, *306*
Hohn, F. E., 130
Horton, Goldie, 442
Howard, A., *404*
Hoyt, J. P., *317, 353*
Huff, W. N., 356
Hummel, P. M., *252*
Hunt, Burrowes, *170*
Huntington, E. V., *144, 316*
Hutchinson, C. A., *144*
Hyers, D. H., *170*

Irwin, F., *402*
Isbell, J. R., *438*
Ivanoff, V. F., *169*

Jackson, D., *207, 305, 411*
Jackson, Dunham, 254, 409
Jacobson, R. A., *306*
Jacobson, Richard A., 219
James, G., *402*
James, Glenn, *99, 159*
James, R. C., *404*
Jewett, J. W., *132*
Johnson, Marie M., 154
Johnson, R. A., *305, 359*, 354
Johnson, Roger A., 266
Joseph, J. A., *169*
Jungck, G., *160*

Kac, Mark, 185
Kammerer, H. M., *252*
Kaplansky, Irving, 272
Karapetoff, V., *211*
Kazarinoff, D. K., 417
Kearns, D. A., *129*
Kefalas, C. N., *359*
Kelman, R. B., *317*
Kemeny, John G., *144*
Kempner, A. J., *402*
Kennedy, E. C., *132, 376*
Kenyon, Hewitt, *147*
Kirchner, Roger Burr, 83
Klamkin, M. S., *144, 359*, 372, *403*
Klee, V. L., Jr., *146*

Kline, M., *179*
Knebelman, M. S., 113
Konhauser, Joseph D. E., 352
Krall, H. L., *228*
Kreith, Kurt, 101

Lan, Chih-Chin, *228*
Lange, L., *169*
Lange, L. H., *160*
Lange, Luise, 50
Lariviere, R., 401
Lars, L. S., *179*
Leach, Ernest, 435
Leader, S., 167
Lee, H. L., *353*
Lehmer, D. H., *441*
Leighton, W., *445*
Lennes, N. J., *305*, *403*
Levi, H., *228*
Lewis, D. C., 243
Libby, W. F., *402*
Lightstone, A. H., *133*
Livingston, A. E., *253*, *316*
Llano, A., *144*
Longley, W. R., *159*
Lovitt, W. V., *366*
Lowell, L. I., *376*
Lubin, C. I., *129*
Lunn, Arthur C., 229
Lyon, Richard, 432

MacDonald, J. K. L., *144*, *403*
MacDuffee, C. C., 45
MacNeish, H. F., *341*
Macnie, J., *169*
Mancill, J. D., *129*, 278, *306*, *359*
Manheim, J., *317*
Manheim, J. H., *170*
Marcus, M. D., *159*
Marcus, S., *317*
Maria, Alfred J., *253*
Matlak, R. F., *133*
Matsuoka, Yoshio, 422
Matz, F. P., *358*
May, Kenneth O., 100, 345
Mazkewitsch, D., *170*
McCarthy, John, 156
McKiernan, M., *170*
McShane, E. J., *411*
Mead, D. G., *170*, 310
Mendensohn, N. S., 431
Menger, K., *99*
Menger, Karl, 425
Metcalf, F. T., *160*
Meyer, Burnett, *403*, *404*
Michelow, J., *404*
Milkman, J., *133*

Miller, H. C., Jr., *133*
Miller, N., *252*, *402*
Miller, Norman, 284
Miller, W. G., *353*
Moore, C. N., 377
Moore, E. H., *99*
Morduchow, M., *317*, *336*
Moritz, R. E., *159*, *252*, 352
Morley, R. K., *403*, 367
Morris, W. L., *366*
Moser, Leo, *404*
Mosteller, F., *376*
Mott, T. E., *376*
Moulton, E. J., *99*, *146*
Munro, W. D., *404*
Munroe, M. E., 70
Myers, W. M., Jr., *253*

Nash, Stanley W., *179*
Newton, T. A. *404*
Nicholas, C. P., *252*, 180, 245, 344
Nicholson, J. W., *132*, *358*
Niven, I., *441*
Norris, M. J., *403*
Nowlan, F. S., *99*
Nyberg, J., *144*
Nyberg, J. A., *102*

Oakley, C. O., 275
Ogilvy, C. S., 126, 175, *253*, 294, 297, *306*
Ogura, K., *424*
Olds, C. D., *258*, *353*, 324
Olds, E. G., *179*
Osborn, Roger, *228*, 231

Paige, L. J., 232
Pall, G., *144*
Pandres, D. Jr., *170*
Pang, H-C., *359*
Parker, F. D., 248, 330
Parker, S. T., *403*
Pascual, M. I., 117, 351
Pease, D. J., *129*
Pease, Donald K., 334
Pennington, W. B., *160*
Pennisi, L. L., *353*, *434*, *441*
Perkins, Fred W., 150
Perlin, I. E., *403*
Perry, Gary, 127
Phelps, C. R., *341*
Phipps, C. G., 196, *306*
Poliferno, M. J., 222
Pólya, G., 287
Porges, A., *359*
Porter, G. J., *170*
Porter, M. B., *129*, *132*, *402*

Potts, D. H., *317*
Powderly, Mary, 168
Prenowitz, W., *99*
Price, G. B., *144*
Punga, V., 329
Pursell, L. E., *317*
Putney, T., 212

Rademacher, H., *316*
Radó, T. 48
Raisbeck, Gordon, *160*
Rajagopal, C. T., *403*
Randolph, J. F., 185
Ransom, W. R., *99*, 191
Rao, S. K. Lakshmana, *353*, 444
Read, C. B., 122
Rearick, David, *404*
Reyner, S. W., *404*
Reynolds, J. B., 321, *359*, 391, *424*
Richardson, M., *99*
Richmond, D. E., 84, 93, 114, *376*
Rickert, N. W., *236*
Rietz, H. L., *99*
Rinehart, R. F., *306*
Rivlin, T. J., *317*
Robbins, H., *376*
Robbins, H. E., *316*
Roe, E. D., Jr., *144*
Roever, W. H., *366*
Rosenthal, A., *316*
Rosenthal, Arthur, 27
Rouse, H. R., *133*
Rudinger, G., *252*
Rutledge, G., *353*

Sadowsky, M., *376*
Saltzer, C., *306*
Sandham, H. F., *353*, *438*
Scarborough, J. B., *376*
Schaefer, Paul, *236*
Schaumberger, Norman, 350
Schillo, Paul, *317*
Schmiedel, O., *424*
Schoy, C., *44*
Schwartz, J., *366*
Schwerdtfeger, H., *404*
Seebeck, C. L., Jr., *132*, *133*, 241, *376*
Seeley, R. T., 365
Sensenig, W., *402*
Sewell, W. E., *316*
Shain, J., 228
Shanks, E. B., *404*
Sharpe, F. R., *403*
Sheffer, I. M., *403*
Shieh, P., *170*
Shklov, N., *376*
Shohat, J. A., *316*, 386

Sklar, Abe, *317*
Slobin, H. L., *146*, *316*
Smith, D. E., *44*
Smith, H. W., *353*
Smith, W. B., *144*
Snyder, R. W., *376*
Spiegel, M. R., 69, 119, 161, 225, 234, 327, *341*, *353*, *424*
Sprague, Atherton H., 157
Staib, John H., 340
Stelley, R., *403*
Stanaitis, O. E., *404*
Stein, S. K., *184*
Stewart, B. M., 268, *306*
Sutcliffe, A., *359*
Swartz, William, 332
Swift, Elijah, 265
Swift, W. C., *160*
Sydnor, T. E., *341*
Szasz, O., *411*

Tamarkin, J. D., *159*, *179*
Taylor, A. E., *179*, *236*, *353*, *411*
Taylor, M. E., *404*
Thielman, H. P., *146*, *159*
Thomas, G. B., Jr., 419
Thurston, H. A., *179, 207*
Thurston, Hugh A., 80, 128, 303
Thurston, H. S., 357
Tull, J. P., *404*

Ulam, S. M., *170*
Underwood, R. S., *424*
Uspensky, J. V., *129, 306*

Valentine, F. A., *179*
Vaughan, H. E., *129*
Veblen, Oswald, 103
Verghese, K., *170*
Viertel, William K., 336

Walker, A. W., 178, *353*
Walsh, J. L., *316,* 273
Wang, Chung Lie, *228*
Ward, J. A., 155
Ward, L. E., *129*
Ward, M., *403*
Ward, Morgan, 432
Watson, G. C., *236*
Weiner, L. M., *376*
Weisner, L., *424*
Westlund, J., *359*
Weston, J. D., *211*
Wetzel, J. E., *147*
Whitaker, H. C., *1?*

White, C. E., *424*
Whitman, E. A., 1, 360
Wilansky, A., *404*
Wilansky, Albert, *132,* 169, *170*
Wilczynski, E. J., *424*
Williams, G. T., *424*
Williamson, A. W., *424*
Wilson, E. B., *179*
Wolfe, James, 247
Wolfe, J. M., *376*
Wolinsky, A., *341*
Wren, F. L., *44*
Wrench, J. W., *441*
Wrestler, Ferna, 122
Wright, E. M., *146*
Wylie, C. R., Jr., *353*

Yamabe, Hidehiko, 210
Yates, R. C., *132,* 180, 215, 344, *404*
Yates, Robert C., *179*
Yocom, K. L., *306*
Yong-Jeng, Lee, *404*
Young, G. S., 145
Youse, B. K., *376*

Zatzkis, H., *366*
Zeitlin, D., *359*
Zeitlin, David, 214
Zerr, G. B. M., *352, 358, 423, 424*
Zimmerberg, H. J., *366*

CONTENTS
Part II

PREFACE

1. HISTORY

Who Gave You the Epsilon? Cauchy and the Origins of Rigorous Calculus
  JUDITH V. GRABINER   1
An Application of Geography to Mathematics: History of the Integral of the Secant
  V. FREDERICK RICKEY AND PHILIP M. TUCHINSKY   12
The Changing Concept of Change: The Derivative from Fermat to Weierstrass
  JUDITH V. GRABINER   17
Bibliographic Entries: History   29

2. PEDAGOGY

Calculus as an Experimental Science   R. P. BOAS, JR   30
The Problem of Learning to Teach
  P. R. HALMOS, E. E. MOISE, GEORGE PIRANIAN   34
A Mock Symposium for Your Calculus Class   DENNIS WILDFOGEL   45
Calculus by Mistake   LOUISE S. GRINSTEIN   47
Testing Understanding and Understanding Testing
  JEAN PEDERSON AND PETER ROSS   54
Bibliographic Entries: Pedagogy   61

3. FUNCTIONS

(a) CONCEPTS

On the Notion of "Function."	G. J. Minty	62
Bibliographic Entries: Concepts		63

(b) TRIGONOMETRIC FUNCTIONS

On the Differentiation Formula for sin θ	Donald Hartig	64
π and the Limit of $(\sin \alpha)/\alpha$	Leonard Gillman	65
Graphs and Derivatives of the Inverse Trig Functions	Daniel A. Moran	69
Trigonometric Identities through Calculus	Herb Silverman	70
Bibliographic Entries: Trigonometric Functions		71

(c) LOGARITHMIC FUNCTIONS

The Logarithmic Mean	B. C. Carlson	72
Is ln the Other Shoe?	Byron L. McAllister and J. Eldon Whitesitt	76
The Place of ln x Among the Powers of x	Henry C. Finlayson	79
Bibliographic Entries: Logarithmic Functions		79

(d) EXPONENTIAL AND HYPERBOLIC FUNCTIONS

An Elementary Discussion of the Transcendental Nature of the Elementary Transcendental Functions	R. W. Hamming	80
A Matter of Definition	M. C. Mitchelmore	84
The Relationship Between Hyperbolic and Exponential Functions	Roger B. Nelsen	89
Bibliographic Entries: Exponential and Hyperbolic Functions		91

4. LIMITS AND CONTINUITY

Some Thoughts about Limits	Ray Redheffer	92
Bibliographic Entries: Limits and Continuity		103

5. DIFFERENTIATION

(a) THEORY

An Introduction to Differential Calculus	D. G. Herr	104
An Elementary Proof of a Theorem in Calculus	Donald E. Richmond	107
Inverse Functions and their Derivatives	Ernst Snapper	108
An Elementary Result on Derivatives	David A. Birnbaum and Northrup Fowler III	111
Mapping Diagrams, Continuous Functions and Derivatives	Thomas J. Brieske	113
A Self-contained Derivation of the Formula $d/dx\,(x^r) = rx^{r-1}$ for Rational r	Peter A. Lindstrom	119
$(x^n)' = nx^{n-1}$: Six Proofs	Russell Jay Hendel	121
A Note on Differentiation	Russell Euler	123
Differentials and Elementary Calculus	D. F. Bailey	124
The Differentiability of a^x	J. A. Eidswick	126
Bibliographic Entries: Theory		127

(b) APPLICATIONS TO GEOMETRY

The Width of a Rose Petal	S. C. Althoen and M. F. Wyneken	128
Differentiating Area and Volume	Jay I. Miller	134
Convexity in Elementary Calculus: Some Geometric Equivalences	Victor A. Belfi	136
Does "holds water" Hold Water?	R. P. Boas	141
Transitions	Jeanne L. Agnew and James R. Choike	142
A Note on Parallel Curves	Allan J. Kroopnick	152
Bibliographic Entries: Applications to Geometry		153

(c) APPLICATIONS TO MECHANICS

Velocity Averages	Gerald T. Cargo	154
Travelers' Surprises	R. P. Boas	156
Related Rates and the Speed of Light	S. C. Althoen and J. F. Weidner	163
Intuition Out to Sea	William A. Leonard	167
Bibliographic Entries: Applications to Mechanics		168

(d) DIFFERENTIAL EQUATIONS

A Fact About Falling Bodies	William C. Waterhouse	169
The Homicide Problem Revisited	David A. Smith	170
A Linear Diet Model	Arthur C. Segal	175
Bibiographic Entries: Differential Equations		176

(e) PARTIAL DERIVATIVES

Using the Multivariable Chain Rule	Fred Halpern	177
Bibliographic Entries: Partial Derivatives		178

6. MEAN VALUE THEOREM FOR DERIVATIVES, INDETERMINATE FORMS

(a) MEAN VALUE THEOREM

A Versatile Vector Mean Value Theorem	D. E. Sanderson	179
Who Needs Those Mean-Value Theorems, Anyway?	Ralph P. Boas	182
Bibliographic Entries: Mean Value Theorem for Derivatives		186

(b) INDETERMINATE FORMS

Lhospital's Rule Without Mean-Value Theorems	R. P. Boas, Jr	187
Indeterminate Forms of Exponential Type	John V. Baxley and Elmer K. Hayashi	190
Counterexamples to L'Hôpital's Rule	R. P. Boas	192
L'Hôpital's Rule Via Integration	Donald Hartig	194
Indeterminate Forms Revisited	R. P. Boas	196
L'Hôpital's Rule and the Continuity of the Derivative	J. P. King	202
Some Subtleties in L'Hôpital's Rule	Robert J. Bumcrot	203
Bibliographic Entries: Indeterminate Forms		204

7. TAYLOR POLYNOMIALS, BERNOULLI POLYNOMIALS AND SUMS OF POWERS OF INTEGERS

(a) TAYLOR POLYNOMIALS

From Center of Gravity to Bernstein's Theorem	RAY REDHEFFER	205
A Simple Derivation of the Maclaurin Series for Sine and Cosine	DENG BO	207
Trigonometric Power Series	JOHN STAIB	208
Rediscovering Taylor's Theorem	DAN KALMAN	211
Bibliographic Entries: Taylor Polynomials		216

(b) BERNOULLI POLYNOMIALS AND SUMS OF POWERS OF INTEGERS

(no papers reproduced)

Bibliographic Entries: Bernoulli Polynomials and Sums of Powers of Integers
216

8. MAXIMA AND MINIMA

A Strong Second Derivative Test	J. H. C. CREIGHTON	217
Cutting Certain Minimum Corners	L. H. LANGE	219
An Optimization Problem	RICHARD BASSEIN	224
An Old Max-Min Problem Revisited	MARY EMBRY-WARDROP	229
Maximize $x(a - x)$	L. H. LANGE	232
Construction of an Exercise Involving Minimum Time	ROBERT OWEN ARMSTRONG	236
A Bifurcation Problem in First Semester Calculus	W. L. PERRY	241
To Build a Better Box	KAY DUNDAS	244
A Surprising Max-Min Result	HERBERT BAILEY	251
Hanging a Bird Feeder: Food for Thought	JOHN W. DAWSON, JR.	255

Peaks, Ridge, Passes, Valley, and Pits. A Slide Study of $f(x, y) = Ax^2 + By^2$
                                                        CLIFF LONG   257
A Surface with One Local Minimum
                                    J. MARSHALL ASH AND HARLAN SEXTON   259
"The Only Critical Point in Town" Test
                                    IRA ROSENHOLTZ AND LOWELL SMYLIE   262
Bibliographic Entries: Maxima and Minima   264

9. INTEGRATION

(a) THEORY

A Fundamental Theorem of Calculus that Applies to All Riemann
    Integrable Functions   MICHAEL W. BOTSKO   265
Finding Bounds for Definite Integrals   W. VANCE UNDERHILL   268
Average Values and Linear Functions   DAVID E. DOBBS   271
Riemann Integral of cos x JOHN H. MATHEWS AND HARRIS S. SHULTZ   275
Sums and Differences vs. Integrals and Derivatives   GILBERT STRANG   276
Bibliographic Entries: Theory   284

(b) TECHNIQUES OF INTEGRATION

Inverse Functions and Integration by Parts
                                    R. P. BOAS, JR. AND M. B. MARCUS   285
Proof Without Words: Integration by Parts   ROGER B. NELSEN   287
A Discovery Approach to Integration by Parts
                                    JOHN STAIB AND HOWARD ANTON   288
Formal Integration: Dangers and Suggestions   S. K. STEIN   290
The Evaluation of $\int_a^b x^k \, dx$ N. SCHAUMBERGER   300
Bibliographic Entries: Techniques of Integration   301

(c) SPECIAL INTEGRALS

Bibliographic Entries: Special Integrals   302

(d) APPLICATIONS

Disks and Shells Revisited   WALTER CARLIP   303

Upper Bounds on Arc Length	Richard T. Bumby	305
A Theorem on Arc Length	John Kaucher	308
A Note on Arc Length	John T. White	309
"Mean Distance" in Kepler's Third Law	Sherman K. Stein	310
Some Problems of Utmost Gravity	William C. Stretton	312
A New Look at an Old Work Problem	Bert K. Waits and Jerry L. Silver	317
Some Surprising Volumes of Revolution	G. L. Alexanderson and L. F. Klosinski	321
Surface Area and the Cylinder Area Paradox	Frieda Zames	324
Using Integrals to Evaluate Voting Power	Philip D. Straffin, Jr	329
Area of a Parabolic Region	R. Rozen and A. Sofo	332
Bibliographic Entries: Applications		334

(e) MULTIPLE INTEGRALS AND LINE INTEGRALS

Change of Variables in Multiple Integrals: Euler to Cartan	Victor J. Katz	335
Three Aspects of Fubini's Theorem	J. Chris Fisher and J. Shilleto	345
The Largest Unit Ball in Any Euclidean space	Jeffrey Nunemacher	348
Volumes of Cones, Paraboloids, and Other "Vertex Solids."	Paul B. Massell	350
Interchanging the Order of Integration	Stewart Venit	352
Bibliographic Entries: Multiple Integrals and Line Integrals		353

10. NUMERICAL, GRAPHICAL, AND MECHANICAL METHODS AND APPROXIMATIONS (INCLUDING USE OF COMPUTERS)

Numerical Differentiation for Calculus Students	David A. Smith	354
The Error of the Trapezoidal Method for a Concave Curve	S. K. Stein	358
A Non-Simpsonian Use of Parabolas in Numerical Integration	Arthur Richert	361
Reconsidering Area Approximations	George P. Richardson	362

An Interpolation Question Resolved by Calculus
  Martin D. Landau and William R. Jones 366
Generalized Cycloids: Discovery via Computer Graphics
  Sheldon P. Gordon 370
Behold! The Midpoint Rule is Better Than the Trapezoidal Rule for Concave Functions Frank Burk 376
Applications of Transformations to Numerical Integration
  Chris W. Avery and Frank P. Soler 377
Circumference of a Circle—the Hard Way
  David P. Kraines, Vivian Y. Kraines and David A. Smith 380
Note on Simpson's Rule Anon 383
Bibiographic Entries: Numerical, Graphical, and Mechanical Methods and Approximations, (Including use of Computers) 384

11. INFINITE SEQUENCES AND SERIES

(a) THEORY

Interval of Convergence of Some Power Series Eugene Schenkman 385
An Alternative to the Integral Test for Infinite Series G. J. Porter 387
A Sequence of Convergence Tests J. R. Nurcombe 389
Every Power Series is a Taylor Series Mark D. Meyerson 392
Power Series Without Taylor's Theorem Wells Johnson 394
More–and Moore–Power Series Without Taylor's Theorem
  I. E. Leonard and James Duemmel 396
$N!$ and the Root Test Charles C. Mumma II 398
A Simple Test for the nth Term of a Series to Approach Zero
  Jonathan Lewin and Myrtle Lewin 399
Convergence and Divergence of $\sum_{n=1}^{\infty} 1/n^p$
  Teresa Cohen and William J. Knight 400
A Differentiation Test for Absolute Convergence Yaser S. Abu-Mostafa 401
A Note on Infinite Series Louise S. Grinstein 405
Power Series for Practical Purposes Ralph Boas 407
Bibliographic Entries: Theory 411

(b) SERIES RELATED TO THE HARMONIC SERIES

An Interesting Subseries of the Harmonic Series	A. D. Wadhwa	412
A Proof of the Divergence of $\Sigma 1/p$	Ivan Niven	414
Proofs that $\Sigma 1/p$ Diverges	Charles Vanden Eynden	415
The Sum of the Reciprocals of the Primes	W. G. Leavitt	418
The Bernoullis and the Harmonic Series	William Dunham	419
Rearranging the Alternating Harmonic Series	C. C. Cowen, K. R. Davidson and R. P. Kaufman	425
Bibliographic Entries: Series Related to the Harmonic Series		427

(c) SUMS OF SPECIAL SERIES

A Simple Proof of the Formula $\sum_{k=1}^{\infty} k^{-2} = \pi^2/6$	Ioannis Papadimitriou	428
Application of a Mean Value Theorem for Integrals to Series Summation	Eberhard L. Stark	430
Summing Power Series with Polynomial Coefficients	John Klippert	432
A Geometric Proof of the Formula for ln 2	Frank Kost	434
On Sum-Guessing	Mangho Ahuja	436
The Sum is One	John H. Mathews	441
Bibliographic Entries: Sums of Special Series		442

12. SPECIAL NUMBERS

(a) e

A Discovery Approach to e	J. P. Tull	443
Simple Proofs of Two Estimates for e	R. B. Darst	444
Which is Larger, e^π or π^e?	Ivan Niven	445
Proof Without Words: $\pi^e < e^\pi$	Fouad Nakhli	448
Two More Proofs of a Familiar Inequality	Erwin Just and Norman Schaumberger	449
An Alternate Classroom Proof of the Familiar Limit for e	Norman Schaumberger	450
Bibliographic Entries: e		452

    (b) π

Bibliographic Entries: π 453

    (c) EULER'S CONSTANT

Bibliographic Entries: Euler's Constant   453

13. THE LIGHT TOUCH

The Derivative Song	TOM LEHRER	454
There's a Delta for Every Epsilon (Calypso)	TOM LEHRER	454
The Professor's Song	TOM LEHRER	455
Limerick	ARTHUR WHITE	455
The Versed of Boas	RALPH P. BOAS	456
Area of an Ellipse	ED BARBEAU	457
Cauchy's Negative Definite Integral	ED BARBEAU	457
A Positive Vanishing Integral	ED BARBEAU AND M. BENCZE	458
$\cos x = \sinh x$ and $1 = 0$	ED BARBEAU AND ROBERT WEINSTOCK	458
A Power Series Representation	ED BARBEAU	458
Differentiating the Square of x	ED BARBEAU AND A. W. WALKER	459
More Fun with Series	ED BARBEAU	459
A New Way to Obtain the Logarithm	ED BARBEAU AND LEWIS LUM	459
All Powers of x are Constant	ED BARBEAU AND ALEX KUPERMAN	460
A Natural Way to Differentiate an Exponential	ED BARBEAU AND GERRY MYERSON	460

SOURCES   461

AUTHOR INDEX   469